second edi

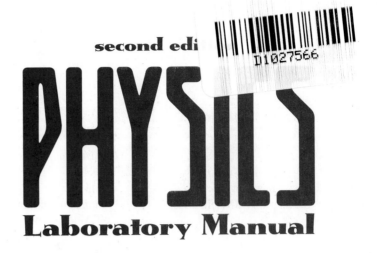

Laboratory Manual

David H. Loyd

Angelo State University

BROOKS/COLE

™

THOMSON LEARNING

Australia • Canada • Mexico • Singapore • Spain • United Kingdom • United States

8   9   10   11   12   05   04   03   02

0-03-024561-3

Asia
Thomson Learning
60 Albert Complex, #15-01
Alpert Complex
Singapore 189969

Australia
Nelson Thomson Learning
102 Dodds Street
South Street
South Melbourne, Victoria 3205
Australia

Canada
Nelson Thomson Learning
1120 Birchmount Road
Toronto, Ontario M1K 5G4
Canada

Europe/Middle East/South Africa
Thomson Learning
Berkshire House
168-173 High Holborn
London WC1 V7AA
United Kingdom

Latin America
Thomson Learning
Seneca, 53
Colonia Polanco
11560 Mexico D.F.
Mexico

Spain
Paraninfo Thomson Learning
Calle/Magallanes, 25
28015 Madrid, Spain

Preface

This laboratory manual is intended for use with a two-semester introductory physics course either calculus based or non-calculus based. For the most part, the manual includes the standard laboratories that have been used by many physics departments for years. The major changes in this edition of the manual from the previous edition include improvements in some of the pre-laboratory questions, improvements in some questions at the end of each laboratory, and better attention to significant figures and units in the data tables and calculations tables. Two completely new laboratories have also been added to this edition, one involving Kirchhoff's rules and the other demonstrating the use of the oscilloscope. Accompanying this new edition of the laboratory manual is a greatly expanded Instructor's Manual which includes actual data for each laboratory complete with analysis and completed questions. The data presented represents essentially the best experimental results that can be expected from these laboratories.

Most of the equipment suggested is of a variety that is available from scientific companies. A few of the laboratories describe homemade equipment that is easily constructed. A somewhat demanding pre-laboratory assignment and the use of rather sophisticated data analysis are major features of this manual. All of the laboratories are written in the same format which is described below. Each of the sections attempts to achieve a specific goal, and these are briefly described.

Prelaboratory... For each laboratory a pre-laboratory assignment is given which is based upon the laboratory description. It is intended that the process of answering a series of questions about the theory and procedure will better prepare the student to perform the laboratory. Some pre-laboratory questions ask for simple restatement of facts discussed in the laboratory instructions. However, there are usually several quantitative questions and problems in the pre-laboratory assignment which require some real understanding of the laboratory principles.

Objectives... Each laboratory has a brief statement of the laboratory goals and a discussion of what concepts will be specifically demonstrated or determined from the measurements.

Equipment List... Each laboratory contains a brief list of the major equipment needed to perform the laboratory.

Theory... The laboratory theory section is intended to be a brief description of the theory underlying the laboratory. The theory discussion is intended to include only what is really needed to understand the basis for the laboratory. The emphasis is to clearly define the relevant variables of the problem and to state the needed equations that relate those variables.

Experimental Procedure... The procedure given is usually very detailed. It attempts to give very explicit instructions on how to perform the measurements. The data tables provided include the units in which the measurements are to be recorded. With the exception of two or three laboratories, SI units are used throughout the manual.

Calculations... Very detailed descriptions of the calculations to be performed are given. In many cases even the intermediate calculated quantities are required. For all but one or two laboratories, any quantity calculated from a measured quantity is recorded in a separate table labeled as a Calculations Table. Most instructors attempt to emphasize the distinction between directly measured quantities and quantities calculated from the measured quantities; yet most laboratory manuals provide a table labeled Data Table where both data and calculated quantities are recorded. The use of separate Calculations Tables helps to emphasize the distinction.

Whenever it is feasible, repeated measurements are performed, and the student is asked to determine the mean and standard error of the measured quantities. For data which is expected to show a linear relationship between two variables, a linear least squares fit to the data is required. Although detailed descriptions are given of the methods to perform these calculations, in the interest of time it is much preferred that the student uses a hand-held calculator capable of performing them automatically.

The repeated use of the mean and standard deviation calculations and the linear least squares fit analyses used throughout the laboratory are considered to be a major feature of the laboratory manual. Thirty-five of the 47 laboratories employ one or both of these types of calculations.

Graphs... Any graphs required are specifically described. All linear data is graphed, and the least squares fit to the data is shown on the graph along with the data.

Laboratory Report... The laboratory report includes the Data Tables, Calculations Tables, a Sample Calculations section, and a list of questions. The attempt of the laboratory questions is to encourage the student to think critically about the conclusions that can be drawn from the data actually taken. Vague general questions are avoided in favor of specific questions about the data actually taken, and what can be learned from the data.

Acknowledgments

I wish to acknowledge the mutual exchange of ideas and techniques for improvement in laboratory instruction that has occurred among Professors H. Ray Dawson, C. Varren Parker and myself for over 25 years. I am grateful both for their specific suggestions and for the influence they have had in the development of my teaching philosophy as expressed in this manual. Professor Varren Parker has been especially helpful with suggestions for improvements in the manual through the years. I also wish to thank Professor Bernie Young for several specific suggestions for improvements in this current edition. I thank the following reviewers whose comments were helpful in improving this new addition:

I am grateful to the Literary Executor of the late Sir Ronald A. Fisher, F. R. S., to Dr. Frank Yates, F. R. S., and to Longman Group Ltd., London, for permission to reprint the table in Appendix I from their book *Statistical Tables for Biological, Agricultural and Medical Research.* (6th edition, 1974)

My most important acknowledgment is to my wife and best friend, Judy. Her patience, good humor, and encouragement continue to be the most important motivation in my life.

David H. Loyd

Contents

For each laboratory listed below the symbol ❖ preceding the lab means that lab requires a calculation of the mean and standard deviation of some repeated measurement. The symbol ✱ preceding the laboratory means that a linear least squares fit to two variables that are presumed to be linear is required in the laboratory.

Contents   **ix**

PURPOSE OF LABORATORY

In the study of physics there are several resources available to students. These include textbooks, lectures, problem solving, and laboratory exercises. The laboratory provides a unique opportunity to validate physical theories in a quantitative manner. Laboratory experience teaches a student the limitations inherent in the application of physical theories to real physical situations. It teaches the role that experimental uncertainty plays in physical measurements and introduces ways to minimize experimental uncertainty. In general, the purpose of these laboratory exercises is both to demonstrate some physical principle and to allow the student to learn and appreciate the techniques of careful measurement.

DATA-TAKING PROCEDURES

It is usually necessary to work in pairs (or sometimes even larger groups) because of space and equipment limitations. This is usually not a disadvantage because many of the exercises require more than one person to perform all phases of the experiment. Avoid the temptation to let one person dominate the process. Before any measurements are made, all partners should agree on exactly what needs to be done and who shall perform each task.

All partners should contribute to the actual process of taking the measurements. If time and other considerations permit, each partner should perform his or her own set of measurements as a check on the procedure. Original data should always be recorded directly in the data tables provided. Avoid the habit of recording the original data on scratch sheets and transferring it to the data tables later. Each partner should separately record data even if only one set of data is taken by the group. Errors of recording are often discovered this way.

SIGNIFICANT FIGURES

The phrase *number of significant figures* tells how many digits are known in some number. This can differ from the total digits in the number because zeros are used as place keepers when digits are not known. For example, in the number 123 there are three significant figures. In the number 1230 one might expect that there are four significant figures, but there are only three because the zero is assumed to be merely keeping a place. Similarly the numbers 0.123 and 0.0123 both have only three significant figures. The rules for determining the number of significant figures in a number are:

1. The most significant digit is the leftmost nonzero digit. In other words, zeros at the left are never significant.

2. If there is no decimal point explicitly given, the rightmost nonzero digit is the least significant digit.

3. If a decimal point is explicitly given, the rightmost digit is the least significant digit, regardless of whether it is zero or nonzero.

4. The number of significant digits is found by counting the places from the most significant to the least significant digit.

As an example consider the following list of numbers: (a) 3456 (b) 135700 (c) 0.003043 (d) 0.01000 (e) 1030. (f) 1.057 (g) 0.0002307. All the numbers in the list have four significant figures for the following reasons: (a) there are four nonzero digits, so they are all significant; (b) the two rightmost zeros are not significant because there is no decimal point; (c) zeros at the left are never significant; (d) the zeros at the left are not significant, but the three zeros at the right are significant because there is a decimal point; (e) there is a decimal point so all four numbers are significant; (f) again, there is a decimal point so all four are significant; (g) zeros at the left are never significant.

READING MEASUREMENT SCALES

For the measurement of any physical quantity such as mass, length, time, temperature, voltage, or current some appropriate measuring device must be chosen. Despite the diverse nature of the devices used to measure the various quantities, they all have in common a measurement scale, and that scale has a smallest-marked-scale-division. All measurements should be done in the following very specific manner. All meters and measuring devices should be read by interpolating between the smallest-marked-scale-division. Generally, the most sensible interpolation is to attempt to estimate 10 divisions between the smallest-marked-scale-division. Consider the section of a meter stick pictured in Figure 1, which shows the region between 2 cm and 5 cm. The smallest-marked-scale-divisions are 1 mm apart. The location of the arrow in the figure is to be determined. It is clearly between 3.4 cm and 3.5 cm, and the correct procedure is to estimate the final place. In this case, a reading of 3.45 cm is estimated. For this measurement, the first two digits are certain, but the last digit is estimated. This measurement is said to contain three significant figures. Much of the data taken in this laboratory will have three significant figures, but occasionally data may contain four or even five significant figures.

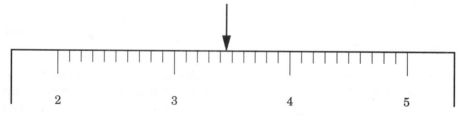

Figure 1

Note that there should be only one estimated place in any measurement. Thus, if a piece of data is to properly contain four significant figures, three must be known and only the fourth estimated. Never attempt to make the estimate between marked scales to more than one place. For example, never attempt to locate the position of the arrow in Figure 1 as 3.451, which would imply four significant figures when only three are justified.

MISTAKES OR PERSONAL ERRORS

All measurements are subject to errors. There are three types of errors, classified as personal, systematic, or random. Random errors are sometimes called "statistical errors." This section deals with personal errors while systematic and random errors will be discussed later. In fact, personal errors are not really errors in the same sense as the other two types of errors. Instead, they are merely mistakes made by the experimenter. Sometimes they are called "illegitimate errors." Mistakes are fundamentally different from the other two types of errors because mistakes can be completely eliminated if the experimenter is careful. These mistakes can be made either in the process of taking the data or later in calculations done with the original data. Either type of mistake is bad, but a mistake made in the data-taking process is probably worse because often it is not discovered until it is too late to correct it.

Typical mistakes made in the data-taking process include misreading of scales and recording data incorrectly. One common mistake in reading scales involves scales that have 10 small divisions between major divisions, and they also have an intermediate division mark halfway between the major divisions. Such a scale is shown in Figure 2. Note that the correct position of the arrow is 3.77 cm. The common mistake occurs when the experimenter is confused by the middle mark and incorrectly reads the position as 3.27 cm. This example is only intended to be representative of the type of mistakes that are made in reading scales. It is very important to understand that there are many different ways for scales to be marked that might lead to possible confusion in reading the scales. However, it is simply not feasible to point out all the specific pitfalls in advance.

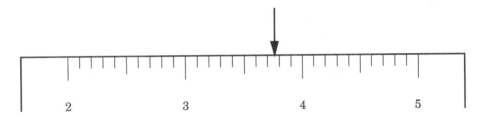

Figure 2

Instead of expecting to have all the potential sources of mistakes pointed out, either by the laboratory instructions or by the instructor, students must develop a specific attitude toward all data-taking processes. The correct attitude is one of skepticism about all the procedures that are carried out in the laboratory. Essentially, this amounts to assuming that things are going to go wrong unless constant attention is given to making sure that no mistakes are made. Unfortunately many students do exactly the opposite. They assume that whatever they are doing is correct without being self-critical of their own understanding of the procedure. For every measurement taken all aspects of the process must be checked and rechecked. Everyone in the group must convince himself or herself that he or she knows exactly what is supposed to be measured, what the correct procedure is to measure it, and that no mistakes are being made by the group in carrying out that procedure.

Exactly the same self-critical approach must be taken to avoid mistakes in calculation of other quantities from the measured quantities. There are at least two kinds of mistakes that can be made in these calculations. The first kind of mistake is to state incorrectly the equation for some derived quantity in terms of the measured quantities. For example, suppose that the distance an object travels and the time it takes that object to travel that distance, are both measured. The task is to calculate the average velocity of the object. The velocity can only be calculated correctly by dividing the distance by the time. Any other calculation using that data will not lead to velocity. Typical mistakes might be to divide the time by the distance or perhaps multiply the distance by the time. Mistakes of this kind are difficult to detect when the first calculated quantity, in this case velocity, is then used to calculate another quantity. For example, if an incorrectly calculated value for velocity is then used to calculate the acceleration, the results obtained for the value of the acceleration will obviously be incorrect. The situation is even more complicated by the fact that often when this kind of mistake is made, a whole series of repeated calculations is performed incorrectly.

The second kind of common mistake occurs when incorrect arithmetic operations are performed on a correct equation. Considering the fact that most students do all their calculations with hand-held electronic calculators, this only amounts to pushing the wrong button, but it is a very common mistake. Again the best defense against such a mistake is to cultivate the habit of continually checking and rechecking all calculations. If all partners do their own independent calculations, mistakes will often be found that would not be discovered if one person's calculations were accepted without question by the group.

A final point to be made about calculations is, to the extent possible, all calculations should be completed before leaving the laboratory. Not only are calculational mistakes more likely to be detected and corrected, but sometimes attempts to perform calculations with the data that has just been taken make it clear that mistakes were made in the data-taking process. If time permits, the original mistake made in taking the data can be corrected by performing the experiment again.

DEVELOPING PROPER LABORATORY HABITS

It is extremely important to appreciate fully the value of systematically attempting to make all measurements with as much care as possible. In many cases the difference between excellent results and complete chaos is the care with which simple measurements are made. No amount of careful calculation after the laboratory period is over can compensate for improperly taken data. If the data are incorrect the point of the laboratory exercise will most likely be completely lost. This leads to frustration for both the student and the instructor.

Sometimes, even if the measurements are performed in a basically correct manner, the point to the exercise may not be readily apparent because of small additional scatter in the data caused by carelessness. For example, if a series of measurements are made like the one illustrated in Figure 1, but only two significant figures are recorded, the effect being measured may change so slightly that the smooth nature of the changes in the measured quantity may not be apparent. The student who strives successfully to develop careful experimental techniques will be rewarded both by success in the laboratory as well as by preparation for future advanced laboratory work.

ACCURACY AND PRECISION

The central point to experimental physical science is the measurement of physical quantities. It is assumed that there exists a true value for any physical quantity, and the measurement process is an attempt to discover that true value. On the other hand, it is not assumed that the process will be perfect and lead to the exact true value. Instead, it is expected that there will be some difference between the true value and the measured value. The terms *accuracy* and *precision* are used to describe different aspects of the difference between the measured value and the true value of some quantity. Note that although most dictionaries make no distinction between them, from a scientific point of view, these have very different meanings.

The *accuracy* of a measurement is determined by how close the result of the measurement is to the true value. For example, in several of the experiments, a value for the acceleration due to gravity will be determined. For this case the accuracy of the result is decided by how close it is to the true value of 9.80 m/s^2. For several of the laboratory experiments the true value of the measured quantity is not known and the accuracy of the experiment cannot be determined from the available data. It is worth noting that even for current research-level experimental work in the physical sciences the determination of the accuracy of a result is extremely difficult in most cases.

The *precision* of a measurement refers essentially to how many significant digits there are in the result. It is an indication also of how reproducible the results are when measurements of some quantity are repeated. When repeated measurements of some quantity are made, the mean of those measurements is considered to be the best estimate of the true value. The smaller the variation of the individual measurements from the mean, the more precise the quoted value of the mean is considered to be. This idea about the relationship between the size of the variations from the mean and the precision of the measurement shall be elaborated on and quantified in a later section on statistical methods.

As an illustration of these ideas about accuracy and precision, consider the following sets of hypothetical measurements of the acceleration due to gravity made by four students named Alf, Beth, Carl, and Dee. Each student made three independent measurements and then took the mean of those three measurements as his or her value for g as shown in Table 1. Also shown in the table is the magnitude of the deviation of each individual measurement from that particular student's mean. In order to be sure of the interpretation of the data table, note that the first set of entries indicates that Alf's three measurements were 7.83, 11.61, and 8.85, and his quoted value for the measurement is 9.43 which is the mean of those three values. Furthermore, the magnitudes of the deviations of each of his individual measurements from his mean are 1.60, 2.18, and 0.58. Of course, all of these values should have units of m/s^2, but the units are omitted here for convenience.

Table 1

	Alf	dev	Beth	dev	Carl	dev	Dee	dev
Measurement 1	7.83	1.60	9.53	0.27	8.70	0.04	9.72	0.04
Measurement 2	11.61	2.18	9.38	0.12	8.75	0.01	9.86	0.10
Measurement 3	8.85	0.58	8.87	0.39	8.77	0.03	9.70	0.06
Mean	9.43		9.26		8.74		9.76	

The accuracy of each student's data is determined by comparing the mean with the true value of 9.80. Doing so indicates that Dee's value of 9.76 is the most accurate, Alf's value of 9.43 is second, Beth's value of 9.26 is third, and Carl's value of 8.74 is the least accurate. Using the size of the deviations from the mean as a criterion for precision, one finds that Carl's value is the most precise, Dee's is second, Beth's is third, and Alf's value is the least precise. Study the data very carefully to be certain that the basis for these conclusions is clear from the ideas that have been given thus far.

In fact, the situation is not quite so simple as has been presented. There is an interplay between the concepts of accuracy and precision that must be considered. If a measurement appears to be very accurate, but the precision is poor, the question arises whether or not the results are really meaningful. Consider Alf's mean of 9.43, which differs from the true value of 9.80 by only 0.37 and thus appears to be quite accurate. However, all of his measurements have deviations greater than 0.37, and two of his deviations are much larger than 0.37. It seems much more likely then that Alf's mean of 9.43 is due to luck than to a careful measurement. It seems likely, however, that Dee's mean of 9.76 is meaningful because the deviations of her individual measurements from the mean are small. In other words, unless a measurement has high precision it cannot really be considered to be accurate. Alf's data cannot really be taken as accurate as his mean implies because his precision is not good enough to warrant faith in that mean.

An examination of the significant figures given in this data leads to essentially the same evaluation of each student's data. Consider Alf's data which indicates by the values stated for the individual measurements that two places to the right of the decimal point are significant. However, that conclusion is not supported by the fact that his deviations occur in the first digit to the left of the decimal point. On the other hand, Dee's results show deviations in the second place to the right of the decimal point in agreement with the fact that two places to the right of the decimal are given as significant in the measured values. Thus from another point of view, Dee's results are seen as meaningful, but Alf's results are questionable.

Carl's results, on the other hand, are an example of a situation that is common in the interplay between accuracy and precision. Carl's precision is extremely high yet his accuracy is not very good. When a measurement has high precision but poor accuracy, it is often the sign of a systematic error. Let us now consider this important class of errors.

SYSTEMATIC ERRORS

Systematic errors are errors that tend to be in the same direction for repeated measurements, giving results that are either consistently above the true value or consistently below the true value. In many cases such errors are caused by some flaw in the experimental apparatus. For example, a voltmeter could be incorrectly calibrated in such a way that it consistently gives a reading that is 80% of the true voltage across its input terminals. It is also possible to have a voltmeter with a zero offset on its scale, which is assumed for this discussion to be 0.50 volt. In the first case the error is a constant fraction of the true value (in this case, 20%), and in the second case the error is a constant absolute voltage. Either one of these is a systematic error, and the answer to the question of which one is worse depends on the magnitude of the voltage to be measured. If the voltage to be measured is 1.00 volt, then the meter with absolute error of 0.50 volt causes an error of 50%, while the meter with relative error causes an error of 20%. On the other hand, if the voltage to be measured is 100 volts, the meter with absolute error of 0.50 volt causes only a 0.5% error, while the other meter still causes a 20% error, or in this case, 20 volts. Also note that if this measured voltage is used to calculate some other quantity, it too will show a systematic error in the results.

It is difficult to deal with systematic errors caused by equipment problems in the time constraints of this laboratory. One cannot afford to spend a great deal of time each week ensuring that all of the equipment has little or no calibration error. On the other hand, it is always a mistake simply to assume that all the equipment in the laboratory works as it should and is well calibrated. The same attitude mentioned previously of healthy skepticism is appropriate for dealing with equipment. A few quick measurements made by two groups on the same object will often serve to at least prove that both group's equipment seems to perform the same or nearly the same. This does not confirm that all is well, but it gives some confidence that the equipment is satisfactory. If there are large differences between the results of the two groups, further investigation is in order. When faulty equipment is suspected ask the instructor for help in either repairing or replacing the equipment.

The general question of the status of equipment in these laboratories is a constant problem apart from the question of systematic errors. It is worth noting here that students often have a very negative feeling about the laboratory unless all the equipment works perfectly. It is a common attitude for students in this laboratory to feel that if their equipment is not perfect, then they are absolved of any responsibility to perform the experiment. Instead of taking that attitude, it is more productive to be aware that dealing with equipment problems is one of the most important aspects of all physical measurements, including those at the research level. This laboratory is a good place to begin learning this valuable skill.

A second common type of systematic error is failure to take into account all of the variables that are important in the experiment. This will often be a problem in the experiments done in this laboratory. In some cases one may be aware that some other factors need to be considered, but may not have the ability to take them into account quantitatively. For example, when using an air table to validate Newton's Laws, as will be done in this laboratory, it is common to ignore friction. This is done because friction is assumed to be small, but also because often there is just no way to determine its contribution easily. It is expected, therefore, that neglecting friction will introduce a systematic error.

Although it may not be possible to eliminate a systematic error in this case, the experiment can be performed in such a way that any systematic error is minimized. For this example, simply choose the forces that are applied to any mass to be as large as possible, consistent with any other constraints of the experiment. If such applied forces are large compared to the frictional force, the smaller is the systematic error.

As previously mentioned, Carl's results for the measurement of g seem very likely to indicate a systematic error. His precision is very high, but all of his individual measurements are about 11% below the true value. In such cases one will very often be asked to state whether or not his or her data show evidence of a systematic error and to suggest possible sources of this error. In attempting to decide if, for example, friction might be the cause of the error, it is important to consider whether the assumed source of the error would tend to produce results that are too high or too low.

If Carl's measurement of g was done by directly measuring some acceleration, friction would tend to cause the results to be too low, and in that case, it is reasonable to suggest friction as a possible cause of the systematic error. On the other hand, if one's measurements for the same experiment showed the same kind of precision as Carl's, but with an average value of 11.37, it is not possible that friction is the problem because the error is in the wrong direction.

For purposes of this laboratory the concern with systematic errors will usually be twofold: (1) to attempt to eliminate any obvious systematic errors to the extent possible; (2) to attempt to identify any data that show systematic error and to suggest possible reasonable causes for such error.

RANDOM ERRORS

The final class of errors are those produced by unpredictable and unknown variations in the total experimental process even when one does the experiment as carefully as is humanly possible. The variations caused by an observer's inability to estimate the last digit the same way every time will definitely be one contribution. Other variations can be caused by fluctuations in line voltage, temperature changes, mechanical vibrations, or any of the many physical variations that may be inherent in the equipment or any other aspect of the measurement process. It is important to realize the following difference between random errors and personal and systematic errors. In principle all personal and systematic errors can be eliminated, but there will always remain some random errors in any measurement. Even in principle the random errors can never be completely eliminated.

This might lead one to conclude falsely that random errors are always the largest contribution because the other kinds can be eliminated in principle. However, as has been said before, in practice systematic errors are not easily eliminated, and in fact, there are many published examples in scientific journals today in which systematic errors may be the largest source of error in the experiment. The statement that systematic errors may be the largest contribution is often all that can be said because only very unreliable estimates can be made of some systematic errors. The main problem with systematic errors is that there is no prescribed process to detect them or to determine their size.

Random errors, on the other hand, can be determined in a prescribed way. This is true because it has been found empirically that random errors often are distributed according to a particular statistical distribution function called the "Gauss distribution function," which is a bell-shaped curve distributed symmetrically about a peak at the mean. The "Gauss distribution function," is also referred to as the normal error function, and random measurement errors are said to be normally distributed when a histogram of the frequency distribution of the results of a large number of repeated measurements produces a bell-shaped curve with a peak at the mean of the measurements. The histogram of the frequency distribution simply means a graph of the number of times the measurements fall within a certain range versus the measured values.

MEAN AND STANDARD DEVIATION

Assume a series of repeated measurements is made in which there are no systematic or personal errors, and thus only random errors are present. Assume there are n measurements made of some quantity x, and the ith value obtained is x_i where i varies from 1 to n. If it is true that the errors are normally distributed, statistical theory says that the mean (of the above n measurements) is the best approximation to the true value. In formal mathematical terms the mean (whose symbol is $\bar{x}$) is given by the equation

$$\bar{x} = \left(\frac{1}{n}\right)\sum_1^n x_i \tag{1}$$

For those not familiar with the notation, the symbol $\sum_1^n$ simply stands for a summation (in other words, addition) process in which the sum is taken from $i = 1$ to $i = n$. For example, assume four measurements are made of some quantity x, and the four results are 18.6, 19.3, 17.7, and 20.4. Equation 1 is simply shorthand notation for the averaging process given by

$$\bar{x} = (1/4)(18.6 + 19.3 + 17.7 + 20.4) = 19.0 \tag{2}$$

The fact that the mean is the best approximation to the true value is not surprising and seems intuitively reasonable. That the mean is indeed the proper choice can be proved mathematically by something called the principle of least squares which can be stated in the following way. The most probable value for some quantity determined from a series of measurements is the value that minimizes the sum of the squares of the deviations between the chosen value and the measured values. It can be shown that the proper choice to produce this minimum sum of deviations is simply the mean of the measurements. The main reason it is mentioned here is because this idea can be usefully generalized later for the case of two variables.

Statistical theory, furthermore, states that the precision of the measurement can be determined by calculation of a quantity called the "standard deviation" from the mean of the measurements. The symbol for standard deviation from the mean is σ_{n-1}, and it is defined by the equation

$$\sigma_{n-1} = \sqrt{\frac{1}{n-1}\sum_1^n [x_i - \bar{x}]^2} \tag{3}$$

For the data given, the standard deviation is calculated from equation 3 to be the following

$$\sigma_{n-1} = \sqrt{\frac{1}{4-1}(18.6 - 19.0)^2 + (19.3 - 19.0)^2 + (17.7 - 19.0)^2 + (20.4 - 19.0)^2} = 1.1$$

The quantity σ_{n-1}, which is also called the "sample standard deviation," is a measure of the precision of the measurement in the following statistical sense. It gives the probability that the measurements fall within a certain range of the measured mean. From the sample standard deviation and tables of the standard error function, it is possible to determine the probability that the measurements fall within any desired range about the mean. The common range to be quoted is the range of one standard deviation as calculated by equation 3. The choice of significant figures used above for the mean (19.0) and σ_{n-1} (1.1) will be addressed later.

Probability theory states that approximately 68.3% of all repeated measurements should fall within a range of plus or minus σ_{n-1} from the mean. Furthermore, 95.5% of all repeated measurements should fall within a range of $2\sigma_{n-1}$ around the mean. For the example given above, 68.3% should fall in the range 19.0 ± 1.1, which is from 17.9 to 20.1, and 95.5% should fall in the range 19.0 ± 2.3, which is from 16.7 to 21.3. Actually, to really make meaningful statements about the expected range, one should do more than four repeated measurements. Something like 10 to 20 measurements would probably be needed to really determine the expected range.

As a final note on the expected distribution for measurements that follow a normal error curve, 99.73% of all measurements should fall within $3\sigma_{n-1}$ of the mean. This implies that if one of the measurements is $3\sigma_{n-1}$ or farther from the mean, it is very unlikely that it is a random error. It is much more likely to be the result of a personal error.

A second issue that can be addressed by these repeated measurements is the precision of the mean. After all, this is what is really of concern, since the mean is the best estimate of the true value. The precision of the mean is indicated by a quantity called the "standard error." The standard error, whose symbol is α, is defined by

$$\alpha = \frac{\sigma_{n-1}}{\sqrt{n}} \tag{4}$$

where σ_{n-1} is the standard deviation from the mean, and n is the number of repeated measurements. For the example given above with $\sigma_{n-1} = 1.1$ and $n = 4$, one gets $\alpha = 0.55$. The significance of α is that if several groups of n measurements are made, each producing a value for the mean, 68.3% of the means should fall in the range 19.0 ± 0.6. Another way to think about this is to say that there is a 68.3%

probability that the true value lies in this range. Of course, all these statements are valid only if there are no errors present other than random errors.

In this laboratory students will often be asked to make repeated measurements of some quantity and to determine the mean. Assuming that α represents the uncertainty in the value of the mean, a crucial question is the appropriate number of significant figures to retain in α. In this laboratory the convention to be followed is to retain *one significant figure* in α and to make the least significant figure in the mean be in the same *decimal place* as α. In this context the appropriate procedure is originally to calculate the mean and σ_{n-1} to more significant figures than it is assumed are needed, and then allow the value of α to determine the significant figures to be retained in the mean. In the example given above, the result should be stated as 19.0 ± 0.6. Notice that as described above, only one significant figure has been retained in α, and the mean has its least significant digit in the same decimal place as α.

PROPAGATION OF ERRORS

Many measurements of interesting quantities involve an indirect process whereby several quantities are measured, and then calculations on them are used to derive a result for some other quantity. For example, speed is often determined by measuring a distance traveled and the time taken to travel that distance, and then dividing the distance by the time. Although it is somewhat indirect, the process as a whole is often referred to as a measurement of speed, when in fact, the speed is derived from the more direct measurement of distance and time.

These measurements of distance and time have some error associated with them. If one wishes to know the error in the speed and how it is related to the errors in the distance and time, there are two cases to be considered. The first is the case in which no repeated measurements have been done, and therefore the number of significant figures in the result is the only guide.

Consider the set of data in Table 2 that was taken by measuring the coordinate position z of some object as a function of time t. From these data the average speed over each time interval can be calculated. The average speed $\overline{v}$ over some time interval Δt during which a distance interval Δz was traveled is given by

$$\overline{v} = \frac{\Delta z}{\Delta t} \tag{5}$$

For the data of Table 2 there are four intervals for the five data points giving for the first two intervals,

$$\overline{v}_1 = \frac{11.97 - 7.57}{2.00 - 1.00} = 4.40 \text{ m/s} \qquad \overline{v}_2 = \frac{16.58 - 11.97}{3.00 - 2.00} = 4.61 \text{ m/s}$$

Table 2

z (m)	t (s)
7.57	1.00
11.97	2.00
16.58	3.00
21.00	4.00
25.49	5.00

The other two intervals give average speeds of 4.42 m/s and 4.49 m/s. A basic question is, on what basis was the decision made to express $\bar{v}_1$, for example, as 4.40 rather than 4.4 or 4.400? Essentially, the answer is given by further extension of the rules for significant figures to include calculations. These calculational rules have been introduced here, rather than in the section on significant figures, in order to emphasize that these rules are related to error propagation.

The following rules should be used to determine the number of significant figures to retain at the end of a calculation:

1. When adding or subtracting, figures to the right of the last column in which all figures are significant should be dropped.

2. When multiplying or dividing, retain only as many significant figures in the result as are contained in the least precise quantity in the calculation.

3. The last significant figure is increased by 1 if the figure beyond it (which is dropped) is 5 or greater.

These rules apply only to the determination of the number of significant figures in the final result. In the intermediate steps of a calculation, one more significant figure should be kept than is kept in the final result.

Consider as an example the addition of the following numbers:

$$
\begin{array}{rr}
753.1 & 753.1 \\
37.08 & 37.1 \\
0.697 & 0.7 \\
\underline{56.3} & \underline{56.3} \\
847.177 & 847.2
\end{array}
$$

Following the above rules strictly implies rewriting each number as shown at the right in which the first digit beyond the decimal is the least significant digit. This is true because that column is the rightmost column in which all digits are significant. Note that one gets the same result if the numbers are added on the calculator (as done at the left), and then it is noted that the first digit beyond the decimal is the last one that can be kept. Therefore 847.177 is rounded off to 847.2.

The same is true for multiplication and division. Consider the calculations below.

$$
\begin{array}{r}
327.23 \\
\times\ \ 36.73 \\
\hline
12019.158
\end{array}
\qquad\qquad
\begin{array}{r}
8.90906 \\
36.73\overline{)327.23}
\end{array}
$$

In each case the result is rounded to four significant figures because the least significant number in each calculation (36.73) has only four significant figures. For the multiplication the result is 12020, and for the division it is 8.909.

Consider again the determination of average speed from the data of Table 2. The reason that the results were given to three significant figures is clear from the application of the above rules. In the subtraction processes to obtain the Δzs and the Δts, there are three significant figures for each of those quantities when the subtraction rules are applied. The average velocity is then calculated by dividing a Δz, having three significant figures, by a Δt, also having three significant figures. Therefore, the result has three significant figures. If, for example, the original data for the time had been given to only two significant figures as 1.0, 2.0, 3.0, 4.0, and 5.0 s, the resulting speeds would have been 4.4, 4.6, 4.4, and 4.5 m/s.

It is difficult to overstate the importance of properly taking significant figures into account in all calculations done in this laboratory. More points on laboratory grades have probably been lost by more students throughout the years over the issue of significant figures than over any other single concern. Most instructors have absolutely no sympathy for failure to adhere to these rules. The wise student will learn these rules for significant figures and apply them without fail.

The second case to be considered concerning the propagation of errors occurs when calculations are to be done on measured quantities to derive another quantity, but now it is assumed that there is specific knowledge of the errors in the original measured quantities.

Suppose variables x, y, and z are measured, and their means $\bar{x}$, $\bar{y}$, and $\bar{z}$ and standard errors α_x, α_y, and α_z are determined. Some quantity $A(\bar{x}, \bar{y}, \bar{z})$ is calculated from the mean values $\bar{x}$, $\bar{y}$, and $\bar{z}$. The question is, What is the standard error, α_A, of the calculated quantity A in terms of α_x, α_y, and α_z? Probability theory states that the appropriate relationship is

$$(\alpha_A)^2 = (\partial A/\partial x)^2 (\alpha_x)^2 + (\partial A/\partial y)^2 (\alpha_y)^2 + (\partial A/\partial z)^2 (\alpha_z)^2 \qquad \textbf{(6)}$$

Obviously, this relationship can be generalized to any number of variables.

Those who have studied calculus recognize the symbol $\partial A/\partial x$ as the partial derivative of A with respect to x. The other symbols are defined in a similar manner. It is understood that some people taking this course have studied calculus, and some have not. Those who have studied calculus should be able to derive an equation for α_A for any function needed using equation 6. For those who have not studied calculus a few examples are given below:

For $A = xy$ $\qquad (\alpha_A)^2 = y^2 (\alpha_x)^2 + x^2 (\alpha_y)^2$ $\qquad \textbf{(7)}$

For $A = x/y$ $\qquad (\alpha_A)^2 = (1/y^4) [y^2 (\alpha_x)^2 + x^2 (\alpha_y)^2]$ $\qquad \textbf{(8)}$

For $A = x + y$ $\qquad (\alpha_A)^2 = (\alpha_x)^2 + (\alpha_y)^2$ $\qquad \textbf{(9)}$

Those who have not studied calculus should not be intimidated by its use here. It is only used to show the origin of these equations for the benefit of those who can use calculus to derive the appropriate equation for other cases of interest. In practice any such equations needed will be given in the laboratory instructions in a form that only requires the use of algebra.

The point to these equations is to give the precise answer to the question of how errors propagate for the case in which the errors in the measured quantities are known. The rules for calculating with significant figures that were given earlier are only an approximate way to handle error propagation in the absence of more complete knowledge implied by equation 6. On the other hand, for many cases in this laboratory, such detailed knowledge of the errors is lacking, and the significant figure calculations are the best way available to estimate the precision of the result.

LINEAR LEAST SQUARES FITS

Often measurements are taken by changing one variable (call it x) and measuring how a second variable (call it y) changes as a function of the first variable. In many cases of interest it is assumed that there exists a linear relationship between the two variables. In mathematical terms one can say that the variables obey an equation of the form

$$y = mx + b \qquad \textbf{(10)}$$

where m and b are constants. This also implies that if a graph is made with x as the abscissa and y as the ordinate, it will be a straight line with m equal to the slope (defined as $\Delta y/\Delta x$) and b equal to the y intercept (the value of y at $x = 0$).

The question is how best to verify that the data do indeed obey equation 10. One way is to make a graph of the data, and then try to draw the best straight line possible through the data points. This will give a qualitative answer to the question, but it is possible to give a quantitative answer to the question by the process described below.

The measurements are repeated measurements in the sense that they are to be considered together in the attempt to determine to what extent the data obey equation 10. It is possible to generalize the idea of minimizing the sum of squares of the deviations described earlier for the mean and standard deviation to the present case. The result of the generalization to two variable linear data is called a "linear least squares fit" to the data. It is also sometimes referred to as a linear regression.

The aim of the process is to determine the value of m and b that produces the best straight line fit to the data. Any choice of values for m and b will produce a straight line, with values of y determined by the choice of x. For any such straight line (determined by a given m and b) there will be a deviation between each of the measured ys and the ys from the straight line fit at the value of the measured xs. The least squares fit is that m and b for which the sum of the squares of these deviations is a minimum. Statistical theory states that the appropriate values of m and b that will produce this minimum sum of squares of the deviations are given by the following equations:

$$m = \frac{n \sum_{1}^{n} y_i - \left(\sum_{1}^{n} x_i\right)\left(\sum_{1}^{n} y_i\right)}{n \sum_{1}^{n} x_i^2 - \left(\sum_{1}^{n} x_i\right)^2} \tag{11}$$

$$b = \frac{\left(\sum_{1}^{n} y_i\right)\left(\sum_{1}^{n} x_i^2\right) - \left(\sum_{1}^{n} x_i y_i\right)\left(\sum_{1}^{n} x_i\right)}{n \sum_{1}^{n} x_i^2 - \left(\sum_{1}^{n} x_i\right)^2} \tag{12}$$

Here n is the number of data points, x_i and y_i are the measured values, and $\sum_{1}^{n}$ stands for the summation from $i = 1$ to $i = n$.

Refer again to the data of Table 2 for coordinate position versus time. The question to be answered is whether or not the data are consistent with constant velocity. It is not assumed that $z = 0$ at $t = 0$. This is the realistic case for much of the data taken in this laboratory. Often the experiments are done in such a manner that there are fixed time intervals (provided by a spark-timer with fixed frequency), but $t = 0$ is arbitrarily determined. For these conditions, if the speed v is constant, the data can be fit by an equation of the form

$$z = vt + z_0 \tag{13}$$

Equation 13 is of the form of equation 10 with z corresponding to y, t corresponding to x, v corresponding to m, and z_0 corresponding to b. Thus v will be the slope of a graph of z versus t, and z_0 will be the intercept, which is the coordinate position at the arbitrarily chosen time $t = 0$. Making these substitutions into equations 11 and 12 leads to

$$v = \frac{n \sum_{1}^{n} t_i z_i - \left(\sum_{1}^{n} t_i\right)\left(\sum_{1}^{n} z_i\right)}{n \sum_{1}^{n} t_i^2 - \left(\sum_{1}^{n} t_i\right)^2} \tag{14}$$

$$z_0 = \frac{\left(\sum_{1}^{n} z_i\right)\left(\sum_{1}^{n} t_i^2\right) - \left(\sum_{1}^{n} t_i z_i\right)\left(\sum_{1}^{n} t_i\right)}{n\sum_{1}^{n} t_i^2 - \left(\sum_{1}^{n} t_i\right)^2} \tag{15}$$

Calculating some of the individual terms gives:

$\Sigma t_i = 1.00 + 2.00 + 3.00 + 4.00 + 5.00 = 15.00$

$\Sigma t_i^2 = (1.00)^2 + (2.00)^2 + (3.00)^2 + (4.00)^2 + (5.00)^2 = 55.00$

$\Sigma z_i = 7.57 + 11.97 + 16.58 + 21.00 + 25.49 = 82.61$

$\Sigma t_i z_i = (1.00 \times 7.57) + (2.00 \times 11.97) + (3.00 \times 16.58) + (4.00 \times 21.00)$
$\qquad + (5.00 \times 25.49) = 292.70$

$\Sigma z_i^2 = (7.57)^2 + (11.97)^2 + (16.58)^2 + (21.00)^2 + (25.49)^2 = 1566.22$

Using these values in equations 14 and 15 gives $v = 4.49$ and $z_0 = 3.06$. Thus, the velocity is determined to be 4.49 m/s, and the coordinate at $t = 0$ is found to be 3.06 m.

At this point the best possible straight line fit to the data has been determined by the least squares fit process. A second goal remains, and that is to determine how well the data actually fit the straight line that has been obtained. Again a qualitative answer to this question is obtained by making a graph of the data and the straight line and qualitatively judging the agreement between the line and the data.

There is, however, a quantitative measure of how well the data follow the straight line obtained by the least squares fit. It is given by the value of a quantity called the "correlation coefficient," or r. This quantity is a measure of the fit of the data to a straight line with $r = 1.000$ exactly signifying a perfect correlation, and $r = 0$ signifying no correlation at all. The equation to calculate r in terms of the general variables x and y is given by

$$r = \frac{n\sum_{1}^{n} x_i y_i - \left(\sum_{1}^{n} x_i\right)\left(\sum_{1}^{n} y_i\right)}{\sqrt{n\sum_{1}^{n} x_i^2 - \left(\sum_{1}^{n} x_i\right)^2}\ \sqrt{n\sum_{1}^{n} y_i^2 - \left(\sum_{1}^{n} y_i\right)^2}} \tag{16}$$

Making the substitutions for the variables of the problem of the fit to the displacement versus time implies substituting t for x and z for y in the above equation to get

$$r = \frac{n\sum_{1}^{n} t_i z_i - \left(\sum_{1}^{n} t_i\right)\left(\sum_{1}^{n} z_i\right)}{\sqrt{n\sum_{1}^{n} t_i^2 - \left(\sum_{1}^{n} t_i\right)^2}\ \sqrt{n\sum_{1}^{n} z_i^2 - \left(\sum_{1}^{n} z_i\right)^2}} \tag{17}$$

Using the appropriate numerical values calculated earlier in equation 17 gives $r = 0.9998$. Thus, the data show an almost perfect linear relationship, since r is so close to 1.000. In calculations of r, keep either three significant figures or enough to ensure the last place is not a 9. *It is extremely important to study this example carefully and to understand completely the process of linear least square fitting of data.*

A very high percentage of the laboratories in this course will involve two variables that are linearly related. For these cases, a least squares fit to the data will usually be required. Although the least squares fit calculations and mean and standard deviation calculations are not difficult in principle, they are tedious and time-consuming. Many hand-held calculators have automatic routines built in that allow the calculation of these quantities by simply inputting the data points one after another. It is advised that students purchase such a calculator if they do not already own one. Be careful to note that many calculators will perform mean and standard deviation calculations automatically, but fewer calculators will also perform an automatic linear least squares fit (or it may be called "linear regression"). Also note that most calculators will calculate two different standard deviations. The one needed is usually denoted σ_{n-1}, and it is the sample standard deviation. Also available on most calculators is a quantity usually denoted as σ_n. It applies to the case when the population is known, and it will never be appropriate for data taken in the laboratory. Always be sure to choose the quantity σ_{n-1} which is the one defined by equation 3.

When performing a least squares fit to data, particularly when a small number of data points are involved, there is some tendency to obtain a surprisingly good value for r even for data that do not appear to be very linear. For those cases the significance of a given value of r can be determined by comparing the obtained value of r with the probability that that value of r would be obtained for n values of two variables that are unrelated. A table for such comparisons is given in Appendix I in a table entitled "Correlation Coefficients."

PERCENTAGE ERROR AND PERCENTAGE DIFFERENCE

In several of the laboratory exercises the true value of the quantity being measured will be considered to be known. In those cases the accuracy of the experiment will be determined by comparing the experimental result with the known value. Normally this will be done by calculating the percentage error of your measurement compared to the given known value. If E stands for the experimental value, and K stands for the known value, then the percentage error is given by,

$$\text{Percentage error} = \frac{|E - K|}{K} \times 100\% \tag{18}$$

In other cases a given quantity will be measured by two different methods. There will then be two different experimental values, E_1 and E_2, but the true value may not be known. For this case the percentage difference between the two experimental values will be calculated. Note that this tells nothing about the accuracy of the experiment, but will be a measure of the precision. The percentage difference between the two measurements is defined as,

$$\text{Percentage difference} = \frac{|E_2 - E_1|}{|E_1 + E_2|/2} \times 100\% \tag{19}$$

Here the absolute difference between the two measurements has been compared to the average of the two measurements. The average is chosen as the basis for comparison when there is no reason to think that one of the measurements is any more reliable than the other. If it is known that there is a significant difference between the magnitude of the errors of the two measured values, it might be appropriate to take a weighted average, weighted inversely as the size of the error in each measurement.

PREPARING GRAPHS

It is helpful to represent data in the form of a graph as an aid to interpreting the overall trend of the data. Most of the graphs for this laboratory will use rectangular Cartesian coordinates. The horizontal axis is called the abscissa, and the vertical axis is the ordinate. Note that it is customary to denote the abscissa as x and the ordinate as y when developing general equations as was done in the development of the equations for a linear least squares fit. However, any two variables can be plotted against one another.

The first thing to do in preparing a graph is to choose a scale for each of the axes. Note that it is not necessary to choose the same scale for both axes. In fact, rarely will it be convenient to have the same scale for both axes. Instead choose the scale for each axis so that the graph will range over as much of the graph paper as possible, consistent with a convenient scale. Choose scales that have the smallest divisions of the graph paper equal to multiples of 2, 5, or 10 units. This makes it much easier to interpolate between the divisions in order to locate the data points when graphing.

The student is expected to bring to each laboratory a supply of good quality linear graph paper. A very good grade of centimeter-by-centimeter graph paper with one division per millimeter is the best choice. Do not, for example, ever use as graph paper 1/4-inch-by-1/4-inch sketch paper or other such coarsely scaled paper. In some cases special graph paper like semilog or log-log graph paper may be required.

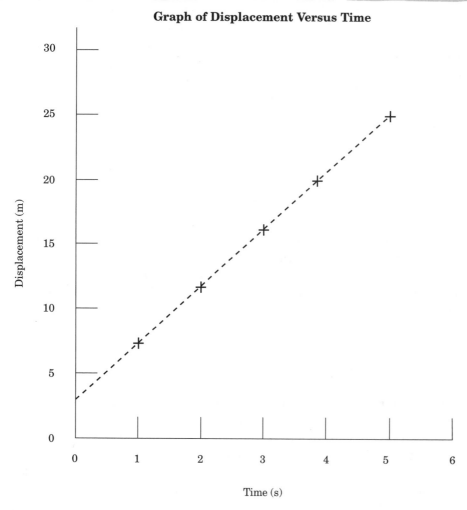

Graph of Displacement Versus Time

Figure 3

Figure 3 is a graph of the data for displacement versus time from Table 2 for which the least squares fit was previously made. Note that scales for each axis have been chosen in order to spread the graph over a reasonable portion of the page. Also note that since the data have been assumed linear, a straight line has been drawn through the data points. The straight line is the one obtained from the least squares fit to the data.

For most experiments the variables will take on only positive values. For that case the scales should range from zero to greater than the largest value for any data point. For example, in Figure 3 the displacement is chosen to range from 0 to 30 because the largest displacement is 25.49, and the time scale has been chosen to range from 0 to 6 because the largest time is 5.00. Also note that the scales should not be suppressed as a means to stretch out the graph. For example, if a set of data contain ordinates that range from 60 to 90, do not choose a scale that shows only that range. Instead a scale from 0 to 100 should be chosen, and there is nothing that can be done in that case to make the graph range over more than about 30% of the graph paper. Scales should always be chosen to increase to the right of the origin and to increase above the origin.

Graphs should always have the scales labeled with the name and units of each variable along each axis. Major scale divisions should be labeled with the appropriate numbers defining the scale. Always include a title for each graph, keeping in mind that it is customary to state the ordinate versus the abscissa.

All graphs should be plotted as points, with no attempt to connect the data with a smooth curve. Do not write the coordinates on the graph next to the data point as is common practice in mathematics classes. The only time it is appropriate to draw any continuous line to represent the trend of the data is when it is assumed the mathematical form of the data is known. In practice, the only time this will be true will be when linearity is assumed, and in that case, it is appropriate to draw the straight line that has been obtained by the least squares fitting procedure.

OBJECTIVES

Physics is an experimental science and, as such, it is largely a science of measurement. The measurement of length is of fundamental importance in scientific work; hence, it is fitting to begin experimental work with this type of measurement. In this experiment, repeated measurements of the dimensions of a laboratory table will be used to accomplish the following objectives:

1. Demonstration of the concept of experimental uncertainty in simple measurements using a meter stick

2. Explanation of the specific knowledge gained by repeated measurements of the length and width of a table

3. Application to these measurements of the statistical concepts of mean, standard deviation from the mean, and standard error

4. Demonstration of the propagation of errors by the determination of the uncertainty in the area calculated from the measured length and width

5. Comparison of the error propagation predicted by statistical theory with the error propagation implied by the simpler concept of significant figures

EQUIPMENT LIST

1. 2-m stick
2. Laboratory table

THEORY

In the section entitled "General Laboratory Information" there is a discussion of the subject of systematic errors. It is possible that there might be small systematic errors associated with this experiment. For example, if the meter sticks used in the experiment were made improperly, or have shrunk or expanded since being manufactured, they would be the source of a systematic error in the results. Any such systematic error will be ignored for the purposes of this experiment. Thus, no statements can be made about the accuracy of the experiment.

There will, of course, be random errors in the measurements associated with the fact that it is not possible to interpolate between the smallest marked scale divisions in exactly the same way every time. Therefore, repeated measurements are expected to show some variation about the mean. If there are no personal errors (mistakes) in the process, this variation will be the basis for the determination of the statistical error or uncertainty in the measurement. If systematic errors are present, the calculated mean will not be a good estimate of the true value, but the precision of the results are still given by the standard error. Note, however, that if personal errors are made, the statistical calculations may not be valid.

A central point of experimental work is the idea that it is necessary not only to determine the value of some measured quantity, but also to attempt to determine the uncertainty in that value. In fact, some would claim that a measurement whose uncertainty is completely unknown is of no use whatsoever! The aim of a measurement is not, therefore, to determine the true value of a quantity (because that cannot be done exactly), but rather to set limits within which it is highly probable that the true value lies. The closer these limits can be set, the more reliable is the measurement.

In this experiment it is assumed that the uncertainty in the measurement of the length and width of the table is due to random errors. If this assumption is valid, then the mean of a series of repeated measurements represents the most probable value for the length or width. For a more complete discussion of this idea, refer to the "General Laboratory Information" section.

Consider the general case in which n measurements of the length and width of the table are made. For this experiment 10 measurements will be made, so $n = 10$ for this case, but the equations will be developed for the case in which n can be any chosen value. If L_i and W_i stand for the individual measurements of the length and width, and $\overline{L}$ and $\overline{W}$ stand for the mean of those measurements, the equations relating them are:

$$\overline{L} = (1/n) \sum_{1}^{n} L_i \qquad \overline{W} = (1/n) \sum_{1}^{n} W_i \qquad (1)$$

To emphasize what was said above, these values of $\overline{L}$ and $\overline{W}$ represent the most probable value for the true values of the length and width assuming no systematic or personal errors are present.

Information about the precision of the measurement is obtained from the variations of the individual measurements using the statistical concept of the standard deviation as described in the section on "General Laboratory Information." The values of the standard deviation from the mean for the length and width of the table, σ_{n-1}^{L} and σ_{n-1}^{W} are given by the equations:

$$\sigma_{n-1}^{L} = \sqrt{1/(n-1) \sum_{1}^{n} (L_i - \overline{L})^2} \qquad \sigma_{n-1}^{W} = \sqrt{1/(n-1) \sum_{1}^{n} (W_i - \overline{W})^2} \qquad (2)$$

If the errors are only random, it should be true that approximately 68.3% of the measurements of length should fall in the range $\overline{L} \pm \sigma_{n-1}^{L}$ and that approximately 68.3% of the measurements of width should fall within the range $\overline{W} \pm \sigma_{n-1}^{W}$.

The precision of the mean for $\overline{L}$ and $\overline{W}$ are given by quantities called the "standard error", α_L and α_W. These quantities are defined by the following equations:

$$\alpha_L = \frac{\sigma_{n-1}^{L}}{\sqrt{n}} \qquad \alpha_W = \frac{\sigma_{n-1}^{W}}{\sqrt{n}} \qquad (3)$$

The meaning of α_L and α_W is that if the errors are only random, there is a 68.3% chance that the true value of the length lies within the range $\overline{L} \pm \alpha_L$, and the true value of the width lies within the range $\overline{W} \pm \alpha_W$.

An important problem in experimental physics is the determination of the uncertainty in some quantity that is derived by calculations from other directly measured quantities. For this experiment, consider the area A of the table as calculated from the measured values of the length and width $\overline{L}$ and $\overline{W}$ by the following:

$$A = \overline{L} \times \overline{W} \qquad (4)$$

This general problem was discussed in the "General Laboratory Information" section, and the case of the product of two measured quantities is given on page 12 by equation 7. Making the appropriate changes of variables for the present case gives

$$\alpha_A = \sqrt{\overline{L}^2\,\alpha_W{}^2 + \overline{W}^2\,\alpha_L{}^2} \qquad (5)$$

This equation gives the standard error of the area of the table, α_A, in terms of the length and width of the table and their associated standard errors. Note that a similar equation applies for standard deviation from the mean:

$$\sigma^A_{n-1} = \sqrt{\overline{L}^2\,(\sigma^W_{n-1})^2 + \overline{W}^2\,(\sigma^L_{n-1})^2} \qquad (6)$$

EXPERIMENTAL PROCEDURE

1. Place the 2-m stick along the length of the table near the middle of the width and parallel to one edge of the length. Do not attempt to line up either edge of the table with one end of the meter stick or with any certain mark on the meter stick.

2. Let x stand for the coordinate position in the length direction. Read the scale on the 2-m stick that is aligned with one end of the table and record that measurement in the Data Table as x_1 (meters). Read the scale that is aligned at the other end of the table and record that measurement in the Data Table as x_2 (meters). Note that the smallest-marked-scale division of the stick is 1 mm. *Therefore, each coordinate should be estimated to the nearest 0.1 mm (nearest 0.0001 m).*

3. Do not perform any subtractions to determine values for the length at this time. Calculations of the length will be performed later from the values of x_1 and x_2, but if calculations are performed now to determine the length, it could bias future readings.

4. Repeat steps 1 and 2 nine more times, for a total of 10 measurements of the length of the table. For each measurement, place the 2-m stick on the table with no attempt to align either end of the stick or any particular mark on the stick with either end of the table. To the extent possible, place the stick along the same line of the table each time. The focus of this experiment is to study the limitation in ability to measure some given length, rather than any additional variation caused by the fact that the table most likely does not have exactly the same length at every line across its width. If the goal was to perform the best possible measurement of the table, such variations should be included. However, the interest here is to examine the statistical variation in the measurement of some fixed length; therefore, choose the same line along the table length each time.

5. Perform steps 1 through 4 for 10 measurements of the width of the table. Let the coordinate for the width be given by y and record the 10 values of y_1 and y_2 (meters) in the Data Table. Again place the stick along the same line each time, but make no attempt to align any particular mark on the stick with either edge of the table.

CALCULATIONS

1. After all measurements are completed, perform the subtractions of the coordinate positions to determine the 10 values of the length L_i, and the 10 values of width W_i. Record the 10 values of L_i and W_i in the Calculations Table.

2. Using equations 1, 2, and 3, calculate the mean length $\overline{L}$, the mean width $\overline{W}$, the standard deviations from the mean σ^L_{n-1} and σ^W_{n-1} and the standard errors α_L and α_W for both the length and width. Initially, calculate $\overline{L}$ and $\overline{W}$ to *seven or eight* significant figures and calculate σ^L_{n-1}, σ^W_{n-1}, α_L, and α_W to *two* significant figures, but do not record those values in the Calculations Table at this time. Do, however, record all these values somewhere for future use in the calculations. Now record the calculated values of σ^L_{n-1}, σ^W_{n-1}, α_L, and α_W to *one* significant figure. Next record the values of $\overline{L}$ and $\overline{W}$ so that they have their most significant digit in the same decimal place as the standard error for that quantity. As an example, suppose that 10 measurements of the length have a value of $\overline{L} = 1.352542$ m, a value of $\sigma^L_{n-1} = 0.0003$ m, and a value of $\alpha_L = 0.00009$ m. Using the above rules for this example, you would record values of $\alpha_L = 0.00009$, $\sigma^L_{n-1} = 0.0003$, and $\overline{L} = 1.35254$. Record your values of $\overline{L}$ and $\overline{W}$ in the Calculations Table to the number of significant figures indicated by the decimal place of the last significant figure in α_L and α_W as described in the example.

3. Using the original values of $\overline{L}$, $\overline{W}$, σ^L_{n-1}, σ^W_{n-1}, α_L, and α_W, in equations 5 and 6, calculate the standard deviation from the mean, σ^A_{n-1} and the standard error, α_A, for the area. Record them in the Calculations Table to *one* significant figure.

4. Using the values of $\overline{L}$ and $\overline{W}$ in equation 4, calculate the area of the table, determining the number of significant figures in the result in two different ways. First, determine the number of significant figures in the area by a procedure similar to the one used above to determine the significant figures in $\overline{L}$ and $\overline{W}$. Let the most significant digit in the calculated area be in the same decimal place as the most significant digit in α_A. Call this value of the area A_1 and record it as A_1 in the Calculations Table.

5. Next, calculate another value for the area, determining the number of significant figures in the result by the following procedure. Use for $\overline{L}$ a value with the number of significant figures contained in each of the values of L_i and use for $\overline{W}$ a value with the number of significant figures contained in each of the values of W_i. Note that this is exactly what the standard rules for significant figures would suggest. Take the product of those values of $\overline{L}$ and $\overline{W}$ using the significant figure rules for multiplication given on page 11 in the "General Laboratory Information" section. Record the result in the Calculations Table as A_2. Note that the only difference between these two values of the area that could result is the number of significant figures, and depending on your precision, there might be no difference at all.

LABORATORY REPORT

Data Table

Trial	x_2 (m)	x_2 (m)	y_2 (m)	y_2 (m)
1				
2				
3				
4				
5				
6				
7				
8				
9				
10				

Calculations Table

Trial	$L_i = x_2 - x_1$ (m)	$W_i = y_2 - y_1$ (m)
1		
2		
3		
4		
5		
6		
7		
8		
9		
10		

$\overline{L} =$ _____ m $\sigma^L_{n-1} =$ _____ m $\alpha_L =$ _____ m

$\overline{W} =$ _____ m $\sigma^W_{n-1} =$ _____ m $\alpha_W =$ _____ m

$A_1 =$ _____ m^2 $\sigma^A_{n-1} =$ _____ m^2 $\alpha_A =$ _____ m^2

$A_2 =$ _____ m^2

QUESTIONS

1. What percentage of your values of L_i fall in the range $\overline{L} \pm \sigma_{n-1}^L$? _____ %
   What percentage fall in the range $\overline{L} \pm 2\sigma_{n-1}^L$? _____ %

2. What percentage of your values of W_i fall in the range $\overline{W} \pm \sigma_{n-1}^W$? _____ %
   What percentage fall in the range $\overline{W} \pm 2\sigma_{n-1}^W$? _____ %

3. According to the theory of random errors, what percentage would be expected for the answers to question 1?

   $\overline{L} \pm \sigma_{n-1}^L$ _____ % $\overline{L} \pm 2\sigma_{n-1}^L$ _____ %

4. Do any of your values of L_i have deviations greater than $3\sigma_{n-1}^L$ from $\overline{L}$? Do any have deviations greater than $3\sigma_{n-1}^W$ from $\overline{W}$? If so, indicate which ones and calculate how many times larger than σ_{n-1}^L or σ_{n-1}^W is the deviation.

5. Based on your answers to questions 1 through 4, are your data reasonably consistent with the assumption that only random errors are present in the experiment? State clearly the basis for your answer.

Using the discussion of standard error in the "General Laboratory Information" section as a guide, state the most probable value of the length, width, and area of the table and their respective uncertainties as determined by statistical theory.

6. length = _____ m   ± _____ m

7. width = _____ m   ± _____ m

8. area = _____ m^2 ± _____ m^2

The two values for the area, A_1 and A_2, represent two different approaches to determination of the uncertainty in a calculated quantity. The value given as A_1 represents the most correct determination of the uncertainty based on the application of statistical theory to assumed random measurement errors. The value given as A_2 assumes that the uncertainty is in the least significant digit for each individual measurement, but does not explicitly state what the uncertainty is in that digit.

9. For your data, does A_1 have more or fewer significant figures than A_2, or do they have the same number of significant figures?

10. Assuming that A_2 is only an approximation to the more correct A_1, state whether A_2 overestimates or underestimates the uncertainty in the area compared to A_1.

OBJECTIVES

The density of an object is defined as its mass per unit volume. Although the standard SI units for density are kg/m^3, it is also very common to use units of g/cm^3. In this laboratory, measurements on several cylinders of different metals will be made to accomplish the following objectives:

1. Determination of the mass of the cylinders
2. Determination of the lengths and diameters of the cylinders
3. Calculation of the density of the cylinders and comparison with the accepted values of the density of the metals
4. Determination of the uncertainty in the value of the calculated density caused by the uncertainties in the measured mass, length, and diameter

EQUIPMENT LIST

1. Three solid cylinders of different metals (aluminum, brass, and iron). (These cylinders should be cut from a metal rod by hand on a band saw in order to introduce some uncertainty in their lengths. It might also be useful if they were ground nonuniformly along their length in order to introduce uncertainty in their diameters.)
2. Vernier calipers
3. Laboratory balance and calibrated masses

THEORY

The goal of this laboratory is to measure the density of three metal cylinders. In fact the process involves measuring directly the mass, length, and diameter of a cylinder and calculating the density from these directly measured quantities. The most general definition of density is mass per unit volume, which can vary throughout the body if the mass is distributed nonuniformly. If the mass in an object is distributed uniformly throughout the object, the density ρ is defined as the total mass M divided by the total volume V of the object. In equation form this is

$$\rho = \frac{M}{V} \tag{1}$$

For a cylinder the volume is given by

$$V = \frac{\pi d^2 L}{4} \tag{2}$$

where d is the cylinder diameter and L is its length. Using equation 2 in equation 1 gives

$$\rho = \frac{4M}{\pi d^2 L} \tag{3}$$

The quantities M, d, and L will be determined by measuring each of them four times independently and calculating the mean and standard error for each quantity. Using the mean of each measured quantity in equation 3 leads to the best value for the measured density ρ. In addition, the uncertainty in the measurement will be determined.

As discussed in the "General Laboratory Information" section, the standard error of ρ is related to the standard error of M, d, and L by

$$(\alpha_\rho)^2 = \left(\frac{\partial \rho}{\partial M}\right)^2 (\alpha_M)^2 + \left(\frac{\partial \rho}{\partial L}\right)^2 (\alpha_L)^2 + \left(\frac{\partial \rho}{\partial d}\right)^2 (\alpha_d)^2 \tag{4}$$

where $\frac{\partial \rho}{\partial M}$ stands for the partial derivative of ρ with respect to M and similarly for the other quantities. The symbol for the standard error for the measurements of M and for the other quantities is α_M. Those who have studied calculus should be able to use equations 3 and 4 to arrive at the following:

$$\alpha_\rho = \rho \sqrt{\left(\frac{\alpha_M}{M}\right)^2 + \left(\frac{\alpha_L}{L}\right)^2 + 4\left(\frac{\alpha_d}{d}\right)^2} \tag{5}$$

If you have no knowledge of calculus do not be intimidated by its use here. Simply accept without proof that equation 5 is the proper relationship between the uncertainty of the measured quantities M, d, and L and the uncertainty in the quantity ρ calculated from them.

The mass of the cylinders will be determined by a laboratory balance. There are several common versions, but the basic principle is essentially the same for all of them. They balance the weight of an unknown mass m against the weight of a known mass m_k. Although the balance is between two forces (the weight of the masses), the scales can be calibrated in terms of mass, assuming that the force per unit mass is the same for both the known and unknown mass. A common form known as the Harvard Trip balance has a calibrated beam along which a permanent sliding weight can be moved in units of 0.1 g up to 10 g. The unknown mass is placed on a pan at the left and balanced against the sum of the permanent sliding weight and additional known masses placed on a pan at the right. The mass of the unknown is the sum of all the known masses placed on the right pan plus the mass equivalent of the permanent sliding mass on the beam when the scales are balanced. Figure 2.3 shows a picture of a Harvard Trip balance.

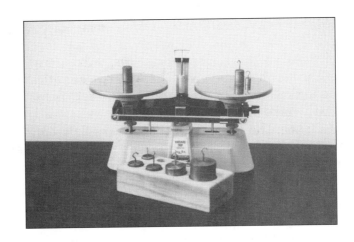

Figure 2.3 Harvard Trip balance.

Before any measurements are made, the balance should be zeroed. Adjustments can be made in the position of a nut on a horizontal screw to ensure that the balance occurs with no mass on either pan when the permanent sliding weight is at the zero position.

The length and diameter of the metal cylinder will be measured with a vernier caliper. Actually, a caliper is any device used to determine thickness, the diameter of an object, or the distance between two surfaces. Often they are in the form of two legs fastened together with a rivet, so they can pivot about the fastened point. The form used in this laboratory consists of a fixed rule containing one jaw and a second jaw with a vernier scale that slides along the fixed-rule scale. Each of the two jaws has two parts pointing in opposite directions. The span between the upper jaws is used to measure the inside diameter between two surfaces. As an example, the upper jaws can be used to measure the inside diameter of a hollow cylinder. The distance between the lower jaws is a measure of the outside diameter of objects over which it is placed.

The caliper has marked on the main scale 1-cm major divisions for which there is both a mark and a number. On the main scale are also marked ten 1-mm divisions between the 1-cm divisions. The 1-mm marks are not labeled with a number. *Vernier* is the name given to any scale that aids in interpolating between marked divisions. This vernier is marked with a scale whose alignment with different marks on the fixed-rule scale allows interpolation between the 1-mm marks on the fixed scale to 0.1-mm accuracy or, in other words, to the nearest 0.01-cm accuracy. A vernier caliper is shown in Figure 2.4.

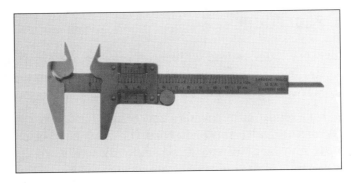

Figure 2.4 Vernier calipers. (Photo courtesy of Sargent-Welch Scientific Co.)

A measurement is made by closing the jaws on some object and noting the position of the zero mark on the vernier and which one of the vernier marks is aligned with some mark on the fixed-rule scale. This is illustrated in Figure 2.5. The position of the zero mark of the vernier scale gives the first two significant figures (2.0 cm in Figure 2.4), and the interpolation between 2.0 cm and 2.1 cm for this case is given by the fact that the sixth mark beyond the vernier zero is best aligned with a mark on the fixed-rule scale. Thus, the reading in this case is 2.06 cm. Study this example carefully to be sure that the instructions on how to read the vernier caliper are clear. If the example is not clear, consult your instructor for additional help in reading these scales.

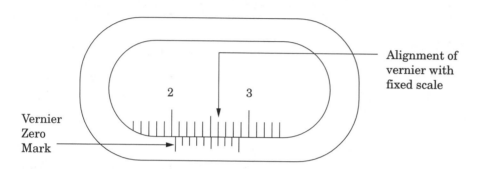

Figure 2.5 Illustration of vernier caliper reading of 2.06 cm.

Before making any measurements it should be determined whether the vernier calipers read zero when the jaws are closed. If the calipers do not read zero when the jaws are closed, they are said to have a zero error. A correction must be made for each measurement made with the calipers. If the vernier zero is to the right of the fixed-scale zero when the jaws are closed, the zero error is negative. Note the mark on the vernier scale that is aligned with the fixed scale, and subtract that number of 0.01-cm units from each measurement. For example, if the third mark to the right of the vernier zero is aligned with the fixed scale when the jaws are closed, then each measurement should have 0.03 cm subtracted from it. On the other hand, if the vernier zero is to the left of the fixed-scale zero, then the zero error is positive. In that case, find which vernier mark is aligned with the fixed scale. Then the number of places to the left of the 10 mark on the vernier scale at which the alignment occurs is the number of 0.01-cm units to be added to the reading. If, for example, the alignment occurs at the 7 mark on the vernier scale, since that is three marks to the left of the 10 mark, then 0.03 cm should be added to the reading.

EXPERIMENTAL PROCEDURE

1. Using the laboratory balance and calibrated masses, determine the mass of each of the three cylinders. Make four independent measurements for each of the cylinders and record the results in the Data Table.

2. Make four separate readings of the zero correction for the vernier calipers. Record the four values in the Data Table. Record the zero correction as positive if the vernier zero is to the right of the fixed-scale zero. Record it as negative if the vernier zero is to the left of the fixed-scale zero.

3. Using the vernier calipers, measure the lengths of the three cylinders. Make four separate trials of the measurement of the length of each cylinder. Measure the length at different places on each cylinder for the four trials in order to sample the variation in length of the cylinders. Record the results in the Data Table.

4. Using the vernier calipers, measure the diameters of the three cylinders. Make four separate trials of the measurement of the diameter of each cylinder. Measure the diameter at four different positions along the length of the cylinders in order to sample the variation in diameter of the cylinders. Record the results in the Data Table.

CALCULATIONS

1. Calculate the mean $\overline{M}$ and the standard error α_M for the four measurements of the mass of each cylinder and record the results in the Calculations Table. Keep only *one* significant figure for all standard errors and then keep the number of *decimal places* in the mean that coincides with the *decimal places* of the standard error.

2. Determine the measured length and diameter for each trial by making the appropriate zero correction to each measurement and then calculating the means $\overline{d}$ and $\overline{L}$ and the standard errors α_d and α_L for each cylinder. Record the results in the Calculations Table. Again keep only *one* significant figure in the standard error and then keep the number of *decimal places* in the mean that coincides with the *decimal places* in the standard error.

3. Using equation 3, calculate the density ρ of each of the cylinders. Use the mean values for the mass, diameter, and length. Calculate the standard error of the density ρ using equation 5. Record the results in the Calculations Table. Again keep only *one significant figure* in the standard error and then keep the number of *decimal places* in the mean that coincides with the *decimal places* in the standard error.

4. For purposes of this laboratory, assume that the density of aluminum is 2.70 g/cm^3, the density of brass is 8.40 g/cm^3, and the density of iron is 7.85 g/cm^3. Calculate the percentage error in your results for the density of each of these metals.

Force Table and Vector Addition of Forces

OBJECTIVES

Physical quantities that can be completely specified by magnitude only are called "scalers." Examples of scalers include temperature, volume, mass, and time intervals. Some physical quantities have both magnitude and direction, and these are called "vectors." Examples of vector quantities include spatial displacement, velocity, and force. In this laboratory, a force table will be used to determine the magnitude and direction of several simultaneously applied forces to accomplish the following objectives:

1. Demonstration of the process of the addition of several vectors to form a resultant vector
2. Demonstration of the relationship between the resultant of several vectors and the equilibrant of those vectors
3. Illustration and practice of graphical solutions for the addition of vectors
4. Illustration and practice of analytical solutions for the addition of vectors

EQUIPMENT LIST

1. Force table with pulleys, ring, and string
2. Mass holders and slotted masses
3. Protractor and compass

THEORY

As an example of the process of vector addition, consider the case of several forces with different magnitudes and directions that act at the same point. It is desired to find the net effect produced by the several forces by finding a single force which is equivalent in its effect to the effect produced by the several applied forces. That single vector is called the resultant vector of the several applied vectors. This resultant vector can be found theoretically by a special addition process known as vector addition.

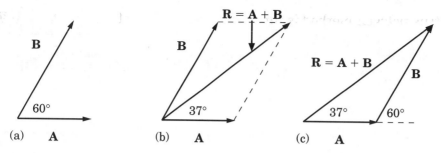

Figure 3.2 Illustration of the parallelogram and triangle addition of vectors.

First, consider the process of vector addition by graphical techniques. Figure 3.2(a) shows the case of two vectors **A** and **B**, which are assumed to represent two forces, with **A** having a magnitude of 20.0 N and **B** having a magnitude of 30.0 N. The scale of 1.00 cm = 10.0 N has been chosen; thus, these vectors are shown as 2.00 cm and 3.00 cm in length, respectively. The vectors are assumed to act at the same point, but 60.0° different in direction. Thus, the tails of the arrows are drawn from the same point, but the heads of the arrows point in directions 60.0° apart. Figure 3.2(b) shows the graphical addition process called the "parallelogram method." The dotted lines shown were constructed with one parallel to **A** and equal in length to **A** drawn from the tip of **B**, and the other parallel to **B** and equal in length to **B** drawn from the tip of **A**. The resultant **R** of the vector addition of **A** and **B** was found by constructing the straight line from the point at the tails of the two vectors to the opposite corner of the parallelogram formed by the original vectors and the constructed dotted lines. A careful measurement of the length of **R** in Figure 3.2(b) shows it to be about 4.35 cm in length, and a measurement of the angle between **R** and **A** shows it to be about 37°. Since the scale is 1.00 cm = 10.0 N, the value of the resultant **R** is 43.5 N, and it acts in a direction 37° with respect to the direction of **A**.

In the graphical vector addition process known as the polygon method, one of the vectors is first drawn to scale. Then each successive vector to be added is drawn with its tail starting at the head of the preceding vector. The resultant vector is then the vector drawn from the tail of the first arrow to the head of the last arrow. Figure 3.2(c) shows this process for the case of only two vectors (for which the polygon method is the triangle method). Note that the second vector, **B**, must be drawn at the proper angle relative to **A** by extending a dotted line in the direction of **A** and constructing **B** relative to that dotted line. In Figure 3.2(c) the length of **R** is again seen to be 4.35 cm, corresponding to 43.5 N, and it acts at 37° with respect to **A**.

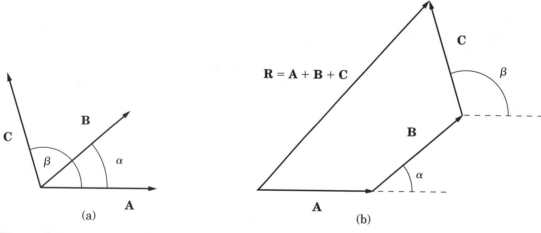

Figure 3.3 Illustration of the polygon method for vector addition.

The true polygon method is illustrated for the case of three vectors in Figure 3.3. The extension of the idea to four or more vectors should be clear. Note carefully that each successive vector must be drawn in the proper relationship to the previous vector. Thus, vector **A** is first drawn, **B** is then drawn at the proper angle α relative to **A**, and **C** is then drawn at the proper angle β relative to **B**. Finally, the resultant **R** is the vector connecting the tail of **A** and the head of **C**.

In each of the graphical methods, the results are exact in principle. In practice they depend greatly on the size of the scale chosen and the care with which the lengths are measured. In general, the larger the scale used, the better the accuracy of the results. In contrast, the analytical process is as exact as the number of significant figures chosen for the calculations. It uses trigonometry to express each vector in terms of its components projected on the axes of a rectangular coordinate system. The process of determining the components of a vector is illustrated in Figure 3.4. In Figure 3.4(a) is shown a vector **A**.

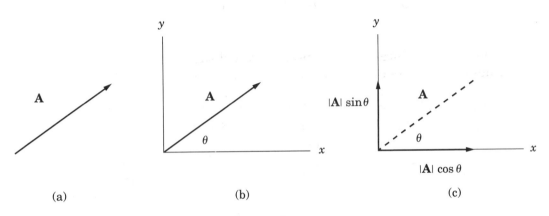

(a)   (b)   (c)

Figure 3.4 Illustration of analytical resolution of a vector.

In Figure 3.4(b) the vector **A** is shown with a rectangular coordinate system superimposed, and in Figure 3.4(c) the components of the vector along the axes are shown to be $|\mathbf{A}|\cos\theta$ along the x-axis and $|\mathbf{A}|\sin\theta$ along the y-axis.

In the process of vector addition, each vector to be added is first resolved into components as shown in Figure 3.4. The components along each axis are then added algebraically to produce the net components of the resultant vector along each axis. Those components are at right angles, and thus the magnitude of the resultant can be found from the Pythagorean theorem. Consider the case of three vectors **A**, **B**, and **C** shown in Figure 3.5. Assume that their magnitudes are 10.0, 15.0, and 12.0, and their directions relative to the x-axis are as shown in that figure. Taking the algebraic sum of each of the components of the three vectors leads to the following:

$$R_x = A_x + B_x + C_x = 10.0\cos(0°) + 15.0\cos(30.0°) - 12.0\cos(45.0°) = 14.5 \quad \textbf{(1)}$$

$$R_y = A_y + B_y + C_y = 10.0\sin(0°) + 15.0\sin(30.0°) + 12.0\sin(45.0°) = 16.0 \quad \textbf{(2)}$$

Note very carefully that the x-component of the **C** vector is negative because it points to the left, which is the negative x-direction. The magnitude of the resultant, R, is found to be the following because the components R_x and R_y are at right angles.

$$|\mathbf{R}| = R = \sqrt{R_x^2 + R_y^2} = \sqrt{(14.5)^2 + (16.0)^2} = 21.6 \qquad \textbf{(3)}$$

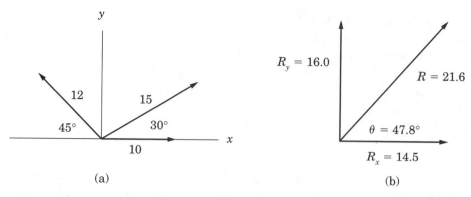

Figure 3.5 Illustration of the analytical addition of vectors.

Furthermore the angle θ that the resultant makes with the x-axis is given by the following:

$$\theta = \text{arc tan}\left(\frac{R_y}{R_x}\right) = \text{arc tan}\left(\frac{16.0}{14.5}\right) = 47.8° \qquad (4)$$

The process of experimentally determining the value of the resultant of several forces is complicated by the fact that when a nonzero resultant force acts on an object it tends to be accelerated. Thus, another force must be applied to produce an equilibrium. For example, if two known forces, F_1 and F_2, are applied to some object, they will have some resultant F_R. In order to keep the object in equilibrium, a force equal in magnitude and opposite in direction of F_R must be applied. The force applied in order to produce equilibrium is called the "equilibrant force," or F_E.

For any number of applied forces ($F_1, F_2, F_3, \ldots$) the equilibrant force F_E is the force that must be applied (along with the original applied forces) in order to keep an object in equilibrium. The magnitude and direction of F_E can be found by trial and error experimentally. The resultant force, F_R, can be found from the knowledge that F_R and F_E have the same magnitude but opposite directions.

The force table (Figure 3.6) used in the experiment is designed to allow application of forces of any chosen magnitude at any chosen angle. The force is provided by the gravitational attraction on masses that are attached to a ring by a string passing over a pulley. Each force is applied over a separate pulley, and the pulley positions can be adjusted to any desired position around a circular plate. It is important to note that the force applied when a mass is hung on the string is equal to the weight of the mass. The weight (N) is equal to the product of the mass (kg) multiplied by the acceleration due to gravity (9.80 m/s²).

Figure 3.6 Force Table. (Photo courtesy of Sargent-Welch Scientific Co.)

EXPERIMENTAL PROCEDURE

Part 1: Two Applied Forces

1. Place a pulley at the 20.0° mark on the force table and place a total of 0.100 kg on the end of the string. Be sure to include the 0.050 kg of the mass holder in the total. Calculate the magnitude of the force (N) produced by the mass. Assume three significant figures for this and for all other calculations of force on the force table. Record the value of this force as F_1 in Data Table 1.

2. Place a second pulley at the 90.0° mark on the force table and place a total of 0.200 kg on the end of the string. Calculate the force produced and record as F_2 in Data Table 1.

3. Determine by trial and error the magnitude of mass needed and the angle at which it must be placed in order to place the ring in equilibrium. The ring is in equilibrium when it is centered on the force table. Jiggle the ring slightly to be sure that the equilibrium conditions are met. Be sure that all the strings are in such a position that they are directed along a line that passes through the center of the ring. This is crucial because all the forces must act through the point at the center of the table.

4. For the experimentally determined mass, calculate the force produced and record the magnitude and direction of this equilibrant force F_{E1} in Data Table 1.

5. From the value of the equilibrant force F_{E1}, determine the magnitude and direction of the resultant force F_{R1} and record them in Data Table 1.

Part 2: Three Applied Forces

1. Place a pulley at 30.0° with 0.150 kg on it, one at 100.0° with 0.200 kg on it, and one at 145.0° with 0.100 kg on it.

2. Calculate the force produced by those masses and record them as F_3, F_4, and F_5 in Data Table 2.

3. Following the procedure outlined steps 3 through 5 in Part 1 above, determine the equilibrant force and the resultant force and record their magnitudes and directions in Data Table 2 as F_{E2} and F_{R2}.

CALCULATIONS

Part 1: Two Applied Forces

1. Find the resultant of these two applied forces by scaled graphical construction using the parallelogram method. Using a ruler and protractor, construct vectors whose scaled length and direction represent F_1 and F_2. For example, a convenient scale might be 1.00 cm = 0.100 N. Be careful to note that all directions are given relative to the force table, and this must be taken into account in the graphical construction to ensure the proper angle of one vector to another. Read the magnitude and direction of the resultant from your graphical solution and record them in the appropriate section of Calculations Table 1.

2. Using trigonometry, calculate the components of F_1 and F_2 and record them in the analytical solution portion of Calculations Table 1. Add the components algebraically and determine the magnitude of the resultant by the Pythagorean theorem. Determine the angle of the resultant from the arc tan of the components. Record those results in Calculations Table 1.

3. Calculate the percentage error of the magnitude of the experimental value of F_R compared to the analytical solution for F_R. Also calculate the percentage error of the magnitude of the graphical solution for F_R compared to the analytical solution for F_R. For each of those comparisons, also calculate the magnitude of the difference in the angle. Record all values in Calculations Table 1.

Part 2: Three Applied Forces

1. Find the resultant of these three applied forces using scaled graphical construction, but for this case, use the polygon method. Read the magnitude and direction of the resultant from your graphical solution and record them in Calculations Table 2.

2. Using trigonometry calculate the components of all three forces, the components of the resultant, and, the magnitude and direction of the resultant. Record them all in Calculations Table 2.

3. Make the same error calculations for this problem as asked for in step 3 above in Part 1.

Laboratory 4
Uniformly Accelerated Motion on the Air Track

PRELABORATORY ASSIGNMENT

Read carefully the entire description of the laboratory and the section entitled "General Laboratory Information," and answer the following questions based on the material contained in the reading assignment. Turn in the completed prelaboratory assignment at the beginning of the laboratory period prior to the performance of the laboratory.

1. A cart on a linear air track has a uniform acceleration of 0.172 m/s^2. What is the velocity of the cart after 4.00 s if it is released from rest?

2. How far does the cart in question 1 travel in 4.00 s if it is released from rest?

3. An air track like the one shown in Figure 4.2 has a block with a height $h = 12.0$ cm under one support. The other support is $d = 3.50$ m away. What is the angle of inclination θ? What is the acceleration of a cart parallel to the air track?

4. The following set of data were taken for the displacement x of a cart on an air track as a function of time t. Calculate t^2 for each of the data points and then perform a linear least squares fit to x versus t^2, where x is the ordinate and t^2 is the abscissa. Record the slope and correlation coefficient below.

x (m)	0	0.150	0.338	0.600	0.930	1.37	2.45
t (s)	0	0.975	1.48	2.04	2.46	3.03	3.95
t^2 (s^2)							

Slope = _____   Correlation Coefficient = _____

5. According to equation 4 in the laboratory theory section, the slope of the fit done in question 4 is equal to $a/2$, where a is the acceleration of the cart. Determine the acceleration of the cart from the slope of the least squares fit.

6. Consider the correlation coefficient r calculated in question 4. Using the table in Appendix I, state the significance of the fit to these data. Remember that including the point (0,0) there are seven data points in this example.

7. Find the angle of inclination θ of the air track in question 4, which gives the acceleration that you found in question 5.

Uniformly Accelerated Motion on the Air Track

OBJECTIVES

In this laboratory, a linear air track will be used to approximate a one-dimensional frictionless surface. The track will be tilted to form an inclined plane, and a cart will be released and allowed to move down the track. Study of the motion of the cart will be used to accomplish the following objectives:

1. Verification that the displacement of the cart down the inclined plane is directly proportional to the square of the elapsed time
2. Determination of the acceleration of the cart from an analysis of the displacement-versus-time data
3. Determination of an experimental value for g, the acceleration due to gravity, by interpretation of the cart's acceleration as a component of g

EQUIPMENT LIST

1. A 5-m linear air track with a built-in 5-m scale (If a shorter air table is used, the suggested distances can be modified to fit the air track used.)
2. Laboratory timer or stopwatch
3. Block to raise the air track to form an inclined plane

THEORY

When an object undergoes one-dimensional uniformly accelerated motion, its velocity increases linearly with time. If it is assumed that the initial velocity of the object is zero at time $t = 0$, then its velocity v at any later time t is given by

$$v = at \tag{1}$$

where a is the acceleration, which is assumed to be constant in magnitude and direction.

Consider a time interval between $t = 0$ and any later time t. The velocity is zero at $t = 0$, and the velocity is v at time t. Therefore, the average velocity $\bar{v}$ during the time interval is

$$\bar{v} = \frac{0 + v}{2} = \frac{v}{2} \tag{2}$$

The displacement x of the object during the time interval t is given by

$$x = \bar{v}t = \frac{vt}{2} \tag{3}$$

Substituting equation 1 for v in equation 3 gives

$$x = \frac{at^2}{2} \qquad \text{(4)}$$

Thus, equation 4 states that if an object is released from rest, its displacement is directly proportional to the square of the elapsed time. Figure 4.1 shows graphs of both x versus t and x versus t^2 for uniformly accelerated motion.

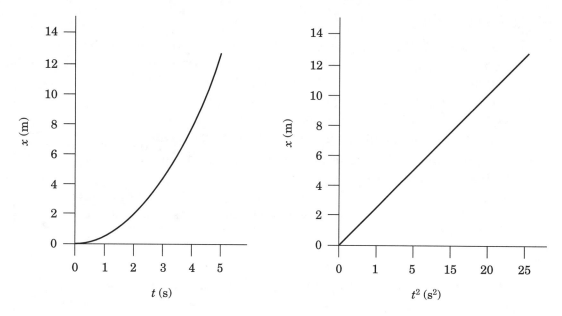

Figure 4.1 Graphs of displacement versus time and displacement versus the square of the time for uniformly accelerated motion. Note that displacement is linear with time squared.

Consider a cart that is placed on an air track that is raised at one end to form an inclined plane whose angle of inclination is θ. The cart moves down the inclined plane with an acceleration a that is simply a component of the acceleration due to gravity g. The acceleration due to gravity points directly downward, but it can be resolved into components that are perpendicular and parallel to the plane. Figure 4.2 shows a cart on an air track.

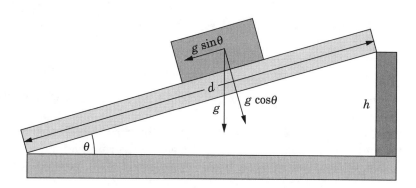

Figure 4.2 Components of g, the acceleration due to gravity, on an air track.

The component perpendicular to the plane is seen to be $g \cos\theta$. The component parallel to the plane is equal to the acceleration a and is given by

$$a = g \sin\theta \qquad \text{(5)}$$

EXPERIMENTAL PROCEDURE

1. Place a block approximately 10 cm in height under the support at one end of the air track. Measure the height h of the block. Attempt to estimate the value of h to the nearest 0.1 mm. Have five members of the class perform this measurement independently and record those five trials in the Data Table.

2. Have five members of the class independently measure the distance d between the points of support of the air track as shown in Figure 4.2. Record the five trials for d in the Data Table.

3. Have one member of the class release the cart from rest at the top of the incline and simultaneously start a timer. Stop the timer when the cart has traveled 0.250 m down the track. Have four other members of the class repeat this measurement, each one taking one trial for a total of five trials at this distance. Record all times in the Data Table.

4. Repeat step 2 for distances of 0.500, 0.750, 1.000, 1.500, 2.000, 3.000, and 4.000 m. At each distance have five different members of the class perform a measurement. Record all times in the Data Table.

CALCULATIONS

1. Calculate the mean $\bar{h}$ and the standard error α_h for the five trials of the block height. Record the results in the Calculations Table.

2. Calculate the mean $\bar{d}$ and the standard error α_d for the five trials of the distance d. Record the results in the Calculations Table.

3. Calculate the mean time $\bar{t}$ and the standard error α_t for the five trials of the time at each distance. Record the results in the Calculations Table.

4. Calculate the square of each of the mean times $(\bar{t})^2$ for the eight distances and record the results in the Calculations Table.

5. Perform a linear least squares fit to the data of x versus $(\bar{t})^2$ with $(\bar{t})^2$ as the abscissa and x as the ordinate. In that fit make (0,0) one of the points that is included. According to equation 4, the slope of this fit should be equal to $a/2$, where a is the cart's acceleration. From the value of the slope obtained in the least squares fit, calculate the acceleration a of the cart and record it in the Calculations Table.

6. Calculate the value of $\sin\theta = h/d$ and record it in the Calculations Table.

7. Using equation 5, calculate an experimental value for the acceleration due to gravity g_{exp} from the measured values of a and $\sin\theta$ and record it in the Calculations Table.

8. Calculate the percentage error in your value of g_{exp} compared to the accepted value of 9.80 m/s^2 and record it in the Calculations Table.

GRAPHS

1. According to equation 4, a graph of x versus $(\bar{t})^2$ should be a straight line. Construct a graph of the x versus $(\bar{t})^2$ data with x as the ordinate and $(\bar{t})^2$ as the abscissa. Also show on the graph the straight line from the linear least squares fit.

Laboratory 4

Uniformly Accelerated Motion on the Air Track

LABORATORY REPORT

Data Table

	Trial 1	Trial 2	Trial 3	Trial 4	Trial 5
x (m)					
d (m)					

x (m)	t_1 (s)	t_2 (s)	t_3 (s)	t_4 (s)	t_5 (s)

Calculations Table

$\overline{h} =$ _____ $\alpha_h =$ _____ $\overline{d} =$ _____ $\alpha_d =$ _____ $\sin\theta = \overline{h}/\overline{d} =$ _____

x (m)								
$\overline{t}$ (s)								
α_t (s)								
t^2 (s^2)								

Slope = _____ $a =$ _____ $g_{exp} =$ _____ % error = _____

SAMPLE CALCULATIONS

QUESTIONS

1. According to the standard error, how many significant figures are there in the values of $\bar{h}$ and $\bar{d}$?

2. According to the standard error, how many significant figures are there in each of the values of $\bar{t}$?

3. Consider the difference between your measured value of g and the true value of 9.80 m/s^2. Would friction produce an error in the direction of the error you observed for your data? In other words, could friction be the cause of the observed difference?

4. Assuming there was a systematic error caused by a constant reaction time in pressing the stopwatch, would it cause an error in the direction of the observed difference between the experimental value and the true value of g?

5. Calculate the correlation coefficient r for the least squares fit. Using the correlation coefficients in Appendix I, state the probability that the level of correlation is accidental.

6. What was the instantaneous velocity v of the cart at $x = 4.00$ m, assuming your value of the acceleration a is correct?

7. State your opinion of how well the objectives of this laboratory were met. Do your data show the expected results? State clearly the basis for your answer.

Laboratory 5

Uniformly Accelerated Motion on the Air Table

PRELABORATORY ASSIGNMENT

Read carefully the entire description of the laboratory and answer the following questions based on the material contained in the reading assignment. Turn in the completed prelaboratory assignment at the beginning of the laboratory period prior to the performance of the laboratory.

1. The displacement is (a) always a vector, (b) a vector only if an object is at rest, (c) a vector only if an object is in motion, (d) always a scaler, and (e) a scaler only if an object is in motion.

2. For one-dimensional motion, the instantaneous velocity is always defined as $\Delta x/\Delta t$. (a) true (b) false

3. The velocity of an object is proportional to elapsed time: (a) always, (b) only for positive acceleration, (c) only for negative acceleration, or (d) only for constant acceleration?

4. Suppose that an ideal frictionless inclined plane has an angle of inclination of $5.00°$ with respect to the horizontal. What is the acceleration of an object sliding down that plane? Assume that the acceleration due to gravity is $g = 9.80 \text{ m/s}^2$. Show work.

5 to 7. The following data for the instantaneous velocity v versus the time t was obtained. Perform a linear least squares fit to the data (preferably using a hand-held calculator with automatic routines) and determine the intercept, slope, and regression coefficient of the straight line. (Refer to the discussion of linear least squares fit in the "General Laboratory Information" section. Note that for this case v corresponds to y, t corresponds to x, a corresponds to m, and v_0 corresponds to b.)

v (m/s)	0.352	0.496	0.655	0.808	0.939	1.073
t (s)	0.200	0.400	0.600	0.800	1.000	1.200

5. For the above least squares fit, what is the intercept v_0? v_0 = _____ m/sec

6. For the above least squares fit, what is the slope a? a = _____ m/s^2

7. For the above least squares fit, what is the correlation coefficient r? r = _____

8. Use the table in Appendix I entitled "Correlation Coefficients" to state the significance of this fit to the data.

9. Assume that the data of question 5 is from an inclined plane of angle 4.40°. Use equation 5 in the laboratory description to determine an experimental value for g, substituting the value of the acceleration a determined in question 5. Calculate the percentage error between this experimental value g_{exp} and the true value of 9.80 m/s^2.

$g_{exp} =$ _____ m/s^2 Percentage error = _____ %

Uniformly Accelerated Motion on the Air Table

OBJECTIVES

In this laboratory an air table will be used to approximate a frictionless surface. The table will be tilted to form an inclined plane, and the study of the one-dimensional motion of a uniformly accelerated puck will be used to accomplish the following objectives:

1. Definition of a coordinate system to describe the location of the puck
2. Determination of the average velocity of the puck during various time intervals from measured displacements and time intervals
3. Examination of the relationship between the instantaneous velocity of the puck and the average velocity of the puck for the special case of constant acceleration
4. Demonstration of the linear increase in the instantaneous velocity with time for the special case of constant acceleration
5. Determination of the acceleration and initial velocity of the puck by a least squares fit to the velocity versus time data
6. Determination of an experimental value for g the acceleration due to gravity by interpretation of the puck's acceleration as a component of g

EQUIPMENT LIST

1. Air table with sparktimer, airpump, and foot switch
2. Level
3. Recording paper and carbon paper (to fit air table)
4. Meter stick or measuring tape
5. Block to insert under air table leg to form inclined plane
6. Balance and calibrated masses

THEORY

In order to study the motion of an object that moves in a straight line, it is helpful to define a coordinate axis along the direction of motion. An origin is chosen, and motion in one direction (usually to the right) is chosen to be positive. The position is specified by a number with a sign. The number represents the distance of the object from the origin of the coordinate system, and the sign describes on which side of the origin the object is located. For example, as shown in Figure 5.1, the coordinate for object **A** would be $+3$, indicating its position 3 units to the right of the origin, and the coordinate for object **B** would be -2, indicating its position 2 units to the left of the origin.

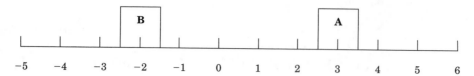

Figure 5.1 Coordinate axis used to locate the position of objects A and B.

As an object moves, the value of its coordinate changes with time. Consider an object whose initial coordinate is x_i at some time t_i, and whose final coordinate is x_f at some later time t_f. The displacements of the object from the origin to each of those coordinate positions are vectors from the origin to the point. Let $\mathbf{x}_i$ and $\mathbf{x}_f$ stand for the displacement vectors from the origin to the coordinate points. The displacement between those two coordinates $\Delta\mathbf{x}$ is defined by

$$\Delta\mathbf{x} = \mathbf{x}_f - \mathbf{x}_i \tag{1}$$

Note that since $\Delta\mathbf{x}$ is the difference between two vectors it is also a vector. Its sign determines its direction. For this one-dimensional case if $|\mathbf{x}_f > \mathbf{x}_i|$ the displacement is positive and thus to the right.

The average velocity $\overline{\mathbf{v}}$ is also a vector quantity and is equal to the time rate of change of the displacement vector $\Delta\mathbf{x}$. It is defined by the following equation:

$$\overline{\mathbf{v}} = \frac{\Delta\mathbf{x}}{\Delta t} = \frac{\mathbf{x}_f - \mathbf{x}_i}{t_f - t_i} \tag{2}$$

Note that the average velocity has the same sign as the displacement because $t_f > t_i$ always. Thus, the direction of the average velocity is the same as the direction of the displacement. The average velocity is defined over some time interval Δt that is finite. The instantaneous velocity, which is the velocity at some particular instant of time, is simply the limit of the average velocity as the time interval Δt approaches zero.

The average acceleration $\overline{\mathbf{a}}$ is also a vector, and it is equal to the time rate of change of the instantaneous velocity. It is defined by the equation

$$\overline{\mathbf{a}} = \frac{\Delta\mathbf{v}}{\Delta t} = \frac{\mathbf{v}_f - \mathbf{v}_i}{t_f - t_i} \tag{3}$$

where $\mathbf{v}_f$ and $\mathbf{v}_i$ are the final instantaneous velocity and initial instantaneous velocity at the times t_f and t_i. The acceleration is positive if the velocity is increasing with time and is negative if the velocity is decreasing with time.

This can be written in a simpler form for the special case in which the acceleration is constant. Assume that $a = |a|$ is constant and also let the initial time $t_i = 0$. Call the final time t_f simply t, which stands for any later time after $t = 0$. For these assumptions equation 3 becomes

$$\overline{a} = \frac{v - v_0}{t} \quad \text{or} \quad v = v_0 + at \tag{4}$$

This equation states that for the special case of constant acceleration, the velocity increases linearly with time from its initial value of v_0 at $t = 0$ to later values of v at any later time t.

It is not difficult to measure the average velocity of some object for almost any kind of motion. It is merely necessary to measure the displacement during a given time interval and divide the displacement by the time interval. In general, however, it is difficult to relate the average velocity to the instantaneous velocity. For the special case of constant acceleration described above, there is a simple relationship between the average velocity and the instantaneous velocity. Because the velocity increases linearly with time for the case of constant acceleration, the average velocity during each time interval is equal to the instantaneous velocity at the middle of the time interval. For example, assume that the coordinate position of some object is measured at fixed time intervals of 0.100 s. The average velocity during the interval from $t = 0$ to $t = 0.100$ s would equal the instantaneous velocity at $t = 0.050$ s (the middle of the time interval between $t = 0$ and $t = 0.100$ s). Similarly, the average velocity during any arbitrarily chosen interval (say the interval between $t = 0.700$ s and $t = 0.800$ s) would be the same as the instantaneous velocity at the middle of that interval (in this case at $t = 0.750$ s).

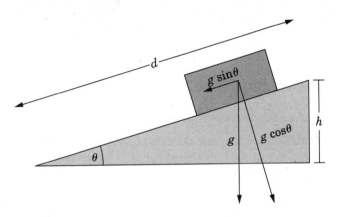

Figure 5.2 Components of g the acceleration due to gravity.

Thus, for the special case of constant acceleration, a determination of the average velocity as a function of time leads to a knowledge of the instantaneous velocity as a function of time. Note, however, that care must be taken to determine the average velocity over a specific time interval, and then determine from it the instantaneous velocity at the time corresponding to the middle of the time interval. When the instantaneous velocity is known as a function of time, that data can be fit to equation 4 by a linear least squares fit to obtain the initial velocity v_0 and the acceleration a.

The acceleration of an object down an inclined plane is a component of the acceleration due to gravity g as shown in Figure 5.2. As seen from the figure, the component of g pointing parallel to the plane causing the acceleration a is given by

$$a = g \sin \theta \tag{5}$$

If θ is known and a is experimentally determined, equation 5 can be solved to provide an experimental value for g. Conversely if θ and g are assumed known, the expected value for the acceleration a can be determined from equation 5.

EXPERIMENTAL PROCEDURE

1. Read the instruction manual for the air table (Figure 5.3) or obtain instructions from your laboratory instructor about the operation of the air table. In particular become familiar with the operation of the airpump, the sparktimer, and the foot switch. Be careful not to touch the conducting portions of the puck when the sparktimer is in operation.

2. Following either the air table instructions manual or instructions from your laboratory instructor level the air table.

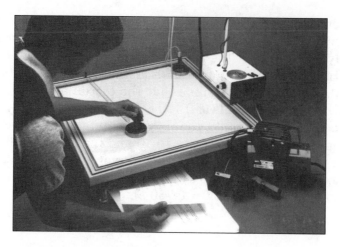

Figure 5.3 Precision Air Table. (Photo courtesy of Central Scientific Co.)

3. Place a small block under the air table single leg, which will make the table act as an inclined plane. The angle of the inclined plane will be determined by the distance between the air table legs d and the height h of the block placed under one leg. Choose h to produce an angle of about 5°. Measure h and d to the nearest 0.0001 m and record them in the Data Table.

4. Set the sparktimer to a spark rate of 10.0 Hz. This will produce a spark every 0.100 s; and thus, the Δt between data points will be 0.100 s for all the data. Record the value of Δt in the Data Table.

5. Place a piece of carbon paper on the air table. Place a sheet of recording paper on the table and place the two pucks on the table. In order to complete the circuit both pucks must be attached to the sparktimer, but one puck can be left at rest at the bottom of the inclined plane. Simultaneously, start the sparktimer and release the other puck from rest at the top of the inclined plane. Allow the puck to accelerate down the incline while the sparktimer is in operation, but stop the sparktimer as the puck hits the bottom rail. Do not let the sparktimer run when the puck rebounds from the rail because the track it makes on the rebound will overlap the spark track it made on the way down. Try to release the puck with no initial velocity but be especially careful not to impart any sideways motion to the puck when releasing it.

6. Once a good spark record has been obtained, choose a point near the beginning of the track as the origin and label it as $x = 0$. Call the corresponding time $t = 0$. Measure to the nearest 0.0001 m the displacement from the origin to each data point. Record in the Data Table the coordinate x and the time t for about 12 data points. Note very carefully that for each data point the displacement is to be measured from the origin to that point.

7. Determine the mass (kg) of the puck and record it in the Data Table.

CALCULATIONS

1. Calculate the magnitude of the displacement Δx for each successive set of sparks by taking the differences between the coordinate values for successive data points. Record the values in the Calculations Table.

2. Using equation 2, calculate the average velocity during each time interval from the values of Δx and the value of Δt (which is the same for all intervals and is

$\Delta t = 0.100$ s). Remember that this average velocity for a given interval is equal to the instantaneous velocity at the middle of that interval. Record the instantaneous velocity and the time at which that velocity occurs in the Calculations Table for each of the calculated velocities.

3. Perform a linear least squares fit to the data of instantaneous velocity v versus time t with v as the ordinate and t as the abscissa. The slope obtained by that fit is equal to the acceleration a, and the intercept obtained is equal to the initial velocity v_0. Record the values of the acceleration, initial velocity, and the regression coefficient r in the Calculations Table. (It is actually helpful to construct the graph described below prior to doing this calculation in order to help find any errors of measurement.)

4. Calculate the angle of the incline θ from the measured values of h and d using the equation $\sin \theta = h/d$. Record the value of θ in the Calculations Table.

5. Using the value of the acceleration a determined by the least squares fit and the measured value of θ, solve equation 5 for g to produce an experimental value for g. Record this value of g as g_{exp} in the Calculations Table. Calculate the percent error for g_{exp} compared to the true value of $g = 9.80$ m/s^2 and record it in the Calculations Table.

GRAPHS

Make a graph of the data for instantaneous velocity versus time with velocity as the ordinate and time as the abscissa. Show the data as points. Also draw on the graph the straight line that was obtained by the least squares fit procedure.

Laboratory 5

Uniformly Accelerated Motion on the Air Table

LABORATORY REPORT

Data Table			Calculations Table		

Data Table

Point	t (s)	x (m)
0		
1		
2		
3		
4		
5		
6		
7		
8		
9		
10		
11		
12		

Calculations Table

Δx (m)	v (m/s)	t (s)

$h =$ _____ m

$d =$ _____ m

$\Delta t =$ _____ s

Mass = _____ kg

$a =$ _____ m/s^2

$v_0 =$ _____ m/s

$\theta =$ _____ degrees

$g_{exp} =$ _____ m/s^2

% Error = _____

QUESTIONS

1. Suppose a different point had been chosen as the origin for the analysis of the data. Would this change significantly the value of a, the acceleration?

2. Would choosing a different point as the origin for the analysis change significantly the value of v_0, the initial velocity?

3. How long before the time you chose as zero for your analysis was the puck actually released? (Hint: Consider the time at which $v = 0$ in equation 4.)

4. Would friction tend to produce an error in the measured value of g that is larger than the true value or smaller than the true value? Is your error in the same direction that friction would produce?

5. Regardless of whether or not your error is in the right direction, calculate the magnitude of the force (N) that would account for the observed difference between your value of g and the true value.

6. Using the table of correlation coefficients in Appendix I state the significance of your value for r, the correlation coefficient. In other words, what is the probability that this level of correlation is accidental?

7. If friction is assumed negligible, and the difference between your measured value g_{exp} and the true value of g is assumed to be caused completely by an error in the frequency of the sparktimer, what is the direction in the error of the sparktimer frequency that would account for the error in your data? In other words, would the frequency that accounts for your error be less than 10 Hz or greater than 10 Hz? Explain your answer carefully.

8. State in your own words whether or not the objectives of this experiment were met. Do your data show the expected results? Can you identify the sources of error in the data?

Laboratory 6

Kinematics in Two Dimensions on the Air Table

PRELABORATORY ASSIGNMENT

Read carefully the entire description of the laboratory and answer the following questions based on the material contained in the reading assignment. Turn in the completed prelaboratory assignment at the beginning of the laboratory period prior to the performance of the laboratory.

A particle is launched with a velocity of 25.0 m/s in a direction $\theta = 35.0°$ above the x-axis as shown in Figure 6.1. There is no acceleration of the particle in the x direction, but there is a downward acceleration in the negative y direction of magnitude 5.00 m/s². Answer questions 1 through 5 about that motion.

1. What is the initial value of v_x at $t = 0$?

2. What is the initial value of v_y at $t = 0$?

3. What is the value of v_x at $t = 2.00$ s?

4. What is the value of v_y at $t = 2.00$ s? (Note that only two significant figures are allowed.)

5. What is the magnitude of the velocity of the particle at $t = 2.00$ s?

(Note that because square root is taken, three significant figures are allowed.)

A particle moves in such a way that its coordinate positions x and y as a function of time are given by the following table.

x (m)	0.00	0.550	1.100	1.650	2.200	2.750	3.300	3.850	4.400
y (m)	0.00	1.425	2.700	3.825	4.800	5.625	6.300	6.825	7.200
t (s)	0.00	0.100	0.200	0.300	0.400	0.500	0.600	0.700	0.800

Answer questions 6 through 8 with respect to these data.

6. What is the particle's average velocity in the y direction, $\overline{v_y}$, between $t = 0.200$ s and $t = 0.300$ s? What is $\overline{v_y}$ between $t = 0.400$ s and $t = 0.500$ s? What is $\overline{v_y}$ between $t = 0.700$ s and $t = 0.800$ s?

7. What is the particle's average velocity in the x direction, $\overline{v_x}$, between $t = 0.200$ s and $t = 0.300$ s? What is $\overline{v_x}$ between $t = 0.400$ s and $t = 0.500$ s? What is $\overline{v_x}$ between $t = 0.700$ s and $t = 0.800$ s?

8. What is the particle's instantaneous velocity in the y direction, v_y, at $t = 0.250$ s? What is v_y at $t = 0.450$ s? What is v_y at $t = 0.750$ s? (*Hint:* Assume that the average velocity during a time interval is equal to the instantaneous velocity at the center of the time interval.)

OBJECTIVES

In this laboratory, an air table will be used to approximate a frictionless two-dimensional surface. The table will be tilted to form an inclined plane to produce an acceleration in the negative y direction of a coordinate system as shown in Figure 6.1. A puck will be launched with an initial velocity having both x and y components. Study of the motion of the puck will be used to accomplish the following objectives:

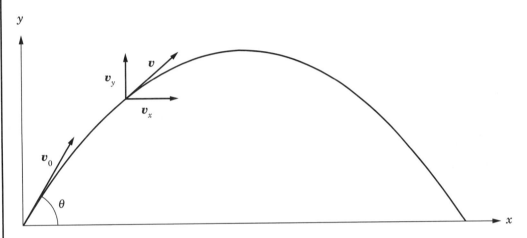

Figure 6.1 Two dimensional motion with acceleration in the negative y direction.

1. Demonstration that the puck follows a projectile-like trajectory
2. Demonstration that the puck is accelerated in the negative y direction
3. Determination of the acceleration in the y direction, a_y, by taking the slope of a graph of the y component of velocity v_y versus time
4. Demonstration that the x component of velocity v_x is very nearly constant in time
5. Determination of the very slight negative acceleration in the x direction, a_x, due to friction by determination of the very slight negative slope of v_x versus time

EQUIPMENT LIST

1. Air table with sparktimer, airpump, foot switch, and level (Figure 6.2)
2. Carbon paper and recording paper
3. Ruler, square, and protractor
4. Block of height about 5 cm to tilt air table

THEORY

Consider an object that is launched at an angle θ relative to the horizontal (x-axis) and moves in a two-dimensional path as shown in Figure 6.1. There is no acceleration in the x direction, but there is a downward acceleration in the negative y direction. Since they are at 90.0° to each other these motions are independent. If v_0 is the magnitude of the velocity with which the object is launched, then its initial velocity in the x direction is $v_0 \cos\theta$, and its initial velocity in the y direction is $v_0 \sin\theta$. There is no acceleration in the x direction, and thus the x component of velocity is constant in time. The y component of acceleration is opposite in direction to the initial component of velocity in the y direction. The two equations that describe the x and y components of velocity v_x and v_y as a function of time are

$$v_x = v_0 \cos\theta \tag{1}$$

$$v_y = v_0 \sin\theta - at \tag{2}$$

Note in equation 2 that the direction of the acceleration has been shown by the negative sign, and a stands for the magnitude of the acceleration. Equations 1 and 2 are valid only for this very special case in which there is no acceleration in the x direction and there is constant acceleration in the y direction.

Consider the case in which the values of the x and y coordinates are known for some general two-dimensional motion. If x_i and x_f stand for the x coordinate positions at the initial time t_i and final time t_f and y_i and y_f stand for the y coordinate positions at these same times, then the displacements Δx and Δy during the time interval Δt are given by

$$\Delta x = x_f - x_i \qquad \Delta y = y_f - y_i \qquad \Delta t = t_f - t_i \tag{3}$$

By definition, the x and y components of the average velocity during a time interval Δt are given by the following equations:

$$\overline{v_x} = \frac{\Delta x}{\Delta t} \quad \text{and} \quad \overline{v_y} = \frac{\Delta y}{\Delta t} \tag{4}$$

Note that the velocities in equations 4 are average velocities during a finite time interval Δt, and these are not the same velocities in equations 1 and 2, which are instantaneous velocities at some particular time t. By definition the components of the instantaneous velocities are the limits of equations 4 as Δt approaches zero. For the special case of *constant acceleration,* there is a relationship between the average velocity during some time interval Δt and the instantaneous velocity at a particular time during the time interval. For the case of *constant acceleration,* the average velocity during the time interval Δt is equal to the instantaneous velocity at a time corresponding to the instant of time at the center of the interval Δt.

For the other special case of *constant velocity,* the average velocity over the interval Δt is equal to the instantaneous velocity at any instant of time throughout the time interval including the time at the center of the interval. Therefore, for the case being considered (constant velocity in the x direction and constant acceleration in the y direction) the instantaneous velocities are given by

$$v_x = \frac{\Delta x}{\Delta t} \quad \text{and} \quad v_y = \frac{\Delta y}{\Delta t} \tag{5}$$

Here v_x and v_y stand for the instantaneous velocities at the instant of time at the center of the interval of time Δt.

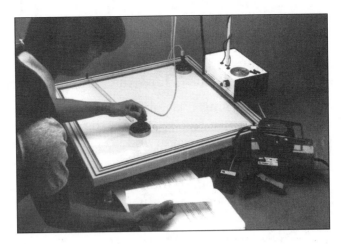

Figure 6.2 Precision Air Table. (Photo Courtesy of Central Scientific Co., Inc.)

EXPERIMENTAL PROCEDURE

1. Read the instruction manual for the air table to become familiar with its operation, including the airpump, sparktimer, and foot switch. Be very careful not to touch the pucks when the sparktimer is in operation, except by the insulated tubes.

2. Level the table by means of the three adjustable legs until a puck placed near the center of the table is essentially motionless.

3. Place a block, with a height of about 5 cm, under the single leg to tilt the table. Place a sheet of carbon paper on the table. Place a sheet of recording paper on top of the carbon paper.

4. Set a spark rate of 10.0 Hz. This will produce a spark every 0.100 s, and thus the Δt between data points will be 0.100 s for all the data. Record the value of Δt in the Data Table.

5. Place the two pucks on the air table. Place one of them at the lower right-hand corner of the table and leave it there for all of the procedure. Launch the other puck by hand from a point near the lower left-hand corner of the table. The puck should move upward and to the right and approach near the top center of the tilted table and then move down to the vicinity of the lower right-hand corner. Several practice launches will probably be needed in order to determine the correct launch speed and direction needed to achieve the desired trajectory.

6. When enough practice has been gained to ensure that the desired path can be achieved, launch the puck and start the sparktimer simultaneously. Stop the sparktimer just before the puck arrives at the lower right-hand corner. Be sure that there are about 20 or so good data points on the record. <u>Do not remove the sheet of recording paper after this measurement has been made! Instead, record a second trace on this same paper in order to experimentally determine the vertical (y direction)</u>. This vertical trace is obtained by carefully releasing the puck from a point near the top of the table on the same side of the air table from which the puck is launched.

7. Remove the recording sheet for data analysis. Choose as the origin the lowest point on the trajectory that comes out clearly and call the corresponding time zero. Label that first point as 0 and then label the next 15 consecutive data points. Note that the times corresponding to these points are already given in the Data Table.

8. Draw a straight line through the vertical line formed by the sparktimer trace for the puck falling vertically. This defines the vertical direction.

9. Construct a straight line passing through the origin chosen in step 7 and parallel to the vertical line drawn in step 8. This is the y-axis of the coordinate system. Construct a line through the origin and perpendicular to the y-axis. This is the x-axis of the coordinate system.

10. Measure the perpendicular distance of each data point from the y-axis. Record these values in the Data Table as the values of the coordinate x.

11. Measure the perpendicular distance of each data point from the x-axis. Record these values in the Data Table as the values of the coordinate y.

CALCULATIONS

1. Calculate the differences between the successive values of x and y and record them in the Calculations Table as the values of Δx and Δy. Note that Δy will change in magnitude, and that it will become negative after the puck reaches its maximum height and starts back down.

2. Using equations 5, calculate the components of the instantaneous velocity v_x and v_y and record them in the Calculations Table. Remember that all the values of Δt are equal to 0.100 s for all the intervals. Also note that the time at which the instantaneous velocities apply are at the center of each interval, and those times have already been recorded in the Calculations Table for each interval.

3. Perform a linear least squares fit to the v_x versus time data with v_x as the ordinate and time as the abscissa. The slope of this fit should be small and negative and is equal to the acceleration caused by friction, and the intercept is the initial velocity in the x direction $(v_x)_0$. Record the values of the acceleration, initial velocity, and the correlation coefficient r in the Calculations Table.

4. Perform a linear least squares fit to the v_y versus time data with v_y as the ordinate and time as the abscissa. The slope of this fit is equal to the acceleration in the y direction, and the intercept is the initial velocity in the y direction $(v_y)_0$. Record the values of the acceleration, initial velocity, and the correlation coefficient r in the Calculations Table.

GRAPHS

1. On one sheet of graph paper, plot both v_x and v_y as a function of time. Use different symbols for v_x and v_y. Since v_y can have both positive and negative values, choose the origin of the velocity scale in the center of the graph paper. Also include on the graph the straight lines obtained by the least squares fit for each case.

Laboratory 6

Kinematics in Two Dimensions on the Air Table

LABORATORY REPORT

Data Table

t (s)	x (m)	y (m)
0.000		
0.100		
0.200		
0.300		
0.400		
0.500		
0.600		
0.700		
0.800		
0.900		
1.000		
1.100		
1.200		
1.300		
1.400		
1.500		

$\Delta_t =$ _____

Calculations Table

Δx (m)	Δy (m)	v_x (m/s)	v_y (m/s)	t (s)
				0.050
				0.150
				0.250
				0.350
				0.450
				0.550
				0.650
				0.750
				0.850
				0.950
				1.050
				1.150
				1.250
				1.350
				1.450

$a_x =$ _____ $a_y =$ _____

$(v_x)_0 =$ _____ $(v_y)_0 =$ _____

$r =$ _____ $r =$ _____

QUESTIONS

1. Suppose a different point had been chosen as the origin for the analysis of the data. In principle, would this change the value of the acceleration a_y?

2. Would choosing a different point as the origin for the analysis change significantly the value of $(v_y)_0$, the initial velocity?

3. What is the magnitude of the initial velocity of the puck?

4. What is the ratio of the magnitude of a_y to the magnitude of a_x?

5. Assuming that a_x is due to friction, state what the ratio calculated in question 5 indicates about the importance of friction in this laboratory.

6. Friction should give a very small negative acceleration in the x direction. However, friction should be so small that the actual values of v_x should vary very little and thus be almost constant. Calculate the mean and the standard deviation of the values of the velocity v_x. Express the standard deviation as a percentage of the mean.

7. Based on your answers to question 6, is the expectation that v_x is approximately constant valid for your data?

8. Using the values of the initial velocity $(v_y)_0$ and the acceleration a_y, calculate how long it should take for the puck to get to the highest point in its motion. Compare that time with the time at which the maximum value of y is recorded in the data.

PRELABORATORY ASSIGNMENT

Read carefully the entire description of the laboratory and answer the following questions based on the material contained in the reading assignment. Turn in the completed prelaboratory assignment at the beginning of the laboratory period prior to the performance of the laboratory.

1. The direction of a frictional force is always in a direction (a) opposite to, or (b) in the same direction as, the motion or the tendency toward motion of an object.

2. State what two kinds of coefficients of friction exist. Describe the conditions under which each kind is appropriate. State the relationship that is generally true that exists between their magnitudes.

3. Suppose a block of mass 25.0 kg rests on a horizontal plane, and the coefficient of static friction between the surfaces is 0.220. (a) What is the maximum possible static frictional force that could act on the block? _____ N (b) What is the actual static frictional force that acts on the block if an external force of 25.0 N acts horizontally on the block? _____ N. Show your work.

4. In the experiment to measure the coefficient of kinetic friction by sliding a block down an inclined plane, it is required that the block move at constant velocity. Why is this requirement necessary?

5. Frictional forces on a block on an inclined plane are always directed down the plane. (a) true (b) false

6. A 5.00-kg block rests on a horizontal plane. A force of 10.0 N applied horizontally causes the block to move horizontally at constant velocity. What is the coefficient of kinetic friction between the block and the plane? Assume $g = 9.80$ m/s². Show your work.

7. What are the units of the coefficients of friction?

8. How does the force of friction depend on the apparent area of the two surfaces in contact?

9. How does the coefficient of kinetic friction depend on the speed of a moving object?

Coefficient of Friction

OBJECTIVES

In this experiment, the nature of the frictional forces between a smooth wooden block and a smooth flat wooden board will be investigated as an example of the general properties of friction. The board will be used in the horizontal position and as an inclined plane to accomplish the following objectives:

1. Determination of μ_s, the static coefficient of friction, and μ_k, the kinetic coefficient of friction, by sliding the block down the board, which acts as an inclined plane
2. Determination of μ_s and μ_k using the board, with a pulley mounted on it, placing the board in the horizontal position, and applying known forces to the block
3. Comparison of the two different values for μ_s and μ_k obtained by the two different techniques
4. Demonstration that the coefficients of friction are independent of the normal force
5. Demonstration that $\mu_s > \mu_k$

EQUIPMENT LIST

1. Smooth wooden board with pulley attached to one end
2. Smooth wooden block (6-inch-long 2 × 4 for example) with hook attached
3. String
4. Balance and calibrated masses
5. Mass holder and slotted masses
6. Protractor

THEORY

Friction is a resisting force that acts along the tangent to two surfaces in contact when one body slides or attempts to slide across another. The direction of the frictional force on each body is to oppose that body's motion. It is an experimental observation that frictional forces depend on the nature of the materials in contact, including their composition and roughness, and the normal force between the surfaces. Normal force is the force that each body exerts on the other body, and it acts 90° to each surface. The frictional force is directly proportional to the normal force. To a good approximation, the frictional force seems to be independent of the apparent area of contact of the two surfaces.

There are two different kinds of friction. Static friction occurs when two surfaces are still at rest with respect to each other, but an attempt is being made to cause one of them to slide over the other. Static friction arises to oppose any force trying to cause motion tangent to the surfaces. It increases in response to such applied forces up to some maximum value that is determined by a constant characteristic of the two surfaces. This is called the "coefficient of static friction," or μ_s. The frictional force f_s for static friction is given by

$$f_s \leq \mu_s N \qquad (1)$$

where N stands for the normal force between the two surfaces. The meaning of equation 1 is that the static frictional force varies in response to applied forces from zero up to a maximum value given by the equality in that equation. If the applied force is less than the maximum given by equation 1, then the frictional force that arises is simply equal to the applied force, and there is no motion. If the applied force is greater than the maximum given by equation 1, the object will begin to move, and static friction conditions are no longer valid.

The other kind of friction occurs when two surfaces are actually moving with respect to each other. It is called "kinetic friction," and it is characterized by a constant called the "coefficient of kinetic friction," or μ_k. The kinetic frictional force f_k is given by

$$f_k = \mu_k N \qquad (2)$$

where N is again the normal force. Equation 2 states that the kinetic frictional force is a constant value any time the object is in motion. Thus, the kinetic frictional force is independent of the speed of motion.

Note that both equations 1 and 2 refer only to the magnitudes of the forces involved because the frictional forces are at 90.0° to the normal force in every case. It is also true that in general $\mu_s > \mu_k$. This means that when enough force is exerted to overcome static frictional forces, that same force is sufficient to accelerate an object because once it is moving, the kinetic frictional force is less than the applied force.

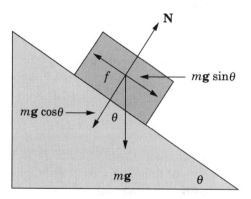

Figure 7.1 Forces acting on a block on an inclined plane.

Consider an inclined plane as shown in Figure 7.1 whose angle θ can be adjusted. If a block is placed on the plane, and the angle is slowly increased, a point will be reached at which the block begins to slip. An examination of the forces on the block shows that the normal force N acts perpendicularly to the plane, and a component of the weight of the block $mg \cos\theta$ acts in the opposite direction. Since the block is in equilibrium as far as motion perpendicular to the plane, these forces are equal and thus

$$N = mg \cos\theta_s \qquad (3)$$

where θ_s is the angle at which the block just begins to slip on the inclined plane. Parallel to the plane there are also two forces. A component of the weight of the block $mg \sin\theta_s$ acts down the plane, and the frictional force f_s acts up the plane. At the point at which the block just slips, the maximum frictional force is exerted, and these two forces must be equal. In equation form this is

$$f_s = \mu_s N = mg \sin\theta_s \qquad (4)$$

Combining equations 3 and 4 leads to:

$$\mu_s = \frac{mg \sin\theta_s}{N} = \frac{mg \sin\theta_s}{mg \cos\theta_s} = \tan\theta_s \qquad (5)$$

Equation 5 can be used to determine μ_s by measuring the angle θ_s at which the block just begins to slip on the inclined plane.

In a similar way, the same inclined plane can be used to determine μ_k. By giving the block a slight push to get it started, an angle θ_k can be determined at which the block slides down the plane at constant velocity. It is necessary to give it a push to get it started because $\mu_s > \mu_k$, and thus the static frictional force is greater than the kinetic frictional force. When the block is moving down the plane at constant velocity, the vector sum of forces on the block is zero. In other words, the frictional force acting up the plane must equal the component of the weight of the block acting down the plane, and the normal force is equal to the component of the weight acting perpendicular to the plane. In equation form this is

$$N = mg \cos\theta_k \quad \text{and} \quad f_k = mg \sin\theta_k = \mu_k N \qquad (6)$$

Performing some algebra and combining these equations leads to the following:

$$\mu_k = \frac{mg \sin\theta_k}{N} = \frac{mg \sin\theta_k}{mg \cos\theta_k} = \tan\theta_k \qquad (7)$$

Equation 7 can be used to determine μ_k by measuring the angle θ_k at which the block slides down the inclined plane at constant velocity after it has been given a slight push to get it started.

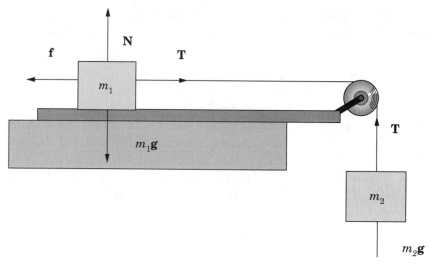

Figure 7.2 Force applied to a block on a horizontal plane.

If the inclined board is lowered to the horizontal position, a force can be applied to the block by means of a string running over a pulley mounted to the board and down to a mass as shown in Figure 7.2. For a given block mass m_1 it is possible to slowly add mass to m_2 until m_1 moves. The point at which the block just moves is the last point at which the system is in equilibrium, and it occurs when the maximum static friction is acting. As shown in Figure 7.2, at that point the following conditions are met:

$$T = f_s \quad T = m_2 g \quad N = m_1 g \quad f_s = \mu_s N \tag{8}$$

In these equations, T is the tension in the string, and the other symbols are the same as previously defined. Combining these four equations leads to

$$f_s = m_2 g = \mu_s N = \mu_s m_1 g \tag{9}$$

Using the second and fourth terms above and cancelling the common factor of g gives

$$m_2 = \mu_s m_1 \tag{10}$$

Thus, equation 10 can be used to determine μ_s by determining the minimum mass m_2 needed to cause the block of mass m_1 to just move. Note carefully that the basic relationship is given by equation 9, and it relates a force $m_2 g$ to another force $\mu_s m_1 g$, but because g is a common factor, it can be cancelled to give equation 10, relating the masses.

The same procedure can also be used to determine the coefficient of kinetic friction. Again refer to Figure 7.2 and imagine f_s is replaced by f_k for the case when the block is in motion. Note that when the system is moving at constant velocity it is also in equilibrium, and the following conditions are met:

$$T = f_k \quad T = m_2 g \quad N = m_1 g \quad f_k = \mu_k N \tag{11}$$

Combining these equations as before leads to

$$m_2 = \mu_k m_1 \qquad (12)$$

Thus, equation 12 can be used to measure μ_k by determining the value of m_2 needed to cause m_1 to move at constant velocity.

EXPERIMENTAL PROCEDURE

Part 1: Board as an Inclined Plane

1. Place the block with its large surface on the board and incline the board until the block just begins to slide on its own. Record the angle at which it slips as θ_s in the Data Table. Repeat the procedure three more times for a total of four trials. Each time, place the block in a slightly different place on the plane in order to average the effects of nonuniformity of the surfaces. Note carefully which side of the block and which side of the board is used, and continue to use the same surfaces for all the measurements made in this experiment.

2. Again place the block on the board and incline the board until the block moves down the plane at constant speed when given a slight push to start it in motion. Record the angle at which this occurs as θ_k. Repeat this process three more times for a total of four trials. Again try to use different parts of the board in order to average the effects of nonuniformity of the surface.

Part 2: Horizontal Board with Pulley

1. Determine the mass of the block using the balance. Record it in the Data Table (Static Friction) as m_1 in the space labeled 0 kg added.

2. Place the board in a horizontal position on the laboratory table with the pulley beyond the edge of the table as shown in Figure 7.2.

3. Attach a piece of string to the hook in the block. Place it over the pulley and attach the mass holder to the other end of the string. Be sure that the same surface of the block and board are in contact as were in the first procedure. Carefully add mass to the mass holder to find the minimum mass needed to just cause the block to move. Record the value of mass so determined as m_2 in the Data Table (Static Friction). Be sure to include the 0.050-kg mass of the holder in the total for m_2. Repeat the procedure two more times for a total of three trials using different parts of the board each time.

4. Repeat step 3 but add 0.200 kg to the top of the block. Record the value of the mass of the block plus 0.200 kg as m_1. Again determine the minimum mass needed to just cause the mass m_1 to move. Do three trials and record each as m_2 in the Data Table (Static Friction).

5. Continue this process, adding 0.400, 0.600, and finally 0.800 kg to the top of the block. In each case take three trials.

6. Perform a similar set of measurements as just described in steps 1 through 5, but this time determine the mass m_2 needed to keep the block moving at constant velocity after it has been started with a small push. Again take three trials for

each case and use values of m_1 beginning with the mass of the block and incrementing in steps of 0.200 kg up to a total added mass of 0.800 kg. Record the values of m_2 and m_1 for all cases in the Data Table (Kinetic Friction).

CALCULATIONS

1. For each θ_s, calculate $\tan\theta_s$ (which by equation 5 is equal to μ_s) and record the values in the Calculations Table. Calculate and record the mean $\overline{\mu_s}$ and standard error α_{μ_s} for the four measurements.

2. For each θ_k calculate $\tan\theta_k$ (which by equation 7 is equal to μ_k) and record the values in the Calculations Table. Calculate and record the mean $\overline{\mu_k}$ and standard error α_{μ_k} for the four measurements.

3. Calculate the mean $\overline{m_2}$ for the three trials of m_2 for each of the values of m_1 for both the static and kinetic friction cases. Record these values in the Calculations Table along with the value of m_1 for each value of $\overline{m_2}$.

4. According to equation 10 there is a linear relationship between m_2 and m_1. A linear least squares fit to the static friction values of $\overline{m_2}$ versus m_1 should produce a straight line whose slope is μ_s. Perform such a fit with $\overline{m_2}$ as the ordinate and m_1 as the abscissa and determine the slope and intercept. Record the value of the slope as μ_s and the value of the intercept as b_s in the Calculations Table.

5. Equation 12 states that there should also be a linear relationship between m_2 and m_1 for the kinetic friction data. Perform a linear least squares fit to the data with $\overline{m_2}$ as the ordinate and m_1 as the abscissa and determine the slope and intercept. Record the slope as μ_k and record the intercept as b_k in the Calculations Table.

6. For a least squares fit with y as the ordinate and x as the abscissa, the standard deviation of the slope m is given by

$$\sigma_{\mu\,\text{stat}} \text{ and } \sigma_{\mu\,\text{kin}} = \sqrt{\frac{\sum\limits_{1}^{N} (mx_i + b - y_i)^2}{N \sum\limits_{1}^{N} x_i^2 - (\sum\limits_{1}^{N} x_i)^2}} \tag{13}$$

For the least squares fits that have been done to determine the static and kinetic coefficients of friction, use equation 13 to calculate the standard deviation of the values of the slopes, keeping in mind that $\overline{m_2}$ corresponds to y and m_1 corresponds to x. Call these values for the standard deviations of the slope $\sigma_{\mu\,\text{stat}}$ and $\sigma_{\mu\,\text{kin}}$ and record them in the Calculations Table.

GRAPHS

1. Graph the static friction data for the horizontal plane case with $\overline{m_2}$ as the ordinate and m_1 as the abscissa. Show on the graph the straight line obtained from the linear least squares fit.

2. Graph the kinetic friction data for the horizontal plane case with $\overline{m_2}$ as the ordinate and m_1 as the abscissa. Show on the graph the straight line obtained from the linear least squares fit.

LABORATORY REPORT

Data Table

Part 1: Inclined Plane

Static Friction

Trial	1	2	3	4
θ_s				

Kinetic Friction

Trial	1	2	3	4
θ_K				

Part 2: Horizontal Board

Static Friction

Added Mass	0.000 kg	0.200 kg	0.400 kg	0.600 kg	0.800 kg
m_1					
m_2 Trial 1					
m_2 Trial 2					
m_2 Trial 3					

Kinetic Friction

Added Mass	0.000 kg	0.200 kg	0.400 kg	0.600 kg	0.800 kg
m_1					
m_2 Trial 1					
m_2 Trial 2					
m_2 Trial 3					

Calculations Table

Part 1: Inclined Plane

$\tan \theta_s = \mu_s$	$\overline{\mu_2}$	α_{μ_s}

$\tan \theta_k = \mu_k$	$\overline{\mu_k}$	α_{μ_k}

$(\mu_{\rm s} m_1 + b_{\rm s} - \overline{m}_2)^2$	m_1 (kg)	$\overline{m}_2$ (kg)	$m_1{}^2$ (kg)^2
$\sum$'s			

Static Friction

$\mu_{\rm s} = \underline{\hphantom{xxxxxxxx}}$

$b_{\rm s} = \underline{\hphantom{xxxxxxxx}}$

$\sigma_{\mu\,{\rm stat}} = \underline{\hphantom{xxxxxxxx}}$

$r = \underline{\hphantom{xxxxxxxx}}$

$(\mu_{\rm k} m_1 + b_{\rm k} - \overline{m}_2)^2$	m_1 (kg)	$\overline{m}_2$ (kg)	$m_1{}^2$ (kg)^2
$\sum$'s			

Kinetic Friction

$\mu_{\rm k} = \underline{\hphantom{xxxxxxxx}}$

$b_{\rm k} = \underline{\hphantom{xxxxxxxx}}$

$\sigma_{\mu\,{\rm kin}} = \underline{\hphantom{xxxxxxxx}}$

$r = \underline{\hphantom{xxxxxxxx}}$

SAMPLE CALCULATIONS

QUESTIONS

1. Discuss the agreement between the two different measured values of μ_s. Calculate the percentage difference between them. Percentage difference = _____ %. Do they agree within the combined uncertainties of the two measurements? Which measurement do you think is the more reliable, and why do you think so?

2. Answer the same question as asked in question 1, but now consider the kinetic friction data. Percentage difference = _____ %.

3. To what extent do your data confirm the expectation that the coefficients of friction, both static and kinetic, are independent of the normal force? State the evidence for your opinion.

4. Do your data confirm the expected relationship between μ_s and μ_k? State clearly what is expected and what your data indicate.

5. State clearly in your own words what was to be accomplished in this laboratory. To what extent did your performance of the laboratory accomplish those goals?

Laboratory 8
_____

Newton's Second Law on the Air Table

PRELABORATORY ASSIGNMENT

Read carefully the entire description of the laboratory and answer the following questions based on the material contained in the reading assignment. Turn in the completed prelaboratory assignment at the beginning of the laboratory period prior to the performance of the laboratory.

1. A net force of 25.0 N acts on a 6.50-kg object. What is the acceleration of the object?

2. See Figure 8.1. What are the two forces acting on mass m_1 that must be considered in determining the acceleration of the mass? Which forces acting on m_1 do not contribute to the acceleration?

3. What are the two forces that act on mass m_2 in Figure 8.1?

4. Suppose that a 0.450-kg puck (m_1) is attached to a 0.0400-kg mass (m_2) as shown in Figure 8.1. A constant frictional force $f = 0.100$ N acts on the puck. What is the acceleration of the system?

5. Suppose that the puck in question 4 starts from rest at $t = 0$. What is the velocity of the puck at $t = 0.500$ s?

6. What is the weight in Newtons of a mass of 0.0500 kg?

The following data for coordinate position x versus time t were taken for a puck that is uniformly accelerated.

t (s)	0.000	0.100	0.200	0.300	0.400	0.500	0.600
x (m)	0.000	0.0102	0.0399	0.0896	0.1603	0.2498	0.3607
Δx (m)							
v (m/s)							
t (s)	0.050	0.150	0.250	0.350	0.450	0.550	

7. Calculate the displacement Δx during each time interval and record each of them in the table above. Calculate the instantaneous velocity at the center of each time interval and record each of them in the table above. Show sample calculations.

8. Perform a linear least squares fit to the velocity versus time data that you have just generated in the table above. The slope of this fit is the acceleration of the puck. Record the acceleration obtained by this fit:

$a =$ _____ m/s^2

OBJECTIVES

When a net force of F acts on a mass of m an acceleration a is produced. The relationship between these quantities is given by Newton's second law as $F = ma$. In this laboratory a mass on an air table will be subjected to a variable applied force to accomplish the following objectives:

1. Determination of the acceleration produced by a series of different forces applied to a given fixed mass
2. Demonstration that the acceleration of the mass is proportional to the magnitude of the applied force
3. Demonstration that the constant of proportionality between the acceleration and the applied force is the mass to which the force is applied
4. Determination of the frictional force that acts on the system

EQUIPMENT LIST

1. Air table with pucks, sparktimer, airpump, foot switch, and level
2. Carbon paper and recording paper
3. Pulley and string
4. Laboratory balance and calibrated masses
5. Meter stick or tape

THEORY

The relationship between the *net* force F exerted on a body of mass m and the acceleration a of the body was stated by Newton to be

$$F = ma \tag{1}$$

It is important to note that equation 1 is a vector equation, and thus it states that each component of the acceleration is determined by the net force in the direction of that component.

Consider a mass m_1 placed on a horizontal surface and connected to a second mass m_2 by a string running over a pulley as shown in Figure 8.1. There will exist a tension T in the string connecting the masses. Each of the masses is subject to several forces, which are shown in a vector free-body diagram in Figure 8.2.

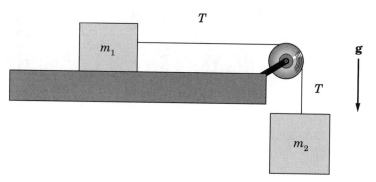

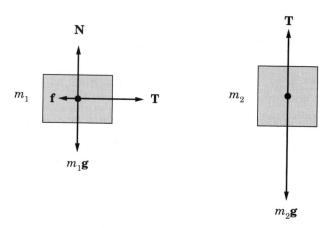

Figure 8.1 Force applied to mass m_1 by weight of mass m_2.

Figure 8.2 Free-body diagram of the forces on masses m_1 and m_2.

There are four forces acting on the mass m_1, but only two of them lie along the direction of motion of the mass. The weight of m_1 and the normal force from the table are equal in magnitude and opposite in direction and thus sum to zero. The net force acting on m_1 is therefore equal to $T - f$ and applying equation 1 gives

$$T - f = m_1 a \qquad (2)$$

The only two forces acting on mass m_2 are its weight $m_2 g$ and the tension T. Because it is tied to mass m_1 it must move with m_1, and so mass m_2 has an acceleration whose magnitude is the same as that of m_1. In equation form

$$m_2 g - T = m_2 a \qquad (3)$$

If equations 2 and 3 are combined and a term f added to both sides, the result is

$$m_2 g = (m_1 + m_2)a + f \qquad (4)$$

The derivation leading to equation 4 actually ignores two effects. It assumes that there is no friction in the pulley, and it also assumes that the pulley is massless with no moment of inertia. To a first approximation the friction in the pulley can be accounted for by assuming that f in equation 4 includes all the friction in the system. It will be assumed that the mass of the pulley is small enough that the moment of inertia of the pulley is negligible.

In this laboratory a series of measurements will be made in which the value of the applied force $m_2 g$ will be varied while the mass $m_1 + m_2$ will be kept constant. The resulting acceleration a of equation 4 will be measured as a function of the force $m_2 g$.

The actual measurements will determine the coordinate position x of the puck as a function of time t. A sparktimer will mark the x coordinate at time intervals of 0.100 s. The average velocity $\bar{v}$ during any time interval Δt is given by

$$\bar{v} = \frac{\Delta x}{\Delta t} \tag{5}$$

where Δx is the displacement during the time interval Δt. Because the acceleration is constant, this average velocity $\bar{v}$ during the time interval Δt can be related to the instantaneous velocity v. In fact, the equation

$$v = \frac{\Delta x}{\Delta t} \tag{6}$$

gives the instantaneous velocity v at the instant of time at the center of the time interval Δt. From the resulting values of v as a function of time the acceleration a will be determined.

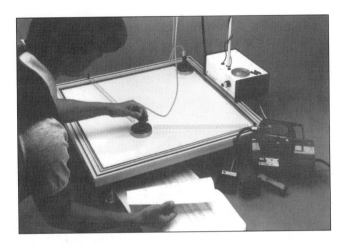

Figure 8.3 Precision Air Table. (Photo courtesy of Central Scientific Co., Inc.)

EXPERIMENTAL PROCEDURE

1. Read the instruction manual for the air table (Figure 8.3) to become familiar with its operation, including the airpump, sparktimer, and foot switch. Be very careful not to touch the pucks except by the insulated tubes when the sparktimer is in operation.

2. Level the table by means of the three adjustable legs until a puck placed near the center of the table is essentially motionless.

3. Place one of the pucks in the corner of the table and leave it there for the rest of the procedure. Place the other puck on a laboratory balance and determine its mass m_p. Record the value of m_p in kg in Data-Calculations Table 1.

4. Place the puck whose mass has been determined on the air table and attach a string to the puck. Run the string over the pulley as shown in Figure 8.1 and attach a mass of 0.0100 kg on the end of the string to serve as mass m_2.

5. Place mass totaling 0.0400 kg on top of the puck. The total mass $m_1 + m_2$ is equal to the puck mass m_p plus the 0.0400 kg plus the 0.0100 kg on the end of the string. Record this value (m_p + 0.0500 kg) in Data-Calculations Table 1 as $m_1 + m_2$.

6. Set a spark rate of 20.0 Hz. Release the puck near the edge of the table with zero velocity and simultaneously start the sparktimer. Stop the sparktimer just before the puck reaches the other side of the table. It is extremely important to release the puck with no initial velocity.

7. Remove 0.0100 kg from the puck and add it to the 0.0100 kg on the end of the string so that m_2 is now 0.0200 kg, but $m_1 + m_2$ remains the same value. Slide the recording paper over slightly, and again release the puck with no initial velocity and record this trace with the sparktimer.

8. Continue this process, each time shifting 0.0100 kg from the puck to the end of the string to produce traces with values of m_2 equal to 0.0300, 0.0400, and 0.0500 kg while the value of $m_1 + m_2$ remains fixed. Record all five traces on the same sheet of paper.

9. Choose an origin for each of the five traces and label every other point. This will produce data points with time intervals of $\Delta t = 0.100$ s. Measure the values of the coordinate position x as a function of time from $t = 0$ to $t = 0.600$ s for each of the traces. Record the values of x in Data-Calculations Table 1.

CALCULATIONS

1. Calculate the values of the displacement Δx between successive data points for each of the five traces. Record these values of Δx in Data-Calculations Table 1.

2. Using equation 6, calculate the instantaneous velocity at the times corresponding to the center of the measured time intervals. Remember that all the Δts are equal to 0.100 s. Record these values of v at the appropriate instantaneous times in Data-Calculations Table 1.

3. From the values of m_2 and g, calculate the values of the applied force $m_2 g$ for each of the applied forces. Use a value of 9.80 m/s² for the acceleration due to gravity. Record the results in Data-Calculations Table 2.

4. For constant acceleration, the velocity v should be linear with time t. Perform a linear least squares fit to each of the five sets of v versus t data with v as the ordinate and t as the abscissa. Calculate the slope of each of these fits as the acceleration a. Record these values of acceleration in Data-Calculations Table 2.

5. According to equation 4, the acceleration a should be proportional to the applied force $m_2 g$ with the total mass $m_1 + m_2$ as the constant of proportionality. Perform a linear least squares fit to the data with the applied force $m_2 g$ as the ordinate and the acceleration a as the abscissa. Calculate the slope of the fit as $(m_1 + m_2)_{exp}$, the experimental value for the total mass, and record it in Data-Calculations Table 2. The intercept of the fit according to equation 4 should be equal to the frictional force f. Calculate the intercept of the fit and record it in Data-Calculations Table 2 as f.

6. Calculate the percentage error in the value of $(m_1 + m_2)_{exp}$ compared to the known value of $m_1 + m_2$ and record it in Data-Calculations Table 2.

GRAPHS

1. On a single sheet of graph paper, plot the data for the velocity v versus time t for each of the five cases. Also show on the graph the straight line obtained by the linear least squares fit to the data. Use different symbols or in some way label the five different sets of data.

2. Make a graph of the data for the applied force $m_2 g$ versus the acceleration a. Also show on the graph the straight line obtained by the least squares fit to the data.

Laboratory 8

Newton's Second Law on the Air Table

LABORATORY REPORT

Data-Calculations

Table 1

$\Delta t = 0.100$ s $m_p =$ _____ kg $(m_1 + m_2) =$ _____ kg

t (s)		0.000	0.100	0.200	0.300	0.400	0.500	0.600
x (m)								
m_2	Δx (m)							
0.0100	v (m/s)							
kg	t (s)	0.050	0.150	0.250	0.350	0.450	0.550	

t (s)		0.000	0.100	0.200	0.300	0.400	0.500	0.600
x (m)								
m_2	Δx (m)							
0.0200	v (m/s)							
kg	t (s)	0.050	0.150	0.250	0.350	0.450	0.550	

t (s)		0.000	0.100	0.200	0.300	0.400	0.500	0.600
x (m)								
m_2	Δx (m)							
0.0300	v (m/s)							
kg	t (s)	0.050	0.150	0.250	0.350	0.450	0.550	

t (s)		0.000	0.100	0.200	0.300	0.400	0.500	0.600
x (m)								
m_2	Δx (m)							
0.0400	v (m/s)							
kg	t (s)	0.050	0.150	0.250	0.350	0.450	0.550	

t (s)		0.000	0.100	0.200	0.300	0.400	0.500	0.600
x (m)								
m_2	Δx (m)							
0.0500	v (m/s)							
kg	t (s)	0.050	0.150	0.250	0.350	0.450	0.550	

Table 2

Applied Force = $m_2 g$ (N)					
Acceleration (m/s^2)					
Initial velocity (m/s)					
Regression coefficient					

$f =$ _____ N $\qquad (m_1 + m_2)_{exp} =$ _____ kg $\qquad r =$ _____ $\qquad$ % err $=$ _____

SAMPLE CALCULATIONS

QUESTIONS

1. Do your five graphs of velocity versus time show the linear relationship between v and t that is expected for constant acceleration?

2. Does your graph of applied force versus acceleration show the linear relationship that is predicted by Newton's second law?

3. What is the ratio of the frictional force f to the applied force $m_2 g$ for each case? Is f small compared to the applied force in every case?

4. It can be shown that if the pulley is assumed to be a uniform disk, the effect of its moment of inertia would be equivalent to adding a mass equal to one-half the pulley mass to $m_1 + m_2$. Is your experimental result for $m_1 + m_2$ larger than the known value of $m_1 + m_2$? If so, what value does your result imply for the mass of the pulley? (The actual pulley mass is probably less than 10 g.)

5. Comment on the agreement between the actual value of $m_1 + m_2$ and your experimental value obtained from the slope of applied force versus acceleration.

6. Calculate the tension T in the string for each of the five cases.

PRELABORATORY ASSIGNMENT

Read carefully the entire description of the laboratory and answer the following questions based on the material contained in the reading assignment. Turn in the completed prelaboratory assignment at the beginning of the laboratory period prior to the performance of the laboratory.

1. A net force of 3.50 N acts on a 2.75-kg object. What is the acceleration of the object?

2. Describe the basic concept of the Atwood machine. What is the net applied force? What is the mass to which this net force is applied?

3. An Atwood machine consists of a 1.0600-kg mass and a 1.0000-kg mass connected by a string over a massless and frictionless pulley. What is the acceleration of the system? Assume g is 9.80 m/s². Show your work.

4. Suppose the system in question 3 has a frictional force of 0.056 N. What is the acceleration of the system? Show your work.

5. Suppose the frictionless system described in question 3 is released from rest at $t = 0$. How long does it take for the large mass to fall 1.200 m? Show your work.

6. Suppose the system with friction described in question 4 is released from rest at $t = 0$. How long does it take for the large mass to fall 1.200 m? Show your work.

The following data were taken with an Atwood machine whose total mass $m_1 + m_2$ is kept constant. For each of the values of mass difference $m_2 - m_1$ shown in the table the time for the system to move $x = 1.000$ m was determined.

$m_2 - m_1$ (kg)	0.0100	0.0200	0.0300	0.0400	0.0500
t (s)	8.30	5.06	3.97	3.37	2.98
a (m/s²)					
$(m_2 - m_1)g$ (N)					

7. From the data above for x and time t, calculate the acceleration for each of the applied forces and record them in the table above. Show a sample calculation.

8. From the mass differences $m_2 - m_1$ calculate the applied forces $(m_2 - m_1)g$ and record them in the table above. Use a value of 9.80 m/s² for g. Show a sample calculation.

9. Perform a linear least squares fit to the data with the applied force as the ordinate and the acceleration as the abscissa. Calculate the slope of the fit that is equal to the total mass $m_1 + m_2$, the intercept of the fit that is the frictional force f, and the regression coefficient of the fit r. Record the results below.

$m_1 + m_2 = $ _____ kg $f = $ _____ N $r = $ _____

Newton's Second Law on the Atwood Machine

OBJECTIVES

When a net force F acts on a mass of m an acceleration a is produced. The relation between these quantities is given by Newton's second law as $F = ma$. In this laboratory an Atwood machine will be used to apply different forces to a fixed total mass to accomplish the following objectives:

1. Determination of the acceleration produced by a series of different forces applied to a given fixed mass
2. Demonstration that the acceleration of the mass is proportional to the magnitude of the applied force
3. Demonstration that the constant of proportionality between the acceleration and the applied force is the mass to which the force is applied
4. Determination of the frictional force that acts on the system

EQUIPMENT LIST

1. Atwood machine pulley (very-low-friction ball-bearing pulley)
2. Calibrated slotted masses and slotted-mass holders
3. Strong, thin nylon string
4. Laboratory timer, laboratory balance, and meter stick

THEORY

The relationship between the *net* force F exerted on a body of mass m and the acceleration a of the body was stated by Newton to be

$$F = ma \tag{1}$$

It is important to note that equation 1 is a vector equation, and thus it states that each component of the acceleration is determined by the net force in the direction of that component.

Consider the system shown in Figure 9.1, which is called an Atwood machine. It consists of two masses at the ends of a string passing over a pulley.

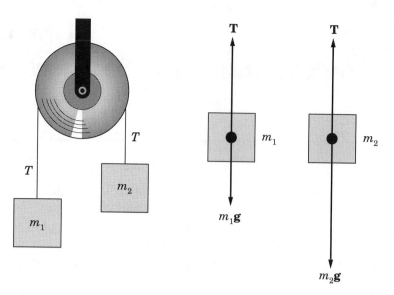

Figure 9.1 Atwood machine and free-body diagram of the forces on each mass.

Also shown in the figure is a free-body diagram of the forces on each mass. Assuming that $m_2 > m_1$, equation 1 applied to each of the masses gives

$$T - m_1 g = m_1 a \quad \text{and} \quad m_2 g - T = m_2 a \tag{2}$$

where T is the tension in the string and a is the magnitude of the acceleration of either mass. Combining the two equations 2 leads to the following:

$$(m_2 - m_1)g = (m_1 + m_2)a \tag{3}$$

Essentially, equation 3 states that a force $(m_2 - m_1)g$ equal to the difference in the weight of the two masses acts on the sum of the masses $m_1 + m_2$ to produce an acceleration a of the system. In fact, there will be some frictional force f in the system that will tend to subtract from the applied force $(m_2 - m_1)g$. Including the frictional force but moving it to the other side of the equation gives

$$(m_2 - m_1)g = (m_1 + m_2)a + f \tag{4}$$

Consider the Atwood machine shown in Figure 9.2, where $m_2 > m_1$, the mass m_1 is initially on the floor, and m_2 is released from rest a distance x above the floor at $t = 0$. Successive positions of the two masses are shown in Figure 9.2 at later times until the final picture, which shows m_2 as it strikes the floor at some time t after its release.

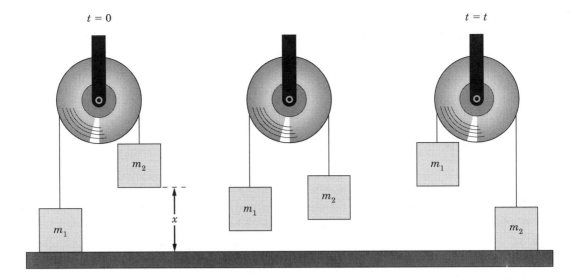

$t = 0$ $t = t$

Figure 9.2 Atwood machine as mass m_2 falls a distance x in time t.

The relationship between the distance x, the acceleration a of the system, and the time t is given by

$$x = \frac{at^2}{2} \tag{5}$$

Solving equation 5 for a in terms of the measured quantities x and t gives

$$a = \frac{2x}{t^2} \tag{6}$$

This laboratory will measure the acceleration for the Atwood machine for several different values of the applied force $(m_2 - m_1)g$ using a fixed total mass $m_1 + m_2$.

EXPERIMENTAL PROCEDURE

1. Using the laboratory balance, place 1.000 kg of slotted masses (including a 0.050-kg-mass holder) on each pan of the balance. Include five 0.002-kg masses on one of the pans. If all the masses are accurate, the scale should be balanced. If, however, the scales are not exactly balanced, add 0.001-kg slotted masses as necessary to produce as perfect a balance as can be obtained to the nearest 0.001 kg. At most, one or two 0.001-kg masses should be needed. The purpose of this procedure is to produce two masses that are as nearly equal as possible. Record in the Data Table the sum of all mass on the balance as $m_1 + m_2$.

2. Place these two collections of slotted masses on their mass holders at each end of a string over the pulley. Place the group of masses that contain the collection of 0.0020-kg masses on the left side (m_1) of the pulley and the other masses on the right (m_2).

3. Transfer a 0.0020-kg mass from the left side (m_1) to the right side (m_2) while holding the system fixed. This produces a mass difference ($m_2 - m_1$) of 0.0040 kg.

4. While one partner holds mass m_1 on the floor, another partner should measure the distance x from the bottom of mass m_2 to the floor as shown in Figure 9.2. The height of the pulley above the floor should be chosen so that x is at least 1.000 m. Record this distance as x in the Data Table. *Use this same distance for all measurements.*

5. Release mass m_1 and simultaneously start the timer. Stop the timer when mass m_2 strikes the floor. Repeat this measurement four more times, for a total of five trials with a mass difference $m_2 - m_1$ of 0.0040 kg. Record all times in the Data Table.

6. Repeat steps 4 and 5 using mass differences $m_2 - m_1$ of 0.0080, 0.0120, 0.0160, and 0.0200 kg by transferring an additional 0.0020 kg each time. Make a total of five trials for each mass difference and record the measured times in the Data Table.

CALCULATIONS

1. Calculate the applied force $(m_2 - m_1)g$ for each of the mass differences and record the results in the Calculations Table. Use a value of 9.80 m/s^2 for g.

2. Calculate the mean time $\bar{t}$ and the standard error α_t for the five measurements of time at each of the mass differences $m_2 - m_1$. Record those values in the Calculations Table.

3. Using equation 6, calculate the acceleration a from the measured values of x and $\bar{t}$ for each value of applied force. Record these values of a in the Calculations Table.

4. According to equation 4, the applied force $(m_2 - m_1)g$ should be proportional to the acceleration a with the total mass $m_1 + m_2$ as the constant of proportionality and the frictional force f as the intercept. Perform a linear least squares fit to the data with the applied force as the ordinate and the acceleration as the abscissa. Calculate the slope of the fit and record it as $(m_1 + m_2)_{exp}$ in the Calculations Table. Calculate the intercept of the fit and record it as f in the Calculations Table.

5. Calculate the percentage error in the value of $(m_1 + m_2)_{exp}$ compared to the known value of $m_1 + m_2$ and record it in the Calculations Table.

GRAPHS

Make a graph of the data for the applied force versus the acceleration. Also show on the graph the straight line obtained by the least squares fit to the data.

LABORATORY REPORT

Data Table

$m_1 + m_2 = $ _____ kg $x = $ _____ m

$m_2 - m_1$ (kg)	t_1 (s)	t_2 (s)	t_3 (s)	t_4 (s)	t_5 (s)
0.0040					
0.0080					
0.0120					
0.0160					
0.0200					

Calculations Table

$(m_2 - m_1)\,g$ (N)	$\bar{t}$(s)	α_t (s)	a (m/s^2)

$(m_1 - m_2)_{\text{exp}} = $ _____ kg $f = $ _____ N $r = $ _____ $\% \; err = $ _____

SAMPLE CALCULATIONS

QUESTIONS

1. Do your data and graph of the applied force versus the acceleration show the linear relationship that is predicted by Newton's second law? Consider the value of r, the regression coefficient, and the table in Appendix I in your answer.

2. Is your experimental value for $m_1 + m_2$ obtained from the slope of applied force versus acceleration larger or smaller than the actual value of $m_1 + m_2$?

3. If friction is truly a constant for all applied forces it will not affect the slope of the data. If, however, the frictional force is a function of the applied force (f either increases or decreases with applied force), that would tend to change the slope of the data and thus the experimental mass. Which way would the force have to change (i.e., get larger with the applied force or get smaller with the applied force) in order to account for the results stated in question 2?

4. Express the frictional force as a percentage of the applied force, $(m_1 - m_2)g$, for each applied force. Based on these percentages, does friction appear to be in the right direction and of the approximate size to account for the difference in the measured mass?

5. It can be shown that if the pulley is assumed to be a uniform disk, the effect of its moment of inertia would be equivalent to adding a mass equal to one-half the pulley mass to $m_1 + m_2$. What value does the error in your results imply for the mass of the pulley?

6. Measure the mass of the pulley or obtain its value from your instructor. Based on this value and your answer to question 5, could the pulley mass account for a significant portion of your error in the experimental value of $m_1 + m_2$?

7. Calculate the tension T in the string for each of the five applied forces.

Torques and Rotational Equilibrium of a Rigid Body

OBJECTIVES

Forces acting on a body of finite size tend to both translate and rotate the body. If the body is to be in equilibrium, it must be in equilibrium both with respect to translation and to rotation. In this laboratory, a meter stick pivoted on a support whose position is adjustable will be subjected to various forces by hanging weights on the meter stick. Measurements of the magnitude and position of forces on the meter stick will be used to accomplish the following objectives:

1. Application of the complete conditions for equilibrium of a rigid body to a meter stick pivoted on a knife edge

2. Experimental determination of the center of gravity of the meter stick

3. Determination of the mass of the meter stick by the application of known torques to the meter stick

4. Comparison of the experimentally determined location for a given applied force to produce rotational equilibrium with the location predicted theoretically

5. Determination of the mass of an unknown object

EQUIPMENT LIST

1. Meter stick with adjustable knife-edge clamp and support stand
2. Laboratory balance and calibrated hooked masses
3. Thin nylon thread and unknown mass with hook

THEORY

If a force F acts on a rigid body that is pivoted about some axis, the body tends to rotate about that axis. The tendency of a force to cause a body to rotate about some axis is measured by a quantity called torque τ. The torque τ is defined by

$$\tau = Fd_\perp \tag{1}$$

where F is the magnitude of the force, and $d_\perp$ stands for the lever arm of the force that is the perpendicular distance from the force to the location of the axis of rotation. The units of torque are Nm. The implication is that the torque due to a given force must be defined relative to a specific axis of rotation. Figure 10.4 shows two forces F_1 and F_2 acting on an arbitrarily shaped body.

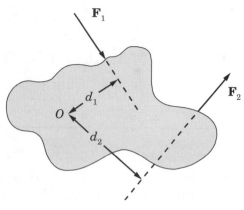

Figure 10.4 Lever arms about the point O for two forces acting on a body.

The axis of rotation of the body is labeled as O and is along a line through O perpendicular to the page. For each force, the direction of the line of action of the force is shown by a dotted line extended in either direction along the force vector. The lever arm for each force is shown as the perpendicular distance from O to the line of action of the force. In this case there are two torques, τ_1 and τ_2, acting on the body, where

$$\tau_1 = F_1\, d_1 \quad \text{and} \quad \tau_2 = F_2\, d_2 \tag{2}$$

Torques tend to rotate the body in either a clockwise or a counterclockwise direction about the axis. The convention used in this laboratory will be to let counterclockwise torques be positive and clockwise torques be negative. Using that convention, the net torque due to forces F_1 and F_2 about an axis through O is given by

$$\tau_{\text{net}} = F_2\, d_2 - F_1\, d_1 \tag{3}$$

The net torque τ_{net} could be either counterclockwise or clockwise depending on whether $F_2\, d_2$ or $F_1\, d_1$ is greater in magnitude. If the magnitudes of $F_2\, d_2$ and $F_1\, d_1$ are the same, the net torque is zero.

   In this laboratory, a meter stick will be the rigid body to which forces will be applied to produce mechanical equilibrium of the body. There are two conditions that must be satisfied for complete equilibrium of a rigid body. They are as follows:
   (1) The vector sum of all the forces acting on the body must be zero. This ensures translational equilibrium.
   (2) The net torque about any axis of the body must be zero. This means the magnitude of $\Sigma \tau_{\text{ccw}}$ (the sum of the counterclockwise torques) must be equal to the magnitude of $\Sigma \tau_{\text{cw}}$ (the sum of the clockwise torques). This ensures rotational equilibrium.

   The center of gravity of a body is defined as that point through which the sum of all the torques due to all the differential elements of mass of the body is zero. If it is true that the gravitational field is uniform throughout the body, then the center of gravity and the center of mass are the same point. Effectively, the entire mass of the body can be assumed to act at the center of gravity. A body that is uniform and symmetric has its center of gravity at the center of symmetry. For example, a meter stick that is uniform and symmetric has its center of gravity at the 0.500-m mark. Any real meter stick will probably be very close to uniform and symmetric, so its center of gravity would probably be very close to the 0.500-m mark.

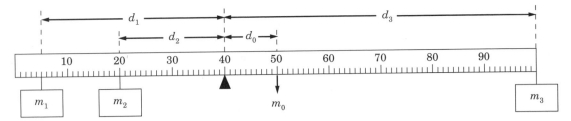

Figure 10.5 Meter stick balanced at point away from the center of gravity. The stick has torques from the three applied masses and from the meter stick mass at the center of gravity.

Consider a meter stick with a knife-edge clamp on a support stand as shown in Figure 10.5. The mass of the meter stick is m_0, and the three other masses m_1, m_2, and m_3 are shown hung from the meter stick at various places. The masses m_1, m_2, and m_3 produce forces equal to the weight of the masses $m_1 g$, $m_2 g$, and $m_3 g$ at the point where they are placed. The support exerts a force F_S directed upward at the point of the support. The weight of the meter stick $m_0 g$ is a force exerted at the center of gravity of the meter stick, labeled as x_g.

Assume that the meter stick in Figure 10.5 is in complete equilibrium. Forces in the upward direction are considered positive, and forces in the downward direction are considered negative. Counterclockwise torques are considered positive, and clockwise torques are considered negative. Torques are taken about an axis perpendicular to the page through the point of support, which is labeled as the position x_0. The lever arms for each mass d_i are calculated from the equation $d_i = |x_i - x_0|$ where x_i is the position of the ith mass. The application of the two conditions for complete equilibrium of a rigid body to this example leads to the following equations:

$$\Sigma \mathbf{F} = 0 \text{ leads to } F_S - m_1 g - m_2 g - m_3 g - m_0 g = 0 \qquad (3)$$

$$\Sigma \tau_{ccw} - \Sigma \tau_{cw} = 0 \text{ leads to } m_2 g d_2 + m_1 g d_1 - m_0 g d_0 - m_3 g d_3 = 0 \qquad (4)$$

Consider the following numerical example of the situation described in Figure 10.5. The mass of the meter stick is $m_0 = 0.1200$ kg, and masses $m_1 = 0.1500$ kg and $m_2 = 0.2000$ kg are placed as shown in the figure. The point of support is at the 0.4000-m mark. What value of mass m_3 must be placed at the 1.0000 m mark in order to put the system in equilibrium? What is the resulting force F_S with which the support pushes upward on the meter stick? *Solution:*

$$\Sigma \tau_{ccw} - \Sigma \tau_{ccw} = 0 \text{ is}$$

$$(0.1500)g(0.3500) + (0.2000)g(0.2000) - (0.1200)g(0.1000) - m_3 g(0.6000) = 0$$

$$0.05250 + 0.04000 = 0.01200 + m_3(0.6000)$$

$$m_3 = \frac{0.08050}{0.6000} = 0.1342 \text{ kg}$$

$$\Sigma \mathbf{F} = 0 \text{ is}$$

$$F_S = (m_1 + m_2 + m_3 + m_0)g = (0.1500 + 0.2000 + 0.1342 + 0.1200)(9.800) = 5.921 \text{ N}$$

Note that since the acceleration due to gravity was a common factor in the torque equation, it was cancelled from every term and was not needed in the calculation of the mass m_3. On the other hand, the factor g must be included in the calculation of the force F_S that the support exerts.

In the numerical example, the value of m_3 was determined for which the sum of torques about an axis through the point of support was zero. When the conditions of complete equilibrium have been met for some specific axis, the sum of torques about any axis is then ensured to be zero. In the case of the numerical example given above, consider the torques about an axis perpendicular to the page through the left end of the meter stick. For that axis the masses m_1, m_2, m_0, and m_3 exert clockwise torques, and the force F_S exerts the only counterclockwise torque. Calculating those values gives

$$\tau_{cw} = [(0.1500)(0.0500)+(0.2000)(0.2000)+(0.1200)(0.5000)+(0.1342)(1.000)][9.800]$$
$$= 2.368 \text{ Nm}$$

$$\tau_{ccw} = [(F_S)0.4000] = (5.921)(0.4000) = 2.368 \text{ Nm}$$

Thus, the net torque about an axis perpendicular to the page through a point at the left end of the meter stick is equal to zero because the magnitude of the counter-clockwise and clockwise torques are equal. Once equilibrium is established, the torque will be zero about any axis considered.

In the experimental arrangements to be considered, attention will be directed to satisfying the conditions of rotational equilibrium. The support force F_S will always act through the support position. If the support position is chosen as the axis for torques, F_S will not contribute to the torque because it will then have a zero lever arm. When the rotational equilibrium conditions are met, the value of the support force F_S will be such to ensure translational equilibrium as well.

In all of the experimental procedures described below, use a small loop of nylon thread to hang the hooked masses at a given position. It may prove helpful to use a very small piece of tape to hold the thread at whatever position it is desired to place the mass.

EXPERIMENTAL PROCEDURE

Part I: Torque Due to Two Known Forces

1. Remove the knife-edge clamp from the meter stick. Determine the mass of the meter stick using the laboratory balance, and record the value in the Meter Stick Data Table.

2. Place the knife-edge clamp on the meter stick and place it on the support. Adjust the position of the clamp until the best balance is achieved. Be sure the clamp is tight at balance. The position of the knife-edge clamp is now the position of the center of gravity of the meter stick. Record this position as x_g in the Meter Stick Data Table.

3. With the meter stick supported at the position x_g, place a mass $m_1 = 0.1000$ kg at the 0.1000-m mark. Experimentally determine to the nearest 0.1 mm the position x_2 at which a mass $m_2 = 0.2000$ kg must be placed in order to balance the meter stick. Record the value of x_2 in Data Table 1.

4. Calculate the lever arm d_i for each force. For this case $d_i = |x_g - x_i|$, where x_i is the position of the ith mass. Note that since the support is at the position of the center of gravity x_g, the meter stick mass has zero lever arm and thus will not contribute to the torque. Record the values of d_1 and d_2 in Calculations Table 1.

5. Calculate and record in Calculations Table 1 the value for the sum of the counterclockwise torques $\Sigma\tau_{ccw}$. In this case, the only counterclockwise torque is due to m_1, and $\Sigma\tau_{ccw} = m_1gd_1$. Calculate and record in Calculations Table 1 the value for the sum of the clockwise torques $\Sigma\tau_{cw}$. In this case the only clockwise torque is due to m_2, and $\Sigma\tau_{cw} = m_2gd_2$. Use a value of 9.800 m/s^2 for g.

6. According to theory, the magnitude of $\Sigma\tau_{ccw}$ and $\Sigma\tau_{cw}$ should be equal since the system is in equilibrium. Calculate the percentage difference between $\Sigma\tau_{ccw}$ and $\Sigma\tau_{cw}$ and record the result in Calculations Table 1.

Part 2: Torque Due to Three Known Forces

1. Support the meter stick at x_g, the center of gravity of the meter stick. Place $m_1 = 0.1000$ kg at the 0.1000-m mark and $m_2 = 0.2000$ kg at the 0.7500-m mark. Experimentally determine to the nearest 0.1 mm the position x_3 at which a mass $m_3 = 0.0500$ kg must be placed in order to balance the system. Record the value of x_3 in Data Table 2.

2. For this experimental arrangement, the meter-stick mass m_0 again makes no contribution to the torque. Calculate the lever arm for each of the other masses and record the values in Calculations Table 2 ($d_i = |x_g - x_i|$).

3. Calculate the value of $\Sigma\tau_{ccw}$ and record it in Calculations Table 2. Calculate the value of $\Sigma\tau_{cw}$ and record it in Calculations Table 2. Note that you must decide for each mass whether it contributes to the counterclockwise or clockwise torque. Use a value of 9.800 m/s^2 for g.

4. Calculate the percentage difference between $\Sigma\tau_{ccw}$ and $\Sigma\tau_{cw}$ and record it in Calculations Table 2.

Part 3: Determination of the Meter-Stick Mass by Torques

1. Place a mass $m_1 = 0.2000$ kg at the 0.1000-m position. Loosen the knife-edge clamp and slide the meter stick in the clamp until the torque exerted by the weight m_1g is balanced by the torque due to the weight of the meter stick acting at the point of the center of gravity x_g. Tighten the clamp back when the best balance is achieved. The position at which the meter stick is now supported is x_0. Record in Data Table 3 the value of x_0 to the nearest 0.1 mm.

2. The values of the lever arms are given by $d_1 = |x_1 - x_0|$ and $d_0 = |x_g - x_0|$. Calculate and record the values of d_1 and d_0 in Calculations Table 3.

3. For these conditions $\Sigma\tau_{ccw} = m_1gd_1$ and $\Sigma\tau_{cw} = m_0gd_0$, where m_0 stands for the mass of the meter stick, which is assumed to be unknown. Equating the two torques gives $m_1gd_1 = m_0gd_0$, and solving that equation gives $m_0 = m_1(d_1/d_0)$. Calculate the value of the meter stick from this equation and record it in Calculations Table 3 as the experimental value of the meter-stick mass $(m_0)_{exp}$.

4. Calculate and record in Calculations Table 3 the percentage error in this experimental value $(m_0)_{exp}$ for the meter-stick mass compared to the meter-stick mass recorded in Data Table 1 as determined by the laboratory balance.

Part 4: Comparison of Experimental and Theoretical Determinations of the Location of an Applied Force

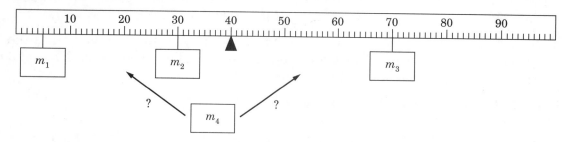

Figure 10.6 The location of mass m_4 needed to place the system into equilibrium is to be determined both experimentally and theoretically and then compared.

1. Adjust the knife-edge clamp to the 0.4000-m mark on the meter stick. Place $m_1 = 0.0500$ kg at the 0.0500-m mark, $m_2 = 0.3000$ kg at the 0.3000-m mark, and $m_3 = 0.2000$ kg at the 0.7000-m mark as shown in Figure 10.6. *With the meter stick supported at the 0.4000-m mark*, experimentally determine to the nearest 0.1 mm the position at which mass $m_4 = 0.1000$ kg must be placed to balance the meter stick. Record this position as x_4 in Data Table 4. Using this value of x_4 calculate the experimental value of the lever arm d_4. Record this value of d_4 in Calculations Table 4 as $(d_4)_{exp}$.

2. In the space provided in Data Table 4, write an equation for the rotational equilibrium in which the counterclockwise torques are positive and the clockwise torques are negative. Use m_1, m_2, m_3, and m_4 as the symbols for the appropriate mass, and d_1, d_2, d_3, and d_4 as the symbols for the lever arms. *Be sure to also include the contribution from the meter stick mass* m_0 *acting at the center of gravity of the meter stick* x_g.

3. In the equation, treat the lever arm d_4 of mass m_4 as unknown and all the other quantities as known. Solve the equation to obtain a value for d_4 and record that result in Calculations Table 4 as $(d_4)_{theo}$.

4. Calculate the percentage error in $(d_4)_{exp}$ compared to $(d_4)_{theo}$ and record it in Calculations Table 4.

Part 5: Determination of an Unknown Mass by Torques

1. Using an experimental arrangement with the meter stick that is similar to those that have been used thus far, devise a method to determine an unknown mass. Describe carefully the procedure that is followed. Use at least one known mass and state its value and location on the meter stick. Write down an equation that describes the equilibrium of the system treating the unknown mass as the unknown in the equation. Include a sketch of the experimental arrangement showing the position of all masses known and unknown. Construct your own Data Table 5 and Calculations Table 5 listing all the relevant quantities.

2. Determine the unknown mass using the laboratory balance. Calculate the percentage error in your measured value of the unknown mass compared to that obtained using the laboratory balance.

Laboratory 11

Conservation of Energy on the Air Table

PRELABORATORY ASSIGNMENT

Read carefully the entire description of the laboratory and answer the following questions based on the material contained in the reading assignment. Turn in the completed prelaboratory assignment at the beginning of the laboratory period prior to the performance of the laboratory.

1. What is the definition, both in words and equation form, of the spring constant k? What are the units of k?

2. What is the equation for the spring potential energy of a spring? Define all terms used in the equation.

3. What is the principle of conservation of mechanical energy as applied to a mass m attached to a spring of spring constant k? State it in words and equation form and define all terms used.

4. What kinds of forces conserve mechanical energy? Is mechanical energy conserved if frictional forces are present?

5. A spring has a spring constant $k = 2.75$ N/m. It has an unstretched length of 0.1000 m. What is its spring potential energy when the spring is stretched to a length of 0.1700 m?

6. A spring is stretched by applying mass m over a pulley to the spring as shown in Figure 11.1. The position of the end of the spring as a function of the applied mass is determined, and the data given below is obtained. Find the spring constant k by performing a linear least squares fit to the data of *applied force* versus x.

m (kg)	0.0200	0.0400	0.0600	0.0800	0.1000	0.1200
x	0.0000	0.0490	0.1000	0.1510	0.2010	0.2490
F (N)						

$k =$ _____ N/m   Intercept = _____ N $r =$ _____

7. A 2.00-kg mass moves under the force of a spring whose spring constant is $k = 5.65$ N/m. At one instant of time, the mass has a speed of 4.75 m/s when the spring has a displacement from equilibrium of 1.557 m. What is the speed of the mass at a later time when the spring displacement from equilibrium is 0.857 m? (*Hint:* Use conservation of mechanical energy.)

Conservation of Energy on the Air Table

OBJECTIVES

A puck on an air table subjected to two springs pulling in opposite directions can be made to move in an elliptical path when released under the appropriate initial conditions. The total mechanical energy of the system at any time consists of the kinetic energy of the puck and spring potential energy of the two springs. In this laboratory, measurements will be made of the elongation of each spring as a function of the force applied to the spring. The position of the puck as a function of time when subjected to the two springs will also be determined. These measurements will be used to accomplish the following objectives:

1. Determination of the spring constant k (the elongation per unit force) for each of the springs
2. Determination of the speed of the puck as a function of time from the change in coordinate position of the puck per unit time
3. Determination of the kinetic energy of the puck as a function of the coordinate position of the puck
4. Determination of the total spring potential energy of the two springs as a function of the coordinate position of the puck
5. Evaluation of the extent to which the total mechanical energy (sum of kinetic energy plus spring potential energy) is constant as a function of position of the puck

EQUIPMENT LIST

1. Air table, sparktimer, airpump, foot switch, level, carbon paper, and recording paper
2. Two springs, two spring clamps, and a double hook for the puck
3. Pulley, string, calibrated masses, and laboratory balance
4. Meter stick or metal ruler

THEORY

The force F exerted by a spring as it is elongated from its natural length x_0 to some greater length x is given by

$$F = -k(x - x_0) \tag{1}$$

where k is a constant called the "spring constant." The spring constant is defined as the force per unit elongation, and it has units of N/m. The value of k in general will be different for each spring. It depends on the properties of the material from which the spring is made and also on how tightly the spring is wound. The value of k for a particular spring can be determined by applying known forces and measuring the resulting elongation of the spring.

The force exerted by a spring is an example of a class of forces known as conservative forces. A conservative force is one for which a potential energy function can be defined. In the case of the force due to a spring, the potential energy U is given by

$$U = \tfrac{1}{2}k\,(x - x_0)^2 \tag{2}$$

If a mass m is attached to the end of a spring and allowed to move under the influence of the spring, the speed V of the mass will change as the force of the spring on the mass varies. At any particular instant when the mass m has speed V the particle has a kinetic energy K given by

$$K = \tfrac{1}{2}mV^2 \tag{3}$$

The sum of the potential and kinetic energies is equal to the total energy E. In an isolated system, where only conservative forces are present, the total energy E is a constant throughout the motion. This is the principle of conservation of mechanical energy. In equation form the total energy E is given by

$$E = \tfrac{1}{2}mV^2 + \tfrac{1}{2}k\,(x - x_0)^2 \tag{4}$$

Mechanical energy is not conserved when nonconservative forces are present. Nonconservative forces are ones for which it is not possible to define a potential energy function. Friction is a nonconservative force, and mechanical energy is not conserved in the presence of frictional forces.

In this laboratory, a puck on an air table will be caused to move under the influence of two springs arranged in such a way that they pull in opposite directions on the puck. The spring constants k_1 and k_2 of the two springs will be determined. The position of the puck as a function of time will be determined with sparktimer measurements. At each position of the puck along its recorded path, the speed of the puck, as well as the elongation of each spring, will be determined. From the measured speed and mass of the puck, the kinetic energy will be deduced. From the measured spring constants and the elongation of each spring, the total spring potential energy will be deduced. From the kinetic energy and spring, potential energy at each point, the total energy E at each point will be determined. A graph of the total energy E as a function of position will show the extent to which the total energy remains constant.

EXPERIMENTAL PROCEDURE—SPRING CONSTANTS

1. Label one of the springs as spring 1. Using that spring in the arrangement described in Figure 11.1, place enough mass m on the end of the string to elongate the spring slightly and thus place it under some tension. Probably a value for m of 0.0200 kg will be about the right choice for the initial value. Record the value of m in Data-Calculations Table 1. When the puck is at rest with the spring under tension, mark the position of the puck with a short burst of the sparktimer.

2. Continue to add mass to the end of the string over the pulley and record the position of the puck for each value of applied mass by a short burst of the sparktimer. Altogether a total of five measurements of mass versus position are required. In order to find a convenient range of the mass needed for a particular spring, apply enough mass to extend the spring about 0.300 m. This will be a convenient number for the maximum mass needed, and the other trials can be at intermediate values of the maximum. Record the five values of m used in Data-Calculations Table 1 and mark the position corresponding to each mass by a short burst of the sparktimer.

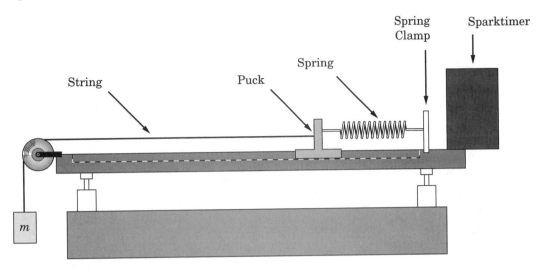

Figure 11.1 Air table arrangement for measuring the spring constant k.

3. Remove the recording paper from the table and draw a straight line through the points obtained from the sparktimer bursts. Arbitrarily choose an origin and measure the distance from the origin to each point. Record these values as x in Data-Calculations Table 1.

4. Label the second spring as spring 2 and repeat steps 1 through 3 above for the second spring.

CALCULATIONS—SPRING CONSTANTS

1. For each applied mass m, calculate the force mg using $g = 9.80$ m/s^2. Record the values of the force in Data-Calculations Table 1.

2. According to equation 1, the applied force mg should be proportional to the coordinate position x. Perform a linear least squares fit to the data with the force as the ordinate and x as the abscissa. Do this for the data from each of the springs. Determine the slope of each of these least squares fit and record those values in Data-Calculations Table 1 as the spring constants k_1 and k_2.

EXPERIMENTAL PROCEDURE—ENERGY CONSERVATION

1. Place the double hook on the puck and measure the mass of the puck plus the hook using the laboratory balance. Record this value as m_p in Data-Calculations Table 2.

2. Measure the unstretched length of each of the two springs and record them in Data-Calculations Table 2 as L_{01} and L_{02}. Also record the values of the spring constants k_1 and k_2 in Data-Calculations Table 2.

3. Level the air table and place a sheet of carbon paper and a sheet of recording paper on the table. Attach the leads to both of the pucks and place one of the pucks in one corner of the table, where it will remain.

4. Place the spring clamps on the table on opposite sides. Attach the two springs to opposite ends of the double hook on the puck. Attach the other end of each of the two springs to the two spring clamps. Make certain that the springs are clearly labeled and that there is no confusion about which is spring 1 and which is spring 2.

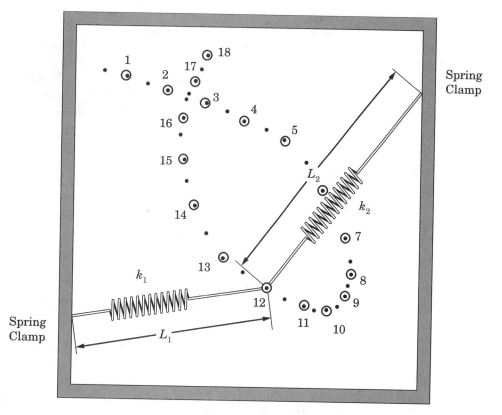

Figure 11.2 Sparktimer record produced by the elliptical motion of puck on an air table under the influence of two springs clamped on opposite sides of the table.

5. The puck is to be released so that it *moves in an elliptical trajectory* as shown in Figure 11.2 A second condition on the motion is that *the two springs must both be under stretched tension at all times during the motion.* Launch the puck on several trial runs to be sure that both of these conditions are satisfied.

6. On the recording paper, mark the position of the two spring clamps and indicate which spring is attached to which clamp. A small piece of paper may need to be taped onto the recording paper in order to extend it in the region of the spring clamps.

7. Set the sparktimer to 20.0 Hz and record the trajectory of the puck. Carefully note the location of the initial position of the puck.

8. Choose a point near the beginning of the puck's motion (but not the very first point of the motion). Circle that point and label it as 1. Continuing in the direction of the puck's motion, circle every other point, and number them consecutively up to 15 as shown in Figure 11.2. Since every other point was chosen, the time interval between the numbered points is 0.100 s.

9. For each circled point, measure the distance from the point to each spring clamp. These are the stretched lengths of the springs at that point. These lengths are illustrated for the point labeled 12 in Figure 11.2. Record in Data-Calculations Table 2 the values of these lengths as L_1 and L_2.

10. At each *numbered point*, measure the distance between the *preceding uncircled point* and the *next uncircled point* as illustrated in Figure 11.3. This gives the average displacement of the puck during the time interval $\Delta t = 0.100$ s that includes 0.050 s before the point and 0.050 s after the point. For each circled point, record this distance as Δs in Data-Calculations Table 2. Note: As shown in Figure 11.3, if the path has sharp curvature, measure Δs as the distance from uncircled to circled to uncircled points as shown in the right-hand side of that figure.

Figure 11.3 Illustrating calculation of Δs when path is sharply curved at the right as opposed to the case when the path is essentially straight as shown at the left.

CALCULATIONS—ENERGY CONSERVATION

1. Calculate the speed V at each point from the relationship $V = \Delta s/\Delta t$. Remember that all the points have $\Delta t = 0.100$ s. Record the results in Data-Calculations Table 2.

2. Calculate the kinetic energy K of the puck at each point, using equation 3. Record the values of K in Data-Calculations Table 2.

3. For each data point, calculate the quantities $|L_1 - L_{01}|$ and $|L_2 - L_{02}|$. These quantities represent the amount that each of the springs are stretched at each point. Calculate the total spring potential energy U at each point due to both springs from the relationship $U = \frac{1}{2} k_1 |L_1 - L_{01}|^2 + \frac{1}{2} k_2 |L_2 - L_{02}|^2$. Record the values of U in Data-Calculations Table 2.

4. Calculate the total energy E from $E = K + U$ at each point. Record the values of E in Data-Calculations Table 2.

GRAPHS

Using a single sheet of linear graph paper, make a graph that has energy units on the vertical scale and position number (1 through 15) on the horizontal scale. Using different symbols for each, graph the kinetic energy K, the total spring potential energy U, and the total mechanical energy E as a function of position number. Do not suppress the energy scale. Show the whole range of energy from zero to slightly greater than E.

Laboratory 11

Conservation of Energy on the Air Table

LABORATORY REPORT

Data-Calculations

Table 1

Spring 1 $k_1 =$ _____ N/m

Point	m (kg)	mg (N)	x (m)
1			
2			
3			
4			
5			

$r =$ _____

Spring 2 $k_2 =$ _____ N/m

Point	m (kg)	mg (N)	x (m)
1			
2			
3			
4			
5			

$r =$ _____

SAMPLE CALCULATIONS

Table 2

$m_p =$ kg	$k_1 =$ N/m	$k_2 =$ N/m	$L_{01} =$ m	$L_{02} =$ m			
Point	L_1 (m)	L_2 (m)	Δs (m)	V (m/sec)	K (J)	U (J)	E (J)
1							
2							
3							
4							
5							
6							
7							
8							
9							
10							
11							
12							
13							
14							
15							

QUESTIONS

1. Calculate the mean and standard error for the 15 values for the total energy E. Divide the standard error by the mean. Express this ratio as a percentage. Considering these calculations, do your data indicate that E is approximately constant in value?

2. Does the energy E tend to decrease from the beginning to the end of the data? If friction were present, such a trend should occur. Calculate the decrease in energy from the start of the motion to the end. Divide that loss in energy by the total length of the path over which the loss occurs. The result is the force of friction. What value of friction results from this procedure for your data?

3. Using your calculated value of f, the frictional force, calculate μ_k, the coefficient of kinetic friction for the puck on the air table.

4. In the determination of the spring constant k, the origin from which the position x was measured was arbitrarily chosen. Why does that not affect the value of the spring constant obtained from the data?

5. Calculate the average value of the potential energy $\overline{U}$ and the average value of the kinetic energy $\overline{K}$. How do these two quantities compare? Is one significantly greater than the other, or are they about the same?

6. Calculate the maximum difference in the observed values of U, the spring potential energy, and call it ΔU. Express $\Delta U/\overline{U}$ as a percentage. Calculate the maximum difference in the observed values of K the kinetic energy and call it ΔK. Express $\Delta K/\overline{K}$ as a percentage. Calculate the maximum difference in the observed values of E the total mechanical energy and call it ΔE. Express $\Delta E/\overline{E}$ as a percentage.

7. Summarize in your own words the physical theory that this laboratory is supposed to demonstrate. Do your results support the theory? Use the results of your calculations in question 6 to be as quantitative in your answer as possible.

Laboratory 12

Conservation of Spring and Gravitational Potential Energy

PRELABORATORY ASSIGNMENT

Read carefully the entire description of the laboratory and answer the following questions based on the material contained in the reading assignment. Turn in the completed prelaboratory assignment at the beginning of the laboratory period prior to the performance of the laboratory.

1. A spring has a spring constant of $k = 7.500$ N/m. If the spring is displaced 0.5500 m from its equilibrium position, what is the force that the spring exerts?

2. A spring of spring constant $k = 8.250$ N/m is displaced from equilibrium by a distance of 0.1500 m. What is the stored energy in the form of spring potential energy?

3. What was the work done in stretching the spring in question 2 from its equilibrium position to a displacement of 0.1500 m?

4. If the spring in question 2 is stretched from the displacement of 0.1500 m to a displacement of 0.3500 m, what is the change in spring potential energy between those two positions?

5. A spring of spring constant $k = 12.50$ N/m is hung vertically. A 0.5000-kg mass is then suspended from the spring. What is the displacement of the end of the spring due to the weight of the 0.5000-kg mass?

6. A mass of 0.4000 kg is raised by a vertical distance of 0.4500 m in the earth's gravitational field. What is the change in its gravitational potential energy? This same mass is lowered by a vertical distance of 0.3500 m in the earth's gravitational field. What is the change in its gravitational potential energy?

7. A spring of spring constant $k = 8.750$ N/m is hung vertically from a rigid support. A mass of 0.5000 kg is placed on the end of the spring and supported by hand at a point so that the displacement of the spring is 0.2500 m. The mass is suddenly released and allowed to fall. At the lowest position of the mass, what is the displacement of the spring from its equilibrium position? (*Hint:* In solving this problem you will have to solve a quadratic equation, and one of the solutions will be the original 0.2500-m displacement.)

8. The same spring described in question 7 again has a mass of 0.5000 kg on the end of the spring. It is pulled down by hand and held so that the displacement of the spring from its equilibrium position is 0.7000 m. If the system is suddenly released and the mass is allowed to rise, what is the displacement of the spring when the mass is at its highest position? (*Hint:* In solving this problem you will have to solve a quadratic equation, and one of the solutions will be the original 0.7000-m displacement.)

Conservation of Spring and Gravitational Potential Energy

OBJECTIVES

In this laboratory, a spring will be suspended from a rigid support with a hooked mass on the end of the spring. The mass will be caused to move vertically in the earth's gravitational field under the influence of the spring. Measurement on the extension of the spring for different amounts of mass on the spring and measurements of the change in position of a given mass as it moves in the earth's gravitational field will be used to accomplish the following objectives:

1. Determination of the spring constant k for the spring

2. Determination of the change in the gravitational energy ΔU_g as a mass moves from an initial displacement x_i, where it is at rest, to a final displacement x_f, where it is again at rest with the change in gravitational energy given by $\Delta U_g = mg(x_f - x_i)$

3. Determination of the change in spring potential energy ΔU_k as the mass moves from the same x_i to the same x_f with the change in spring potential energy given by the equation $\Delta U_k = (\frac{1}{2})k(x_f^2 - x_i^2)$

4. Evaluation of the extent to which the change in gravitational potential energy ΔU_g is equal to the change in spring potential energy ΔU_k

EQUIPMENT LIST

1. Spring in the form of a truncated cone made of spring brass with a spring constant of about 10 N/m. (Available from Central Scientific Co. If another spring is used, appropriate adjustments should be made in the masses and distances used.)

2. Set of calibrated hooked masses

3. Table clamp, right-angle clamps, support rods, meter stick

THEORY

When a spring is stretched or compressed from its equilibrium length, the spring exerts a restoring force proportional to the displacement of the spring from equilibrium. In equation form the relationship between the force F and the displacement x is given by

$$F = -kx \tag{1}$$

where k is a constant called the "spring constant," whose units are N/m. The negative sign in the equation indicates that the restoring force is opposite the displacement.

When a spring is compressed or stretched it has stored energy called "spring potential energy." The change in spring potential energy ΔU_k when the spring is stretched from a displacement of x_1 to a displacement of x_2 is equal to the work done to deform the spring. The work is given by the product of the average force and the change in the displacement. In equation form this can be written as

$$\text{Work} = \Delta U_k = \frac{(kx_1 + kx_2)}{2}(x_2 - x_1) = \tfrac{1}{2}k\,(x_2{}^2 - x_1{}^2) \qquad (2)$$

It should be emphasized that x_1 and x_2 must refer to displacements measured from the equilibrium position of the spring.

Consider a spring of spring constant k supported at the top by a rigid support and allowed to hang vertically as shown in Figure 12.1. The vertical position at which the lower end of the spring hangs is the zero of the coordinate x. Suppose a hooked mass m is placed by hand on the end of the spring, and the mass is slowly lowered by hand. The mass will extend the spring by an amount x_0 when the hand is removed, and the mass will then hang at rest in equilibrium. The value of x_0, which is illustrated in Figure 12.1, is determined by the position at which the weight mg of the mass is equal to the force exerted by the spring. In equation form this is

$$kx_0 = mg \quad \text{or} \quad x_0 = \frac{mg}{k} \qquad (3)$$

Note carefully that x_0 refers to the position of the lower end of the spring.

Suppose that the mass is now raised and supported by hand so that the lower end of the spring is at position x_1 above position x_0 as shown in Figure 12.1. In other words, x_1 represents less of an extension of the spring than does x_0. The mass is now released and allowed to fall under the influence of the spring and the earth's gravitational force. The mass will fall to the lowest point where the displacement of the lower end of the spring is x_2 as shown in Figure 12.1. The mass will then rebound and oscillate.

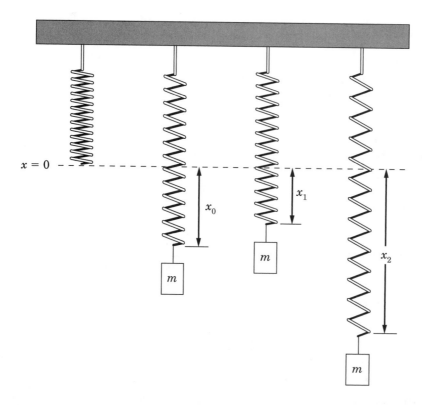

Figure 12.1 Positions of a mass m on the end of a spring in the earth's gravitational field.

The total mechanical energy of the system consists of the sum of the kinetic energy, the spring potential energy, and the gravitational potential energy. The gravitational potential energy relative to any horizontal plane is given by mgx, where x is the distance above the plane.

Consider the total mechanical energy at two particular points in the motion. The first is the point where the mass is originally held before release, and the second is the lowest point to which it falls when released. At each of these two points the kinetic energy is zero. Therefore, for these two points the total mechanical energy of the system is given by the sum of the spring potential energy and the gravitational potential energy. The center of mass of the mass m is a distance d below the lower end of the spring, and the gravitational potential energy zero is taken to be the same zero as the equilibrium point of the spring. With these assumptions, the equation relating the sum of spring potential energy and gravitational potential energy is

$$\tfrac{1}{2}kx_1^2 - mg(x_1 + d) = \tfrac{1}{2}kx_2^2 - mg(x_2 + d) \tag{4}$$

$$\tfrac{1}{2}kx_1^2 - mgx_1 = \tfrac{1}{2}kx_2^2 - mgx_2 \tag{5}$$

Note that both of the gravitational potential energy terms are negative because the mass is below the reference point for both positions. Equation 5 can be rewritten as

$$mg(x_2 - x_1) = \tfrac{1}{2}k(x_2^2 - x_1^2) \tag{6}$$

Equation 6 states that the change in gravitational energy between points 1 and 2 is equal to the change in spring potential energy between those points. This applies only under the experimental conditions described in which the kinetic energy is zero both at points 1 and 2. This laboratory will consist of a series of measurements that will test the validity of equation 6.

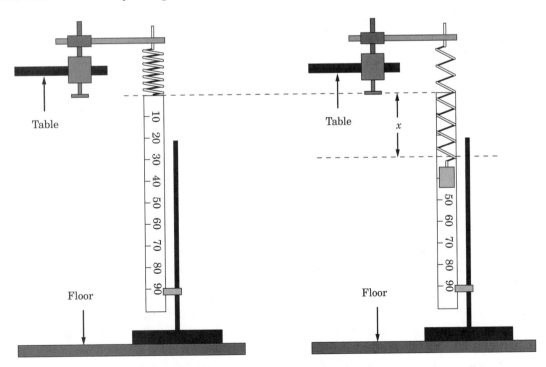

Figure 12.2 Spring supported by table clamp and meter stick aligned with the lower end of the spring. Mass placed on end of the spring caused displacement x of the end of the spring.

EXPERIMENTAL PROCEDURE—SPRING CONSTANT

1. Attach the table clamp to the edge of the laboratory table and screw a threaded rod into the clamp vertically as shown in Figure 12.2. Place a right-angle clamp on the vertical rod and extend a horizontal rod from the right-angle clamp. Hang the spring on the horizontal rod and attach it to the horizontal rod with a piece of tape. Screw a threaded vertical rod into a support stand that rests on the floor. Place a right-angle clamp on the vertical rod and place a meter stick in the clamp so that the meter stick stands vertically. Adjust the height of the clamp on the vertical rod until the zero mark of the meter stick is aligned with the bottom of the hanging spring as shown in Figure 12.2.

2. Place a hooked mass m of 0.1000 kg on the end of the spring. Slowly lower the mass m until it hangs at rest in equilibrium when released. Carefully read the position of the lower end of the spring on the meter-stick scale. Record the value of the mass m and the value of the displacement x in Data Table 1.

3. Repeat step 2 placing in succession 0.2000, 0.3000, 0.4000, and 0.5000 kg on the spring and measuring the displacement x of the spring. Record all values of m and x in Data Table 1. Record x to the nearest 0.1 mm.

CALCULATIONS—SPRING CONSTANT

1. Calculate the force mg for each mass and record the values in Calculations Table 1. Use the value of 9.800 m/s^2 for g.

2. Perform a linear least squares fit to the data with mg as the ordinate and x as the abscissa. Calculate the slope of the fit and record it in Calculation Table 1 as the spring constant k. Also record the value of r, the correlation coefficient of the fit.

EXPERIMENTAL PROCEDURE—ENERGY CONSERVATION

1. Check again and be certain that the lower end of the spring is precisely at the zero mark of the meter stick. Adjust the meter stick if necessary. Hang a 0.5000-kg mass on the end of the spring and support it with your hand so that the lower end of the spring is precisely at the level of the 0.2500-m mark of the meter stick. This is the value of x_1. Record the value of x_1 and the value of the mass in Data Table 2. Release the mass and note how far it falls. Mark the lowest point of the lower end of the spring. Release the mass several times until you have accurately located the lowest point of the motion. It may be easier to note the lowest position of the mass itself, and then place the mass at that position and determine the position of the lower end of the spring. Record this distance as the value of x_2 in Data Table 2.

2. Repeat step 1 for x_1 values of 0.3000, 0.3500, and 0.4000 m. Measure the value of x_2 for each of these values of x_1 and record the values of x_1 and x_2 in Data Table 2.

3. Check again and be certain that the lower end of the spring is precisely at the zero mark of the meter stick. Adjust the meter stick if necessary. Still using a mass of 0.5000 kg on the end of the spring, pull the mass down by hand until the lower end of the spring is precisely at the 0.7500-m mark of the meter stick. This is the value of x_2. Release the mass and determine how high it rises. The position of the lower end of the spring when the mass is at its highest point is x_1. Again

release the mass several times to accurately determine the value of x_1. Record the value of x_1 and the value of the mass in Data Table 3.

4. Repeat step 3 for x_2 values of 0.7000, 0.6500, and 0.6000 m. Measure the value of x_1 for each of these values of x_2 and record the values of x_1 and x_2 in Data Table 3.

CALCULATIONS—ENERGY CONSERVATION

1. For each of the four measurements of the falling mass in Data Table 2, calculate the change in the gravitational potential energy ΔU_g, where $\Delta U_g = mg(x_2 - x_1)$. Calculate the change in spring potential energy ΔU_k, where $\Delta U_k = \frac{1}{2} k(x_2^2 - x_1^2)$. Record the results in Calculations Table 2.

2. Calculate the percentage difference between ΔU_g and ΔU_k for each case of step 1 and record them in Calculations Table 2.

3. For each of the four measurements of the rising mass in Data Table 3, calculate the change in gravitational potential energy ΔU_g and the change in spring potential energy ΔU_k. Record the results in Calculations Table 3.

4. Calculate the percentage difference between ΔU_g and ΔU_k for each case of step 3 and record them in Calculations Table 3.

GRAPHS

Graph the data from Calculations Table 1 for force mg versus displacement x with mg as the ordinate and x as the abscissa. Also show on the graph the straight line obtained from the linear least squares fit to the data.

Laboratory 12

Conservation of Spring and Gravitational Potential Energy

LABORATORY REPORT

Data Table 1

m (kg)	x (m)

Calculations Table 1

mg (N)	k (N/m)	r

Data Table 2

x_1 (m)	x_2 (m)	m (kg)

Calculations Table 2

$mg\,(x_2 - x_1)$(J)	$1/2\,k\,(x_2^2 - x_1^2)$(J)	% Diff

Data Table 3

x_1 (m)	x_2 (m)	m (kg)

Calculations Table 3

$mg\,(x_2 - x_1)$(J)	$1/2\,k\,(x_2^2 - x_1^2)$(J)	% Diff

SAMPLE CALCULATIONS

QUESTIONS

1. Do the data from Data Table 1 for the displacement versus force indicate that the spring constant is indeed constant? State clearly the evidence for your answer.

2. Do the data in Data Table 2 and Data Table 3 indicate that mechanical energy is conserved for this system? State clearly the evidence for your answer.

3. Examine your data in Data Table 2 and in Data Table 3. For the data with the smallest percentage difference in those two tables, perform the following calculations, which will compare the total energy at each point instead of comparing changes in energy. Calculate the sum $U_k + U_g$ at x_1 as $\frac{1}{2}kx_1^2 - mgx_1$. Now calculate that sum at x_2 as $\frac{1}{2}kx_2^2 - mgx_2$. Compare those two quantities and comment on whether or not they agree.

4. Consider the same case as discussed in question 3. Calculate the value of x halfway between x_1 and x_2. Calculate $U_k + U_g = \frac{1}{2}kx^2 - mgx$ for that point. How does it compare with the values calculated in question 3? Explain the reason for this result.

Laboratory **13**

The Ballistic Pendulum and Projectile Motion

PRELABORATORY ASSIGNMENT

Read carefully the entire description of the laboratory and answer the following questions based on the material contained in the reading assignment. Turn in the completed prelaboratory assignment at the beginning of the laboratory period prior to the performance of the laboratory.

1. What are the conditions under which the total momentum of a system of particles is conserved?

2. What kind of collision conserves kinetic energy?

3. What kind of collision does not conserve kinetic energy? What kind of collision results in the maximum loss of kinetic energy?

4. A ball of mass 0.0750 kg is fired horizontally into a ballistic pendulum as shown in Figure 13.1. The pendulum mass is 0.3500 kg. The ball is caught in the pendulum, and the center of mass of the system rises a vertical distance of 0.145 m in the earth's gravitational field. What was the original speed of the ball? Assume $g = 9.80$ m/s^2.

5. How much kinetic energy was lost in the collision of problem 4?

6. A projectile is fired in the earth's gravitational field with a horizontal velocity of $v = 9.00$ m/s. How far does it go in the horizontal direction in 0.550 s?

7. How far does the projectile of question 6 fall in the vertical direction in 0.550 s?

8. A projectile is launched in the horizontal direction. It travels 2.050 m horizontally while it falls 0.450 m vertically, and it then strikes the floor. How long is the projectile in the air?

9. What was the original velocity of the projectile described in question 8?

10. What is the velocity in the horizontal direction of the projectile in question 8 when it strikes the floor? What is its velocity in the vertical direction at this time? What is the magnitude of its velocity as it strikes the floor?

The Ballistic Pendulum and Projectile Motion

OBJECTIVES

In collisions, the forces involved are internal, and therefore momentum is conserved. A collision is called "elastic" if kinetic energy is also conserved. An *inelastic collision* is one in which some kinetic energy is lost. If the colliding particles stick together, the collision is called "completely inelastic," and the maximum possible loss of kinetic energy occurs. A ballistic pendulum is a device used to measure the velocity of a projectile fired into an initially stationary pendulum. The pendulum is designed to catch the ball, causing a completely inelastic collision. Measurements for the ball fired into the pendulum and for the ball fired horizontally as a free projectile will be used to accomplish the following objectives:

1. Determination of the initial velocity of the ball and the initial velocity with which the pendulum plus ball moves after the collision

2. Determination of the kinetic energy loss in the collision of the ball with the pendulum

3. Independent determination of the initial velocity of the ball by firing it as a free projectile

4. Comparison of the velocity of the ball determined by the two different experimental arrangements

EQUIPMENT LIST

1. Ballistic pendulum apparatus with projectile ball
2. Laboratory balance and calibrated masses
3. Meter stick, plain paper, carbon paper, and masking tape

THEORY

According to the principle of conservation of momentum, if there are no external forces acting on a system of several particles, then the total momentum of the system remains constant. Collision processes are particularly good examples of this concept. In this laboratory, a ballistic pendulum will be used to measure the velocity of a ball projected by a spring gun. Consider the process shown in Figure 13.1 in which a ball of mass m moving initially in the horizontal direction with speed v_{x0} strikes a pendulum designed to catch the ball. The pendulum of mass M, upon receiving the ball, swings about a pivot point O to some maximum height y_2 that is greater than its original height y_1. The system of ball plus pendulum rises a vertical distance of $y_2 - y_1$ as a result of the process.

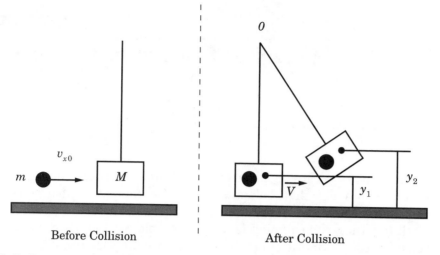

Before Collision

After Collision

Figure 13.1 Ballistic pendulum of mass M before and after collision with ball of mass m.

The analysis of the event is best done in two steps. Momentum is conserved in the collision because the only forces acting on the ball and the pendulum in the direction of motion are the forces of the collision. The collision is completely inelastic because the two particles stick together after the collision and thus move with the same velocity V. The equation for conservation of momentum is

$$mv_{x0} = (m + M)V \tag{1}$$

The collision itself does not conserve mechanical energy. In fact, the original kinetic energy $[\frac{1}{2} mv_{x0}^2]$ is much greater than the kinetic energy immediately after the collision $[\frac{1}{2} (m + M)V^2]$. On the other hand, mechanical energy is conserved in the motion of the combined mass (ball plus pendulum). The combined mass moves with velocity V immediately after the collision. It swings up along the arc of the pendulum. It rises a vertical distance of $y_2 - y_1$ and comes to rest instantaneously at that maximum height as the kinetic energy immediately after the collision is converted into gravitational potential energy. In equation form, $\frac{1}{2} (m + M)V^2 = (m + M)g(y_2 - y_1)$. Solving for V gives

$$V = \sqrt{2g(y_2 - y_1)} \tag{2}$$

If equation 1 is solved for the initial velocity of the ball, the result is

$$v_{x0} = \left(\frac{m + M}{m}\right)V \tag{3}$$

Therefore, a measurement of the height $y_2 - y_1$ leads to a determination of V using equation 2, and with V known, equation 3 can be used to determine v_{x0}.

EXPERIMENTAL PROCEDURE—BALLISTIC PENDULUM

1. Slide the projectile ball (which has a hole in it) onto the rod of the spring gun (Figure 13.2). The ball may have been stored in the pendulum bob. If so, when removing the ball be careful to push up on the spring in the bob that serves as a mechanism to catch the ball. This spring can be easily broken if the ball is forced out

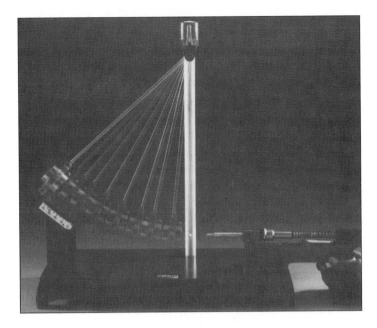

Figure 13.2 Ballistic pendulum apparatus. (Photo courtesy of Central Scientific Co., Inc.)

of the bob without releasing the spring first. When the ball has been placed on the rod, cock the gun by pushing against the ball until the latch catches. *Be very careful not to get your hand caught in the spring gun mechanism.* Fire the gun several times to see how it operates. There are two common problems. If the ball does not catch in the pendulum bob, the spring in the bob should be adjusted or replaced. If the pawl that is designed to catch on the notched track does not engage, the pendulum suspension should be adjusted by means of the screws at the suspension points. The alignment of this suspension is critical and frequently requires adjustment.

2. A sharp curved point on the side of the pendulum (or on some models a dot) marks the center of mass of the pendulum-ball system. Let the pendulum bob hang vertically and measure the distance y_1 of the point marking the center of mass above the base of the gun (Figure 13.1). Record the value of y_1 in Data Table 1.

3. Place the ball on the rod, push against the ball to cock the gun, and fire the ball into the stationary pendulum while it hangs freely at rest. The pendulum will catch the ball, swing up, and then lodge in the notched track. Record in Data Table 1 the position number p at which the pawl on the pendulum catches on the track. Measure the distance y_2 of the center of mass point above the base of the apparatus (Figure 13.1). Record the value of y_2 in Data Table 1. Repeat this procedure four more times for a total of five trials, recording the position p and measuring the distance y_2 for each trial.

4. Loosen the screws holding the pendulum in its support and remove the pendulum consisting of the rod and the bob. Determine the mass of the pendulum (bob and rod) using a laboratory balance. Record the pendulum mass as M in Data Table 1. Determine the projectile ball mass and record it in Data Table 1 as m.

CALCULATIONS—BALLISTIC PENDULUM

1. Calculate the distance $y_2 - y_1$ that the combined mass rises for each trial. Record these values in Calculations Table 1.

2. Using equation 2, calculate the velocity V of the combined mass for each of the five trials and record them in Calculations Table 1.

3. Using equation 3, calculate the initial speed of the projectile ball v_{x0} for the five trials and record those values in Calculations Table 1.

4. Calculate the mean $\overline{v_{x0}}$ and the standard error α_v for the five values of v_{x0}. Record them in Calculations Table 1.

THEORY—PROJECTILE MOTION

The ballistic pendulum apparatus can be used to launch the ball as a free-falling projectile. If the pendulum is raised and the pawl placed in one of the notches on the track before the gun is fired, the ball will travel a path like that shown in Figure 13.3. The ball will travel a horizontal distance X while it falls a vertical distance Y.

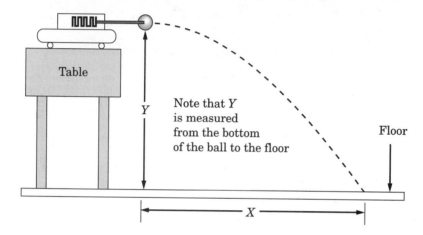

Figure 13.3 Motion of the ball moving horizontal distance X while free-falling height Y.

The original velocity of the ball is completely in the x direction with no y component. The acceleration due to gravity is the only acceleration of the ball, and it is in the y direction with no x component. The equations for the horizontal displacement X and the vertical displacement Y as a function of the time t after the ball is launched are

$$X = v_{x0}\, t \tag{4}$$

$$Y = \tfrac{1}{2}\, gt^2 \tag{5}$$

Note that equation 5 has been written so that a positive displacement is downward in the same direction as g. Equations 4 and 5 can be combined to eliminate the time t and thus express v_{x0} in terms of X and Y as

$$v_{x0} = \frac{X}{\sqrt{2Y/g}} \tag{6}$$

Equation 6 can be used to determine the initial velocity v_{x0} by firing the projectile from a known height Y and measuring the value of X that results.

The velocity in the y direction is initially zero. As the projectile falls under the influence of gravity, it acquires a velocity in the y direction. The magnitude of the velocity in the y direction is given by

$$v_y = gt = \sqrt{g\, 2Y/g} = \sqrt{2gY} \tag{7}$$

EXPERIMENTAL PROCEDURE—PROJECTILE MOTION

In the following procedure, be extremely careful not to fire the ball when anyone is in a position to be struck by it. Serious injury could result.

1. Raise the pendulum and let it catch on the notched track so that the ball can be fired under it.

2. Place the apparatus near the front edge of the laboratory table in a place where there is room for the ball to strike the floor before hitting a wall or any other object. It is crucial that the gun be fired each time from the same position relative to the table. It may be necessary to clamp the apparatus to the table. Place a piece of heavy cardboard or some other object in a position to catch the ball after it strikes the floor but before it strikes a wall. *Do not allow the ball to strike a wall because it will likely damage it.* Make several test firings to locate the approximate place where the ball will land on the floor.

3. Place a sheet of white paper on the floor approximately centered where test firings have landed. Place a piece of carbon paper over the white paper in such a way that the ball striking the carbon paper will leave a dot on the white paper. Tape both of the papers to the floor.

4. Place the ball on the rod of the spring gun. The vertical distance Y that the ball will fall is the distance from the *bottom* of the ball to the floor as shown in Figure 13.3. Measure this distance to the nearest 0.1 mm and record it in Data Table 2 as Y.

5. Fire the ball five times onto the same sheet of paper. Place the ball on the rod and measure the horizontal distance X to the nearest 0.1 mm from the center of the ball to the center of each dot on the paper. Record these five values of X in Data Table 2.

CALCULATIONS—PROJECTILE MOTION

1. Using equation 6, calculate the value of v_{x0} for each of the five values of X. Use a value of $g = 9.80$ m/s^2. Record these values in Calculations Table 2.

2. Calculate the mean $\overline{v_{x0}}$ and standard error α_v for the five values of v_{x0} and record them in Calculations Table 2.

LABORATORY REPORT

Data Table 1

Trial	p	y2 (m)
1		
2		
3		
4		
5		

$m =$ kg	$M =$ kg	$y_1 =$ m

Calculations Table 1

$y_2 - y_1$ (m)	V (m/sec)	v_{x0} (m/sec)

$\overline{v_{x0}} =$ m/sec	$\alpha_v =$ m/sec

Data Table 2

Trial	X (m)
1	
2	
3	
4	
5	

$Y =$ m

Calculations Table 2

v_{x0} (m/sec)

$\overline{v_{x0}} =$ m/sec	$\alpha_v =$ m/sec

SAMPLE CALCULATIONS

QUESTIONS

1. Compare the two different values of $\overline{v_{x0}}$. Calculate the difference and the percentage difference between them. Do the two measurements agree within the combined standard errors of the two measurements?

2. Is either of the two measurements of $\overline{v_{x0}}$ more accurate than the other? Is either of these measurements more precise than the other? State clearly the basis for your answer in each case.

3. Calculate the loss in kinetic energy when the ball collides with the pendulum as the difference between the kinetic energy before and immediately after the collision.

4. What is the fractional loss in kinetic energy? Find by dividing the loss calculated in question 3 by the original kinetic energy.

5. Calculate the ratio $M/(m + M)$ for the values of m and M in Data Table 1. Compare this ratio with the ratio calculated in question 4. Theoretically these two ratios should be the same. State the level of agreement for these two quantities for your data.

6. Consider the ball when fired as a free-falling projectile as it hits the ground. What is the velocity of the projectile in the horizontal direction? What is the velocity of the projectile in the vertical direction? What is the magnitude of its velocity as it hits the ground?

7. We have assumed that the ball when fired as a projectile initially moved exactly in the horizontal direction. If in fact the projectile was fired at some small angle θ to the horizontal, the correct equation relating v_0, θ, X, and Y would then be

$$(4.90X^2/v_0^2)\tan^2\theta - X\tan\theta + (Y + 4.90X^2/v_0^2) = 0$$

Assume that the v_0 determined by the ballistic pendulum measurement is correct. Use that value of v_0 and the measured X and Y in the above equation and solve for the angle. If the ball was fired at that angle relative to the horizontal it would be enough to account for the difference in the two measured values of v_0. (Note the equation is a quadratic in $\tan\theta$ and note that Y is negative.)

Laboratory 14

Conservation of Momentum on the Air Track

PRELABORATORY ASSIGNMENT

Read carefully the entire description of the laboratory and answer the following questions based on the material contained in the reading assignment. Turn in the completed prelaboratory assignment at the beginning of the laboratory period prior to the performance of the laboratory.

1. What is the definition of momentum?

2. What conditions must be satisfied in order for momentum to be conserved?

3. Two pieces of tape are placed a distance of 1.50 m apart on an air track. A 0.350-kg glider on the air track takes a time interval of $\Delta t = 1.30$ s to move between the two pieces of tape. What is the velocity of the glider? What is its momentum?

4. A glider of mass $m_1 = 0.350$ kg moves with a velocity of 0.850 m/s to the right on an air track. It collides with a glider of mass $m_2 = 0.350$ kg at rest. Glider m_1 stops, and m_2 moves in the direction that m_1 was traveling. What is the velocity of m_2? Show your work.

5. An air track glider of mass $m_1 = 0.200$ kg moving at 0.750 m/s to the right collides with a glider of mass $m_2 = 0.400$ kg at rest. If m_1 rebounds and moves to the left with a speed of 0.250 m/s, what is the speed and direction of m_2 after the collision? (*Hint:* Remember that momentum is a vector quantity, and direction is indicated by the sign of the momentum.)

6. For the collision in question 5, calculate the total kinetic energy of the system before the collision. Calculate the total kinetic energy after the collision. Account for any difference in the two.

7. An air track glider of mass $m_1 = 0.300$ kg moving at a speed of 0.800 m/s to the right collides with a glider of mass $m_2 = 0.300$ kg moving at a speed of 0.400 m/s in the opposite direction. After the collision m_1 rebounds at speed 0.200 m/s to the left. After the collision what is the speed and direction of m_2?

8. For the collision in question 7, calculate the total kinetic energy of the system before the collision. Calculate the total kinetic energy after the collision. Account for any difference between the two.

Conservation of Momentum on the Air Track

OBJECTIVES

The momentum of a system of particles is conserved if there are no external forces acting on the system. A collision of two gliders on an essentially frictionless air track should be a good approximation to the ideal case. The forces exerted on each glider by the other is an internal force, and these forces do not change the momentum of the system. Measurements involving a series of collisions between two gliders on a linear air track will be used to accomplish the following objectives:

1. Determination of the velocity and thus momentum of each glider before and after the collision from which the total momentum of the system before and after the collision is obtained

2. Evaluation of the extent to which the total momentum of the system before the collision is equal to the total momentum of the system after the collision

EQUIPMENT LIST

1. Air track, air blower, three gliders (two approximately equal in mass), rubber-band puck launcher

2. Four laboratory timers

3. Laboratory balance and calibrated masses

4. Masking tape

5. Meter stick (if air track does not have marked scale)

THEORY

The momentum of an object of mass m moving with a velocity $\mathbf{v}$ is defined to be

$$\mathbf{p} = m\mathbf{v} \tag{1}$$

Momentum is a vector quantity because it is the product of a scaler (m) with a vector ($\mathbf{v}$). It can be shown for a system of particles that the total momentum of the system is constant if there are no external forces acting on the system. The forces exerted between the particles of the system are called "internal forces," and they cannot change the momentum of the system.

A collision between two objects is an example of a case in which momentum is conserved because the forces that the two objects exert on each other are internal to the system. If $\mathbf{p}_1^i$ and $\mathbf{p}_2^i$ stand for the initial momenta of particles 1 and 2 and $\mathbf{p}_1^f$ and $\mathbf{p}_2^f$ stand for their final momenta after the collision then

$$\mathbf{p}_1^i + \mathbf{p}_2^i = \mathbf{p}_1^f + \mathbf{p}_2^f \tag{2}$$

In the most general case, equation 2 implies that each of the components of the momentum is conserved. For the linear air track, the collision is one dimensional, and the vector nature of the problem is specified by writing momenta to the right as positive and momenta to the left as negative.

Equation 2 is strictly valid only if there are no external forces on the system. Friction on the gliders of an air track will be an external force. Therefore, frictional effects on the gliders will tend to invalidate equation 2 if the frictional forces are comparable to the forces of collision on the gliders.

The measurements will involve determination of the velocities of two gliders before the collision and after the collision. Each of these four velocities is a constant velocity. The value of a velocity will be determined by measuring the elapsed time for the glider to travel some known distance. The collisions will all be arranged to take place in the center of the air track, and thus a fixed path length will be required on either side of the center. A 5-m air track is ideal for performing this experiment, but a shorter air track can be used. Three different collisions will be investigated. They will all be arranged so that any glider that has two different nonzero velocities before and after the collision moves in opposite directions before and after the collision.

EXPERIMENTAL PROCEDURE

1. Using the laboratory balance, determine the mass of the three gliders m_1, m_2, and m_3. The mass of m_1 should be essentially the same as m_2, and m_3 should be greater than m_1. Record the values of the masses in Data Table 1.

2. Using pieces of tape, mark a fixed path length on either side of the center of the track. Leave room in the center of the track for the collision to occur. Leave a distance slightly longer than the sum of the lengths of the gliders. Use as much of the remaining track on either side of the center for the marked paths as is reasonable, but do not extend the marked paths all the way to the end of the track. Figure 14.1 shows how the marker tape should be placed at positions 1, 2, 3, and 4 to define the distances on each side of the collision region for which the time intervals are to be measured. The tape must be placed low enough on the air track for the gliders to clear the tape as they move past. Measure the distance between 1 and 2 and record it in Data Table 1 as d_{12}. Measure the distance between 3 and 4 and record it in Data Table 1 as d_{34}.

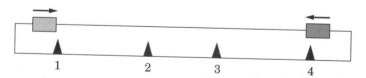

Figure 14.1 Air track with paths 1-2 and 3-4 marked on either side of the center.

3. Two of the cases (Collisions II and III) involve the collision of a moving glider with a second glider at rest. For these cases, an attempt will be made to repeat the collision several times at the same initial velocity. The rubber-band glider launcher (which is just a rubber band stretched across a rigid notched frame as shown in

Figure 14.2) can be used to launch the glider with the same velocity every time with some practice. Launch one of the gliders five or six times starting a timer as the glider is released and stopping the timer when the glider has gone the length of the track. The consistency of the elapsed time in the several trials is a measure of the consistency of the velocity. With practice, the measured time for the glider to go the length of the track should vary by no more than 0.1 s from the average. To achieve this requires both releasing the glider the same way each time and starting the timer on release of the glider. Practice launching the glider until it can be launched with the same velocity every time to the extent that this is possible.

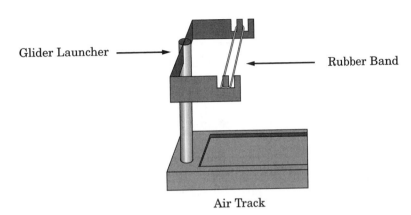

Figure 14.2. Rubber-band glider launcher to assure repeated launches at the same velocity.

4. **Collision I** (Figure 14.3). This collision will be between gliders m_1 and m_2, which have essentially the same mass, with m_2 initially at rest in the center of the track. Glider m_1 will collide with m_2 and essentially stop, and m_2 will move in the direction m_1 was moving before the collision. Launch glider m_1, start a timer as it passes point 1, and stop the timer when it passes point 2. After the collision, start a second timer when glider m_2 passes point 3 and stop this timer when it

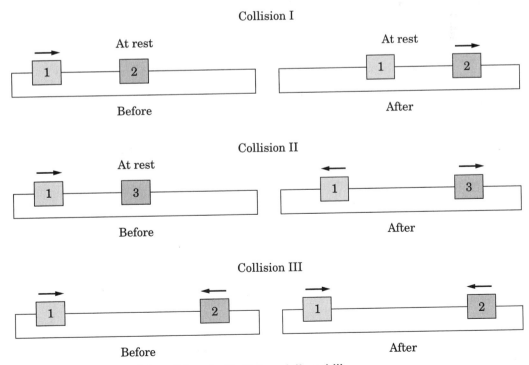

Figure 14.3. Motion of the gliders in Collisions I, II, and III.

passes point 4. Record these time intervals as Δt_{12} and Δt_{34}, respectively, in Data Table 2. Repeat this procedure four more times for a total of five trials. To the extent possible, launch the glider m_1 with the same velocity each time.

5. **Collision II** (Figure 14.3). This collision will be between glider m_1 and glider m_3, with glider m_3 initially at rest. Since $m_3 > m_1$, glider m_1 will rebound after the collision and move back past point 2 and then point 1. Launch glider m_1 and start one timer when m_1 passes point 1 and stop the timer when m_1 passes point 2. This is the time Δt_{12}. After the collision, m_1 will rebound and m_3 will move in the original direction of the motion of m_1. A second timer is used to measure the elapsed time for glider m_3 between points 3 and 4 (time Δt_{34}), and a third timer is used to measure the elapsed time for glider m_1 as it moves back past points 2 and then 1 (time Δt_{21}). Record these three time intervals in Data Table 3. Repeat this procedure four more times for a total of five trials. To the extent possible, launch the glider m_1 with the same velocity every time.

6. **Collision III** (Figure 14.3). This collision will be between gliders m_1 and m_3 launched from opposite ends of the air track with velocities directed toward the center of the air track. Launch m_1 and m_3 with the rubber-band launchers and give m_1 about 30% greater speed than m_3. Thus, m_1 must be launched after m_3 in order for them to collide at the center of the track. Both gliders will rebound after the collision and thus reverse direction. Two timers are used to measure the time intervals Δt_{12} and Δt_{21} for m_1 as it moves toward the center and rebounds, and two other timers are used to measure the time intervals Δt_{43} and Δt_{34} for m_3 as it moves toward the center and then rebounds. Record the values of the four time intervals in Data Table 4. Repeat the procedure four more times for a total of five trials. To the extent possible, launch glider m_1 at approximately the same velocity each time, and launch m_3 at approximately the same velocity each time.

CALCULATIONS

1. Calculate the velocities v from $v = \frac{d}{\Delta t}$ for each of the measured time intervals and distances. Let velocities to the right be positive and velocities to the left be negative. Record the values of the velocities in Calculations Tables 2, 3, and 4.

2. Calculate the momentum for each of the gliders before and after the collision using equation 1. Let momenta to the right be positive and momenta to the left be negative. Record the values of the momenta in Calculations Tables 2, 3, and 4.

3. For cases in which there is more than one glider contributing to the momentum either before or after a collision, calculate the total momentum and record these values in the Calculations Tables.

4. Calculate the percentage difference between the total momentum before the collision and the total momentum after the collision for each trial of each collision. Record the values in the Calculations Tables.

Conservation of Momentum on the Air Track

LABORATORY REPORT

Data Table 1

m_1 (kg)	m_2 (kg)	m_3 (kg)	d_{12} (m)	d_{34} (m)

Data Table 2

		Trial 1	Trial 2	Trial 3	Trial 4	Trial 5
Before	Δt_{12} (s)					
After	Δt_{34} (s)					

Calculations Table 2

		Trial 1	Trial 2	Trial 3	Trial 4	Trial 5
m_1	v_1 (m/s)					
Before	p_1 (kg–m/s)					
m_2	v_2 (m/s)					
After	p_2 (kg–m/s)					
% Difference p_1 & p_2						

SAMPLE CALCULATIONS

Data Table 3

		Trial 1	Trial 2	Trial 3	Trial 4	Trial 5
m_1 Before	Δt_{12} (s)					
m_1 After	Δt_{21} (s)					
m_3 After	Δt_{34} (s)					

Calculations Table 3

		Trial 1	Trial 2	Trial 3	Trial 4	Trial 5
m_1	v_1^i (m/s)					
Before	$\mathbf{p}_1^i$ (kg − m/s)					
m_1	v_1^f (m/s)					
After	$\mathbf{p}_1^f$ (kg − m/s)					
m_3	v_3^f (m/s)					
After	$\mathbf{p}_3^f$ (kg − m/s)					
Total $\mathbf{p}$ After Collision						
% Diff. $\mathbf{p}$ Before & After						

SAMPLE CALCULATIONS

Data Table 4

		Trial 1	Trial 2	Trial 3	Trial 4	Trial 5
m_1 Before	Δt_{12} (s)					
m_3 Before	Δt_{43} (s)					
m_1 After	Δt_{21} (s)					
m_3 After	Δt_{34} (s)					

Calculations Table 4

		Trial 1	Trial 2	Trial 3	Trial 4	Trial 5
m_1	v_1^i (m/s)					
Before	$\mathbf{p}_1^i$ (kg − m/s)					
m_3	v_3^i (m/s)					
Before	$\mathbf{p}_3^i$ (kg − m/s)					
Total $\mathbf{p}$ Before Collision						
m_1	v_1^f (m/s)					
After	$\mathbf{p}_1^f$ (kg − m/s)					
m_3	v_3^f (m/s)					
After	$\mathbf{p}_3^f$ (kg − m/s)					
Total $\mathbf{p}$ After Collision						
% Diff. $\mathbf{p}$ Before & After						

SAMPLE CALCULATIONS

QUESTIONS

1. Consider the percentage differences between the total momentum before the collision and the total momentum after the collision for the various trials of Collision I. Do these data indicate that momentum is conserved for this collision?

2. Consider the percentage differences between the total momentum before the collision and the total momentum after the collision for the various trials of Collision II. Do these data indicate that momentum is conserved for this collision?

3. Consider the percentage differences between the total momentum before the collision and the total momentum after the collision for the various trials of Collision III. Do these data indicate that momentum is conserved for this collision?

4. For each of the Collisions I, II, and III, consider the one trial that has the smallest percentage difference and calculate the total kinetic energy before the collision (K_i) and the total kinetic energy after the collision (K_f) for that trial.

Collision I   Trial _____ $K_i =$ _____ $K_f =$ _____

Collision II   Trial _____ $K_i =$ _____ $K_f =$ _____

Collision III   Trial _____ $K_i =$ _____ $K_f =$ _____

5. What percentage of the original kinetic energy is lost in each collision that you calculated? What happens to the lost kinetic energy?

Collision I   Trial _____ $\Delta K =$ _____   % Lost = _____

Collision II   Trial _____ $\Delta K =$ _____   % Lost = _____

Collision III   Trial _____ $\Delta K =$ _____   % Lost = _____

6. Is the kinetic energy approximately conserved for any of the collisions that you calculated? If so, state which ones.

Laboratory 15

Conservation of Momentum on the Air Table

PRELABORATORY ASSIGNMENT

Read carefully the entire description of the laboratory and answer the following questions based on the material contained in the reading assignment. Turn in the completed prelaboratory assignment at the beginning of the laboratory period prior to the performance of the laboratory.

1. What is the definition of momentum?

2. What conditions must be satisfied for momentum to be conserved?

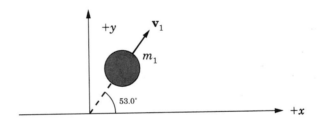

Figure 15.1 Particle of mass m_1 moving at speed v_1 53.0° relative to $+ x$ axis.

3. A particle of mass $m_1 = 0.350$ kg has speed $v_1 = 0.135$ m/s at a direction of 53.0° above the $+x$-axis as shown in Figure 15.1. What is the magnitude of the particle's momentum? Show your work.

4. What is the x component and what is the y component of the particle shown in Figure 15.1? Show your work.

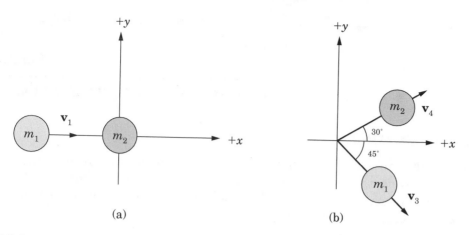

(a)   (b)

Figure 15.2 Particle of mass m_1 at speed v_1 collides with a particle of mass m_2 at rest.

5. A particle of mass $m_1 = 1.000$ kg moves at speed $v_1 = 0.500$ m/s as shown in Figure 15.2(a). It collides with a particle of mass $m_2 = 2.000$ kg at rest at origin. What is the total momentum of the system in the x direction before the collision? What is the total momentum of the system in the y direction before the collision? Show your work.

6. Figure 15.2(b) shows that after the collision, m_1 moves with speed v_3 at an angle $\theta_3 = 45.0°$ below the x-axis, and m_2 moves with a speed v_4 at an angle $\theta_4 = 30.0°$ above the x-axis. Write an expression for the total momentum of the system in the x direction and another expression for the total momentum in the y direction after the collision in terms of the symbols m_1, m_2, v_3, and v_4 and the angles θ_3 and θ_4.

7. Equate the expressions for the x component above to the value of the x component in question 5. Equate the expression for the y component above to the value of the y component in question 5. In the resulting two equations, v_3 and v_4 are the only two unknowns. Solve the two equations for v_3 and v_4. Show your work.

Conservation of Momentum on the Air Table

OBJECTIVES

If two objects moving in a plane undergo a collision of the most general type, each object has components of momentum along two perpendicular coordinates after the collision. The conservation of momentum applied to this case implies that momentum is conserved for each of the components. In this laboratory, collisions between two pucks on an air table will be used to accomplish the following objectives:

1. Determination of the velocities and thus momenta of the pucks before and after several different types of collisions
2. Imposition of an arbitrary coordinate system on each collision in order to separate each of the momenta into components
3. Evaluation of the extent to which the total momentum of the system for each component is the same before and after the collision

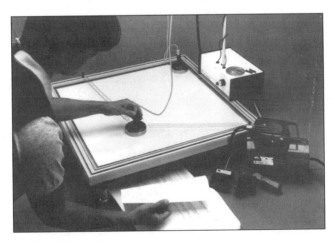

Figure 15.3 Precision air table. (Photo courtesy of Central Scientific Co., Inc.)

EQUIPMENT LIST

1. Air table with pucks, sparktimer, airpump, foot switch, lead weights, Velcro collars, and level
2. Carbon paper and recording paper
3. Meter stick or ruler, protractor, and square

THEORY

The momentum **p** of an object of mass m moving with a velocity **v** is defined to be

$$\mathbf{p} = m\,\boldsymbol{v} \tag{1}$$

Momentum is a vector quantity because it is the product of a scaler (m) with a vector (**v**). It can be shown for a system of particles that the total momentum of the system is constant if there are no external forces acting on the system. The forces exerted between the particles of the system are called "internal forces," and they will not change the total momentum of the system.

A collision between two objects is an example of a case in which momentum is conserved because the forces that the two objects exert on each other are internal to the system. If $\mathbf{p}_1^{\,i}$ and $\mathbf{p}_2^{\,i}$ stand for the initial momenta of particles 1 and 2 before a collision, and $\mathbf{p}_1^{\,f}$ and $\mathbf{p}_2^{\,f}$ stand for their final momenta after the collision then

$$\mathbf{p}_1^{\,i} + \mathbf{p}_2^{\,i} = \mathbf{p}_1^{\,f} + \mathbf{p}_2^{\,f} \tag{2}$$

The most general type of collision between two objects is shown in Figure 15.4. It is an object of mass m_1 moving with a velocity v_1 and an object of mass m_2 moving with a velocity v_2 that collide at the origin of the coordinate system as shown. After the collision the two objects move off with velocities v_3 and v_4. For the collision in Figure 15.4, with coordinates arbitrarily assumed as shown, the equations that represent the conservation of momentum are

x direction $\quad m_1 v_1 \cos\theta_1 - m_2 v_2 \cos\theta_2 = m_2 v_4 \cos\theta_4 - m_1 v_3 \cos\theta_3 \tag{3}$

y direction $\quad m_1 v_1 \sin\theta_1 + m_2 v_2 \sin\theta_2 = m_1 v_3 \sin\theta_3 + m_2 v_4 \sin\theta_4 \tag{4}$

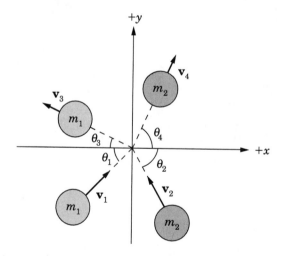

Figure 15.4 Collision between objects mass m_1 and m_2 both having an initial velocity.

This laboratory will investigate the general case shown above as well as several simpler special cases. For example, the collision is simplified if the masses of the colliding objects are equal ($m_1 = m_2$). A simplification results if one of the objects is originally at rest, and another simplification results if the collision is head-on, so that all the motion takes place along a single line. Collisions on the air table between two pucks will be used to study several of these cases.

EXPERIMENTAL PROCEDURE

1. Unless you have prior experience with the air table, read the instruction manual to become familiar with its operation, including the airpump, sparktimer, and footswitch. Be very careful not to touch the pucks except by the insulated tubes when the sparktimer is in operation.

2. Level the air table by means of the three adjustable legs until a puck placed near the center of the table is essentially motionless. *It is especially critical to have the table as level as possible* for these collision events because several of the collisions are to be made with one of the pucks initially at rest.

3. Set the sparktimer to 20.0 Hz. Perform each of the four collisions described in the steps below. For each collision, start the sparktimer just after the pucks are launched and stop it before the pucks collide with the edge of the table. The effects of friction cause disagreement between the experimental results and the theory, and this disagreement is larger the lower the speed of the pucks. For best results the pucks should be moving as fast as is reasonable. However, it is possible to damage the glass tops of the tables if the pucks are thrown too fast. *Be very careful not to damage the air table.* Consult your instructor for advice about the speed of the pucks. Use a clean sheet of recording paper for each collision.

4. **Collision I**. This is a collision between two pucks of approximately the same mass. Using a laboratory balance, determine the mass of each puck. Label one puck 1 and the other 2. Record the masses of each puck in the Data Table. Launch puck 1 from near one corner of the table so that it strikes puck 2, which is at rest near the center of the table. The collision should not be head-on. The pucks should move at some fairly large angle relative to each other after the collision. Before the recording paper is ever removed from the table, label each of the sparktimer tracks with the puck number and the direction that each traveled before and after the collision. Be sure it is clear which tracks are before the collision and which ones are after. Do this labeling for each of the collisions before removing the recording paper from the table. Label this recording as Collision I and put it aside for future analysis.

5. **Collision II**. Use the same pucks as used in Collision I, but this time launch them both at the same time from *adjacent corners* of the table so that they collide at the center of the table at an angle of about 90°. Carefully label the recording as Collision II in the manner described above and set the recording aside for future analysis.

6. **Collision III**. Same as Collision II, except use masses that are not equal. Use puck 1 as it is. Place several of the lead weights that fit over the pucks to increase the mass of puck 2. Using the laboratory balance, determine the mass of puck 2 with the lead on it and record the results in the Data Table. Launch the two pucks from adjacent corners so that they collide near the center of the table. Label the recording paper as above and set it aside for future analysis.

7. **Collision IV**. Remove the lead from puck 2. Place the Velcro collars on each of the two pucks. Determine the mass of each of the pucks and record the values in the Data Table. With the Velcro collars on the pucks, they will stick together on collision; thus, the collision will be completely inelastic. Launch the two pucks of approximately equal mass from adjacent corners of the table so that they collide near the center of the table. It may be necessary to try this several times because when the pucks stick together they may have a tendency to rotate about

a common center of mass as they move off together. This makes it almost impossible to determine the correct direction to assign to the combined mass after the collision. Therefore, it is necessary to achieve a collision in which the combined masses do not rotate as they move off together. A trace showing the combined pucks leaving parallel tracks as they move off together is evidence that no rotation is occurring. It may take several traces before you get one in which the pucks show parallel tracks when stuck together. Label the acceptable recording and set it aside for future analysis.

8. For each of the collisions, determine the displacements along the four tracks that represent the motion of the pucks before and after the collision. Measure the displacements of the puck for a time interval of 0.200 s, which will be four spark intervals, since 20.0 Hz is the sparktimer rate. Be sure that the appropriate displacement is associated with each of the pucks both before and after the collision. Record the displacements Δs in the Data Table.

9. For each of the collisions, perform the following analysis. Draw an x coordinate axis on the recording paper. Remember that this is an arbitrary choice and thus can be chosen in whatever way is most convenient for the analysis. For example, it will usually save some work to choose the $+x$ direction to be the direction of the initial motion of one of the pucks. That puck would then initially have only an x component, thus its y component would be zero. The components for the other puck before the collision, and for both the pucks after the collision, can then be determined by finding the angle that their tracks make with the $+x$-axis. This process is illustrated in Figure 15.5. The $+x$-axis has been chosen to be in the direction of the initial motion of puck 1. The other angles are then determined by the angle that a line drawn through the sparktimer trace for each puck's motion makes with the $+x$-axis. The angles have all been chosen as the angle that the velocity vector makes with either the $+x$-axis or the $-x$-axis. This means that the x components will be given by the cosine and the y components by the sine. The direction of the components must also be taken into account. For example, in Figure 15.5, puck 1 has a negative x component and positive y component after the collision, puck 2 has a negative x component and positive y component before the collision, and both the x and y components of puck 2 are positive after the collision. These signs must be included in the calculation of the various components later in the Calculations section.

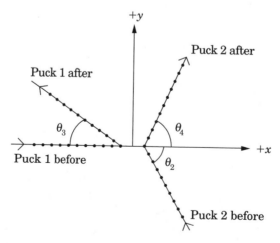

Figure 15.5 Example of how to determine the components of each momentum.

For each of the four collisions, draw in the $+x$-axis through the points of the initial motion of puck 1 in the direction of its motion. Using a protractor, carefully measure the angle of the other three tracks with the x-axis and record the values of those angles in the Data Table. Note that with this choice of $+x$-axis, θ_1 will be zero for all of the collisions.

CALCULATIONS

1. For each of the four collisions, calculate the magnitudes of the velocity for the two pucks before and after the collision from the value of the displacements Δs in the Data Table and time interval $\Delta t = 0.200$ s. The equation is $v = \Delta s/\Delta t$, where Δs stands for each of the displacements (1, 2 before the collision and 3, 4 after the collision). Record the values of these velocities as v_1, v_2, v_3, and v_4 in the Calculations Table.

2. For each of the four collisions, calculate the x component and the y component of the momentum for each puck before and after the collision. The equations for the momentum components are given by

$$p_{1x} = m_1 v_1 \cos\theta_1 \quad \text{and} \quad p_{1y} = m_1 v_1 \sin\theta_1 \tag{5}$$

where the equations have been written using v_1 as an example. The extension of equations 5 for the other velocities should be clear. Using equations 5 and the equations appropriate for the other velocities, calculate the components for each of the collisions and record them in the Calculations Table. Be careful to account for the direction of each component by making it either positive or negative as described earlier. Review the example shown in Figure 15.5 to be sure that you understand how to determine the sign for the calculated components of momentum.

Conservation of Momentum on the Air Table

LABORATORY REPORT

(Record all *masses in kg,* all *displacements in m,* all *velocities in m/s,* and all *momenta in kg-m/s*)

COLLISION I

Data Table

m_1	m_2	Δs_1	Δs_2	Δs_3	Δs_4	θ_1	θ_2	θ_3	θ_4

Calculations Table

v_1	v_2	p_{1x}	p_{1y}	p_{2x}	p_{2y}	$p_{1x} + p_{2x}$	$p_{1y} + p_{2y}$
v_3	v_4	p_{3x}	p_{3y}	p_{4x}	p_{4y}	$p_{3x} + p_{4x}$	$p_{3y} + p_{4y}$

COLLISION II

Data Table

m_1	m_2	Δs_1	Δs_2	Δs_3	Δs_4	θ_1	θ_2	θ_3	θ_4

Calculations Table

v_1	v_2	p_{1x}	p_{1y}	p_{2x}	p_{2y}	$p_{1x} + p_{2x}$	$p_{1y} + p_{2y}$
v_3	v_4	p_{3x}	p_{3y}	p_{4x}	p_{4y}	$p_{3x} + p_{4x}$	$p_{3y} + p_{4y}$

COLLISION III

Data Table

m_1	m_2	Δs_1	Δs_2	Δs_3	Δs_4	θ_1	θ_2	θ_3	θ_4

Calculations Table

v_1	v_2	p_{1x}	p_{1y}	p_{2x}	p_{2y}	$p_{1x} + p_{2x}$	$p_{1y} + p_{2y}$
v_3	v_4	p_{3x}	p_{3y}	p_{4x}	p_{4y}	$p_{3x} + p_{4x}$	$p_{3y} + p_{4y}$

COLLISION IV

Data Table

m_1	m_2	Δs_1	Δs_2	Δs_3	Δs_4	θ_1	θ_2	θ_3	θ_4

Calculations Table

v_1	v_2	p_{1x}	p_{1y}	p_{2x}	p_{2y}	$p_{1x} + p_{2x}$	$p_{1y} + p_{2y}$
v_3	v_4	p_{3x}	p_{3y}	p_{4x}	p_{4y}	$p_{3x} + p_{4x}$	$p_{3y} + p_{4y}$

SAMPLE CALCULATIONS

QUESTIONS

1. In each collision, if momentum is conserved the total x component before the collision $(p_{1x} + p_{2x})$ should equal the total x component after the collision $(p_{3x} + p_{4x})$. Similarly, the y component $p_{1y} + p_{2y}$ should equal $p_{3y} + p_{4y}$. For each collision, calculate the % difference between these quantities. Record the results below.

   Collision I x comp % diff = _____ y comp % diff = _____

   Collision II x comp % diff = _____ y comp % diff = _____

   Collision III x comp % diff = _____ y comp % diff = _____

   Collision IV x comp % diff = _____ y comp % diff = _____

2. Using both the absolute differences and the % differences calculated in question 1 as a guide, evaluate the extent to which your data show that momentum is conserved.

3. For each collision, the magnitude of the total initial momentum is given by the quantity $\sqrt{(p_{1x} + p_{2x})^2 + (p_{1y} + p_{2y})^2}$. Calculate that quantity for each of the four collisions and record the results below.

   Collision I p_{int} = _____

   Collision II p_{int} = _____

   Collision III p_{int} = _____

   Collision IV p_{int} = _____

4. The calculation done in question 1 is not a very sensitive test for cases in which one of the components of the momentum is small compared to the total momentum. As an attempt to compensate for that, consider the following. Divide the difference between each component before and after the collision by the value of p_{int} calculated in question 3 and express it as a percentage. The smaller the value of this quantity the better the data is consistent with momentum conservation.

   Collision I x comp % diff/p_{int} = _____ y comp % diff/p_{int} = _____

   Collision II x comp % diff/p_{int} = _____ y comp % diff/p_{int} = _____

   Collision III x comp % diff/p_{int} = _____ y comp % diff/p_{int} = _____

   Collision IV x comp % diff/p_{int} = _____ y comp % diff/p_{int} = _____

5. Are any of these results significantly better than the results of the calculation in question 1? If so, indicate which ones are better.

6. Calculate the initial kinetic energy, final kinetic energy, and loss in kinetic energy in each collision and record below.

Collision I $K_i =$ _____ $K_f =$ _____ $\Delta K =$ _____

Collision II $K_i =$ _____ $K_f =$ _____ $\Delta K =$ _____

Collision III $K_i =$ _____ $K_f =$ _____ $\Delta K =$ _____

Collision IV $K_i =$ _____ $K_f =$ _____ $\Delta K =$ _____

7. What percentage of the original kinetic energy is lost in each collision? What happens to any lost kinetic energy?

Collision I % Kinetic energy lost = _____

Collision II % Kinetic energy lost = _____

Collision III % Kinetic energy lost = _____

Collision IV % Kinetic energy lost = _____

Centripetal Acceleration of an Object in Circular Motion

OBJECTIVES

A mass M that moves in a circle of radius R at constant speed v has a centripetal acceleration of magnitude v^2/R directed toward the center of the circle. The centripetal force F acting on the mass is therefore given by $F = Mv^2/R$. In this laboratory, independent measurements of the quantities F, M, v, and R will be made to verify that relationship. The following measurements will be made to accomplish this objective.

1. The period T of an object of mass M that rotates at constant speed v in a circle of radius R will be measured.

2. The speed v of the object will be determined from its measured period T and the measured value of R, the radius of the motion.

3. The centripetal force F will be measured directly. It is provided by a spring. The force will be determined by measuring the force required to stretch the spring when the apparatus is not rotating by the same amount it was stretched while it was rotating.

EQUIPMENT LIST

1. Hand-operated centripetal force apparatus (The device described is available from Sargent-Welch Scientific Company. A similar version is available from Central Scientific Company.)

2. Laboratory balance, calibrated slotted masses, and mass holder

3. Laboratory timer, and metal ruler

THEORY

When an object moves in a circle at constant speed the velocity vector of the object's motion is always tangent to the circle. This implies that the direction of the velocity is continuously changing, and thus the object is accelerated because acceleration is by definition a change in velocity per unit time. Figure 16.1 shows the velocity vector at various points around the circle for an object moving in a circle at constant speed. The lengths of the vectors are the same because the speed is constant, and the direction of the vectors indicate the direction of the velocity at that point. Also shown in Figure 16.1 are the velocity vectors $\mathbf{v}_i$ and $\mathbf{v}_f$ at two times t_i and t_f, with a very short time interval between them. In the third part of the figure is shown the vector difference $\Delta\mathbf{v} = \mathbf{v}_f - \mathbf{v}_i$, indicating that the change in velocity $\Delta\mathbf{v}$ always points toward the center of the circle. The acceleration $\mathbf{a}$ is defined by

$$\mathbf{a} = \frac{\Delta\mathbf{v}}{\Delta t} \tag{1}$$

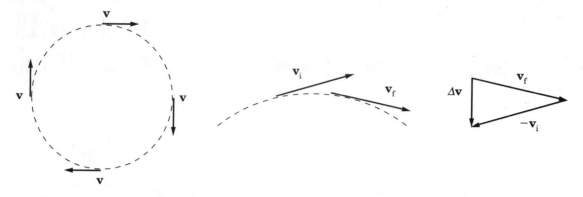

Figure 16.1 Velocity vectors for circular motion at constant speed. Vectors at two times t_i and t_f close together and the change in velocity $\Delta \mathbf{v}$ pointing toward the center of the circle.

Thus, the acceleration is in the direction of $\Delta \mathbf{v}$ and is also always pointed toward the center of the circle. The magnitude of the acceleration $\mathbf{a}$ is given by

$$a = \frac{v^2}{R} \qquad (2)$$

According to Newton's second law, the magnitude of the centripetal force F and the magnitude of the centripetal acceleration a are related by $F = Ma$, where M is the mass of the object moving in a circle at constant speed v. Therefore, using equation 2 for the acceleration gives

$$F = M\frac{v^2}{R} \qquad (3)$$

If the object moves at constant speed v in a circle of radius R, the time for one complete revolution around the circle is the period T. The period T is related to the speed v by the expression

$$v = \frac{2\pi R}{T} \qquad (4)$$

The centripetal force apparatus has a mass bob with a pointed tip at the bottom that is suspended from a horizontal rotating bar. The bob also has a spring hooked between the side of the bob and the central rotating shaft in such a way that the spring provides a horizontal centripetal force when the bob rotates in a horizontal plane. The bob is rotated at a fixed radius R from the central rotating shaft by ensuring that the tip of the bob passes on each revolution precisely over a pointer located a distance R from the central rotating shaft. For a given mass M of the bob and a particular spring, the bob will rotate at a given radius R only for one particular rotation period T. Figure 16.2(a) shows the apparatus when the system is rotating at the period necessary to produce a rotation at radius R. A measurement of T will be made for a given R and M. Equation 4 allows a determination of v, and using that value in equation 3 allows a determination of the centripetal force F. This will be referred to as F_{theo} for the theoretical value of the force.

The force, which the spring exerts on the bob while it is rotating at the distance R from the central shaft, depends on the amount the spring is stretched under those conditions. The value of this force can be measured by determining the force needed to stretch the spring by the same amount when the apparatus is not rotating.

Figure 16.2(b) shows how a string can be attached to the other side of the bob and slotted masses can be applied over the pulley mounted near the end of the base. The weight of the total mass needed to stretch the spring until the tip of the bob is aligned with the pointer is the experimental value of the centripetal force F. This will be referred to as F_{exp}.

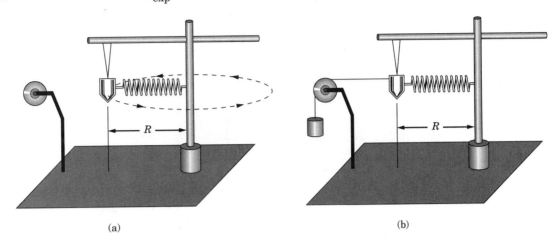

(a)   (b)

Figure 16.2 (a) Centripetal force apparatus rotating. (b) Determination of the centripetal force by measuring the force needed to stretch the spring under static conditions.

EXPERIMENTAL PROCEDURE

1. Detach the bob from its support strings and determine its mass with a laboratory balance. Be sure to remove the spring from the bob before determining its mass. Record the value of the mass of the bob as m_b in the Data Table.

2. Hang the bob from its cross-arm support by the strings that support it (Figure 16.3). Do not attach the spring to the bob but let it hang vertically. Adjust the position of the pointer to its closest position to the rotating shaft for the minimum value of R. Loosen the screw holding the cross arm in the rotating shaft and adjust its position until the tip of the bob is precisely above the tip of the pointer that determines R. The tip of the bob should be about 1 mm above the pointer. Measure with the metal ruler the distance from the center of the pointer to the center of the rotating shaft. Record this value as R in the Data Table.

3. Attach the spring to the bob and to the rotating shaft. Rotate the system as shown in Figure 16.2 (a) by twirling the rotating shaft between your thumb and first finger. The bob will pass over the position of the pointer at radius R only for one particular rotation period T. When this rotation rate has been achieved, measure the total time for 25 complete revolutions of the bob at this radius R and record it in the Data Table as Time 1. Note that you must continue to rotate the apparatus by hand while attempting to keep the rotation speed as constant as possible and at the same time ensuring that the radius of rotation is fixed and the bob passes over the pointer on each rotation. Repeat this process two more times, recording the two other measurements of the time for 25 complete revolutions as Time 2 and Time 3. Record in the Data Table the value of the rotating mass for this part of the procedure as the value of m_b.

Figure 16-3 Centripetal Force Apparatus (Photo courtesy of Sargent Welch Scientific Co.).

4. With the system not rotating, measure directly the centripetal force by attaching a string to the side of the bob opposite the spring. Apply slotted weights over the pulley mounted at the end of the base as shown in Figure 16.2 (b) until the tip of the bob is just above the tip of the pointer. Let m_a stand for the total mass needed to stretch the spring by the proper amount. The experimental value for the centripetal force is then $m_a g$. Record in the Data Table the value of m_a needed to stretch the spring to the position R at which the pointer is located.

5. Repeat steps 3 and 4 above using the same value of R but using two additional values of the rotating mass. First add a 0.0500-kg slotted mass to the bob. Place the slotted mass on the top of the bob with the open end of the slot outward and secure it in place with the knurled nut on the bob. Record the value of the rotating mass in the Data Table as m_b + 0.0500 kg. Perform all the measurements of steps 3 and 4 on this mass and record all the results in the Data Table. Finally, remove the 0.0500-kg mass from the bob and replace it with a 0.1000-kg slotted mass. Record this value of the rotating mass and repeat the measurements of steps 3 and 4. Record all the results in the Data Table.

6. Remove the slotted mass from the bob. Repeat steps 3 and 4 using the bob as the rotating mass but use three different values of R. Remove the spring from the bob and let it hang vertically. Adjust the position of the pointer until it is about 1 cm further from the rotating shaft. Loosen the screw holding the cross arm and adjust it until the bob is just above the pointer at its new position. Measure the new value for R and record it in the Data Table along with the rotating mass, which is again m_b. Reattach the spring and perform the measurements of step 3 and 4 and record those results in the Data Table. Repeat the entire process for two other values of R, increasing the value of R about 1 cm each time if possible. If there is not enough range in the pointer to increase by 1 cm each time, make somewhat smaller increases in R each time. In any case, choose a total of three other values for R within the range of the variation of the pointer. Record all results in the Data Table.

CALCULATIONS

1. Calculate the mean of the three trials of the time for 25 complete revolutions and record it in the Calculations Table as $\overline{\text{Time}}$. Divide the value of $\overline{\text{Time}}$ by 25 and record the result in the Calculations Table as the period T.

2. From the measured values of R and T calculate the speed v for each case using equation 4. Record the results in the Calculations Table.

3. From the values of M, v, and R, calculate the theoretical value for the centripetal force using equation 3. Record the results in the Calculations Table as F_{theo}.

4. From the values of m_a for each case, calculate the experimental value for the centripetal force as $m_a g$. Use a value of 9.80 m/s^2 for g. Record the result in the Calculations Table as F_{exp}.

5. Calculate the percentage difference between the values of F_{theo} and F_{exp}. Record the results in the Calculations Table.

Moment of Inertia and Rotational Motion

OBJECTIVES

For rotation of an object about a fixed axis, the quantity called moment of inertia I is a measure of the tendency of the object to resist a change in its rotational motion. The quantity that produces an angular acceleration α of the objects is called torque τ and the relationship between these quantities is $\tau = I\alpha$. This laboratory will use a solid cylinder with a hub around which a string is wrapped with a weight hung on the string. The weight produces a torque that can be varied by changing the weight. Measurements of the time it takes for the weight on the string to fall a given distance as it causes the cylinder to rotate will be used to achieve the following objectives:

1. Determination of the angular acceleration of the cylinder as a function of the applied torque

2. Determination of the moment of inertia I of the cylinder from the slope of the data for applied torque versus angular acceleration

3. Theoretical calculation of the moment of inertia of the cylinder

4. Comparison of the experimental and theoretical values for the moment of inertia of the cylinder

EQUIPMENT LIST

1. Rotational inertia apparatus (wheel and axle with hub and tripping platform)

2. Meter stick, string, mass holder, and slotted masses

3. Laboratory timer

4. Laboratory balance

5. Vernier calipers and large calipers

THEORY

For linear motion Newton's second law $F = ma$ describes the relationship between the applied force F, the mass m of an object, and its acceleration a. The force is the cause of the acceleration, and the mass is a measure of the tendency of the object to resist a change in its linear translational motion.

For rotational motion of some object about a fixed axis, an equivalent description for the relationship between the applied torque τ, the moment of inertia I, and the angular acceleration α of the object is given by

$$\tau = I\alpha \tag{1}$$

Thus, the torque τ is the cause of the angular acceleration α, and the moment of inertia I is a measure of the tendency of the body to resist a change in its rotational motion.

The moment of inertia I of a rigid body depends on the mass of the body and the way in which the mass is distributed relative to the axis of rotation. A very thin solid cylinder is shown in Figure 17.1. Three arbitrarily chosen elements of mass m_1, m_2, and m_3 are shown located at distances r_1, r_2, and r_3 from the axis of rotation AB. These three elements of mass contribute to the moment of inertia a total equal to $m_1 r_1^2 + m_2 r_2^2 + m_3 r_3^2$. Each of the elements of mass that make up the entire mass of the cylinder contribute in the same way to the total moment of inertia of the cylinder. If $m_i r_i^2$ stands for the contribution to the moment of inertia from an element of mass m_i located a distance r_i from some particular axis, then the total moment of inertia of the cylinder is given by

$$I = \sum_{i=1}^{N} m_i r_i^2 \tag{2}$$

where N s arbitrarily chosen, but must be large enough so that each element of mass ap roximates a point mass. Equation 2 is not applicable just to a cylinder but is in fa t the general definition of the moment of inertia about a particular axis for any object. Note that the moment of inertia of a body is different for different axes of rotation.

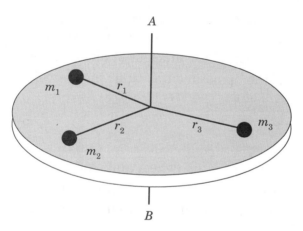

Figure 17.1 Elements of mass contributing to the moment of inertia of a cylinder.

For most objects, the evaluation of equation 2 is best done by taking the limit to produce an infinite number of elements of mass and then using calculus. This has been done for many symmetrically shaped objects. In particular, for a solid cylinder of radius R and total mass M, the results are $I = \frac{1}{2} MR^2$. Note that knowledge of the thickness of the cylinder is not needed in order to calculate the moment of inertia.

In this laboratory a wheel and axle with a small hub will be caused to rotate about its central axis by application of a torque caused by the weight of a mass m on a string wrapped around the hub as shown in Figure 17.2. The mass m moves linearly downward as the cylinder rotates about its fixed axis. The two forces acting on the mass m are the weight mg and the tension T in the string. Applying Newton's second law to the linear motion of the mass gives

$$mg - T = ma \tag{3}$$
or
$$T = m(g - a) \tag{4}$$

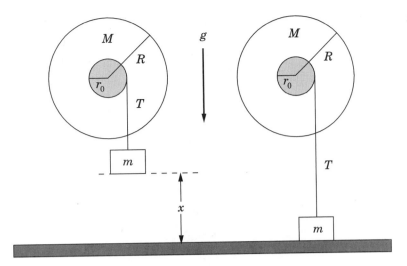

Figure 17.2 Motion of mass m as it moves linearly x and rotates a cylinder.

The wheel of moment of inertia I rotates from the torque applied by the tension in the string T on the hub. The definition of torque is the product of an applied force through a perpendicular lever arm. For the torque due to the tension T this gives

$$\tau = Tr_0 \tag{5}$$

where r_0 is the radius of the hub.

When the mass m is released from rest, it accelerates for a distance x and then strikes the floor as shown in Figure 17.2. It takes some time t to move this distance x. The acceleration a can be determined from rearranging the relationship $x = 1/2\ at^2$ to

$$a = \frac{2x}{t^2} \tag{6}$$

Since a point on the radius r_0 of the hub must travel the same linear distance that mass m travels as it moves the distance x, the angular coordinate, velocity, and acceleration are related to the linear coordinate, velocity, and acceleration by the ratio of the radius of the hub. In particular the linear and angular accelerations are related by

$$\alpha = \frac{a}{r_0} \tag{7}$$

where r_0, the lever arm of the torque, is equal to the radius of the hub.

The relationships described above will be used in this laboratory in the following way. A series of different masses will be placed on the string (Figure 17.3), and the time it takes for each mass to accelerate the distance x will be determined. Equation 6 will then be used to determine the acceleration for each m. Next, equation 7 will be used to determine the angular acceleration α for each mass m. Equation 4 will then be used to determine the tension T, and using T in equation 5, the torque τ will be found. According-ing to equation 1 these data for τ versus α should be linear, with the moment of inertia I as the slope. A linear least squares fit to the data will determine I.

The moment of inertia of the wheel can be determined theoretically by treating the system as a simple cylinder. This is a good approximation because the hub is hollow, made of a lightweight material, and its mass contributes very little to the moment of inertia of the system.

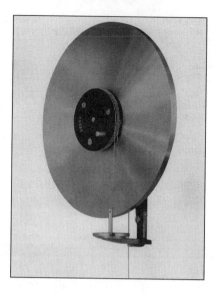

Figure 17-3 Rotational inertia apparatus. (Photo courtesy of Sargent Welch Scientific Co.)

EXPERIMENTAL PROCEDURE

1. Determine the values of the total mass of the wheel M, the radius of the wheel R, and the radius of the small hub r_0. The mass M may be stamped on the wheel, or else it may be available from your instructor. Record the values of M, R, and r_0 in the Data Table.

2. If the wheel-and-axle system is not already mounted on the wall in a permanent position, clamp the system to the laboratory table in such a manner that the wheel hangs over the edge of the table, the plane of the wheel is vertical, and the rotation axis is horizontal. Use thick, very rigid rods to make the system as free from vibration as possible. For best results the wheel should be high enough from the floor in order that the mass placed on the string will travel at least 1 m and preferably 2 m before striking the floor. The larger this distance is, the larger will be the time intervals; and the larger the time interval, the more precision with which it can be measured.

3. The apparatus has a small tripping platform on which the mass holder on the end of the string rests before it is released. Measure the distance from the top of this platform to the floor and record it in the Data Table as x.

4. Make a small loop in each end of a string long enough to wrap around the hub several times and still touch the floor. Place one loop on the peg on the hub of the wheel and wrap the string around the hub. On the loop at the other end of the string place a 0.0500-kg mass holder. Place the mass holder on the platform, which is to be tripped to release the mass.

5. With only the 0.0500-kg mass holder on the end of the string, release the tripping platform and simultaneously start a timer. Stop the timer when the mass holder strikes the floor. Take extreme care to start the timer exactly when the mass is released. Make sure that the mass holder is just resting on the platform with no slack in the string when it is released. Record the value of the time t in the Data Table. Using this value of m, repeat this process three more times for a total of four trials.

6. Repeat the procedure in step 5 for a series of different values of the mass m on the string. Add 0.050 kg each time to produce values of m of 0.1000, 0.1500, 0.2000, 0.2500, and 0.3000 kg. Perform four trials with each value of m. Record all the values of m and t in the Data Table.

CALCULATIONS

1. Using the values of the mass of the wheel M and the radius of the wheel R, calculate the theoretical value of I the moment of inertia from $I = \frac{1}{2} MR^2$. Record this value as I_{theo} in the Calculations Table.

2. Calculate the mean $\bar{t}$ and standard error for the repeated trials of the time for each mass. Record the values of $\bar{t}$ and the values of the standard error in the Calculations Table. (*Note*: In order to avoid confusion with the angular acceleration the standard error is noted in that table as Std Err instead of its usual symbol α.)

3. Using the values of $\bar{t}$ in equation 6, calculate the acceleration a for each value of m. Record the values of a in the Calculations Table.

4. Using the values of acceleration determined above in equation 7, calculate the angular acceleration α for each value of m. Note that the units of α are rad/s^2. Record the values of α in the Calculations Table.

5. Using the values of acceleration determined above in equation 4, calculate the tension in the string T for each value of m. Record these values of T in the Calculations Table.

6. Using these values of T in equation 5, calculate the value of the torque τ for each value of m. Record the values of τ in the Calculations Table.

7. Perform a linear least squares fit to the data, with τ as the ordinate and α as the abscissa. Determine the slope, intercept, and correlation coefficient of the least squares fit. The slope is the experimental value for the moment of inertia of the wheel, and the intercept is τ_f, the frictional torque acting on the wheel. Record these values in the Calculations Table.

GRAPHS

Graph the data for τ versus α with τ as the ordinate and α as the abscissa. Also show on the graph the straight line that was obtained by the linear least squares fit to the data.

Moment of Inertia and Rotational Motion

LABORATORY REPORT

Data Table

$M =$	kg	$R =$	m	$r_0 =$	m	$x =$	m
m (kg)		t_1 (s)		t_2 (s)		t_3 (s)	t_4 (s)

Calculations Table

$\bar{t}$ (s)	Std Err (s)	a (m/s^2)	T (N)	τ (N − m)	α (rad/s^2)
$I_{theo} =$ kg − m^2	$I_{exp} =$ kg − m^2	$\tau_f =$ N − m	$r =$		

QUESTIONS

1. Calculate the percentage error of the experimental value I_{exp} compared to the theoretical value I_{theo}. What does this imply about the accuracy of your results?

2. Using the table in Appendix I for correlation coefficients, determine the significance of the correlation coefficient obtained for the fit to the data.

3. The correlation coefficient is in some sense a measure of the precision of the experimental value I_{exp}. Considering the value for the correlation coefficient that was obtained and the accuracy of the result discussed in question 1, state whether or not there is evidence for a systematic error. State clearly the reasons for your answer.

4. Suggest any possible systematic errors that would produce effects consistent with the observed difference between the experimental and theoretical values for the moment of inertia.

5. What amount of mass placed on the string would produce a torque equal to the value of τ_f?

Laboratory 18

Archimedes' Principle

PRELABORATORY ASSIGNMENT

Read carefully the entire description of the laboratory and answer the following questions based on the material contained in the reading assignment. Turn in the completed prelaboratory assignment at the beginning of the laboratory period prior to the performance of the laboratory.

1. What is the definition of density? What are its units?

2. What is specific gravity? What are its units?

3. State Archimedes' principle.

4. The buoyant force on an object placed in a liquid is (a) always equal to the volume of the liquid displaced, (b) always equal to the weight of the object, (c) always equal to the weight of the liquid displaced, or (d) always less than the volume of the liquid displaced.

5. An object that sinks in water displaces a volume of water (a) equal to the object's weight, (b) equal to the object's volume, (c) less than the object's volume, or (d) greater than the object's weight.

6. An object that sinks in water has a mass in air of 0.0675 kg. Its apparent mass when submerged in water as in Figure 18.1 is 0.424 kg. What is the specific gravity SG of the object? Considering the densities given in Appendix II, of what material is the object probably made?

7. A piece of wood that floats on water has a mass of 0.0175 kg. A lead sinker is tied to the wood, and the apparent mass with the wood in air and the lead sinker submerged in water is 0.0765 kg. The apparent mass with both the wood and the sinker both submerged in water is 0.0452 kg. What is the specific gravity of the wood?

8. An object has a mass in air of 0.0832 kg, apparent mass in water of 0.0673 kg, and apparent mass in another liquid of 0.0718 kg. What is the specific gravity of the other liquid?

OBJECTIVES

Any object in a fluid experiences an upward buoyant force that tends to cause it to float in the liquid. Archimedes' principle states that this force is equal to the weight of the fluid that the object displaces. In this laboratory a series of measurements of the force that objects exert on a laboratory balance when they are in air, in water, and in alcohol will be used to accomplish the following objectives:

1. Determination of the specific gravity of several metal objects that are more dense than water
2. Determination of the specific gravity of a cork that is less dense than water
3. Determination of the specific gravity of alcohol

EQUIPMENT LIST

1. Laboratory balance and calibrated masses
2. 1000-ml beaker
3. String
4. Metal cylinders, and cork or piece of wood (to serve as unknowns)
5. Alcohol

THEORY

The density ρ_0 of an object is defined as the mass m of the object divided by its volume V if the mass is distributed uniformly. In equation form this is

$$\rho_0 = m/V \tag{1}$$

The Système International (SI) units for density are kg/m^3. Other commonly used units for density are g/cm^3, which are in the centimeters-gram-second (CGS) system. A quantity called the "specific gravity" is defined as the ratio of the density of an object to the density of water ρ_w. The equation for the specific gravity is given by

$$\text{Specific gravity} = SG = \rho_0 / \rho_w \tag{2}$$

Note that the specific gravity is a dimensionless quantity because it is the ratio of two densities. Since the density of water in the CGS system is 1.000 g/cm^3, densities in that system are numerically equal to the specific gravity SG. For the SI system, water has a density of 1000 kg/m^3, and densities in the SI system are equal to $SG \times 10^3$.

Archimedes' principle states that an object placed in a fluid has an upward buoyant force on it that is equal to the weight of the fluid displaced by the object. Although the principle applies to all fluids, both liquids and gases, the concern of this laboratory will be with the application of the principle to liquids. An object floats if its density is less than the liquid in which it is placed. It sinks in the liquid to a depth that is sufficient to displace the weight of the liquid equal to its own weight. The object is in equilibrium under the action of two forces of equal magnitude. One is its weight acting downward, and the other is the buoyant force acting upward. If an object has a greater density than the liquid in which it is placed, then it will sink to the bottom of the liquid. It experiences an upward buoyant force equal to the weight of the displaced fluid as before, but that force is less than the weight of the object, and thus it sinks.

For objects whose density is greater than the density of water, Archimedes' principle allows a rather simple determination of the specific gravity of the object. Consider an object that is tied to a string and submerged in water with the string attached beneath the proper arm of a laboratory balance as shown in Figure 18.1. Also shown in the figure is a free-body diagram of the forces acting on the submerged object. If B stands for the upward buoyant force on the submerged object, W is the downward weight of the object, and W_1 is the tension in the string, the following equation is satisfied when the laboratory balance is in equilibrium

$$B + W_1 = W \tag{3}$$

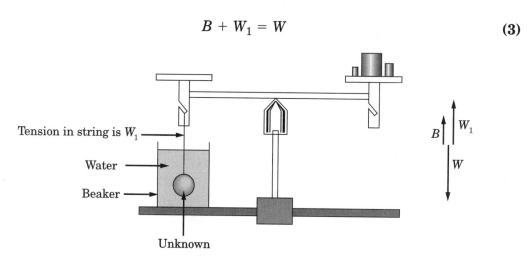

Figure 18.1 Determining the apparent mass of an object with density greater than water.

The quantity W_1 is also equal to the apparent weight read by the laboratory balance when the object is in the water. According to Archimedes' principle the buoyant force is equal to the weight of displaced water. This can be written as

$$B = m_w g = \rho_w V_w g = \rho_w V_o g \tag{4}$$

where m_w is the mass of the displaced water, and V_w and V_o are the volumes of the displaced water and of the object. The last step in equation 4 is valid only because the object sinks and displaces a volume of water equal to its own volume, and thus $V_w = V_o$. Using equations 3 and 4 along with the facts that $W = mg$ and $W_1 = m_1 g$, the following results are obtained:

$$W_1 = W - B = mg - \rho_w V_w g = \rho_o V_o g - \rho_w V_o g \tag{5}$$

$$\frac{W}{W - W_1} = \frac{mg}{mg - m_1 g} = \frac{\rho_o V_o g}{\rho_o V_o g - (\rho_o V_o g - \rho_w V_o g)} = \frac{\rho_o}{\rho_w} \tag{6}$$

$$SG = \frac{\rho_0}{\rho_w} = \frac{m}{m - m_1} \tag{7}$$

Equation 7 implies that the specific gravity of any object that sinks in water can be determined by measuring its mass m and its apparent mass in water m_1. This laboratory will make use of equation 7 to determine the specific gravity of several metals with densities greater than that of water.

In order to use Archimedes' principle to determine the specific gravity of an object that floats, it is necessary to devise some means to submerge the object. This must be done so that the volume of the displaced water is equal to the volume of the object. When this is true the volume is common to every term in the equation. Therefore, it can be cancelled from every term in equation 6. An object that floats can be made to sink if a sinker of high density material is attached to the object. Consider the two situations shown in Figure 18.2. In the first case the object of interest is in the air, and the sinker is in the water. In the second case both the object of interest and the sinker are in the water.

For the case shown in Figure 18.2(a) there are four forces acting on the system. They are the weight of the object W, the weight of the sinker W_s, the buoyant force on the sinker B_s, and the apparent weight W_1, which is equal to the tension in the string tied to the laboratory balance. The equation relating those forces is

$$W_1 + B_s = W + W_s \tag{8}$$

For the case shown in Figure 18.2(b) there are five forces acting on the system. The forces W, W_s, and B_s are the same as before; W_2 refers to the apparent weight when both the object and sinker are in water, and the buoyant force on the object of interest is called B. Note that for an object that is held submerged, its buoyant force B is greater than its weight. The equation relating these five forces is

$$W_2 + B + B_s = W + W_s \tag{9}$$

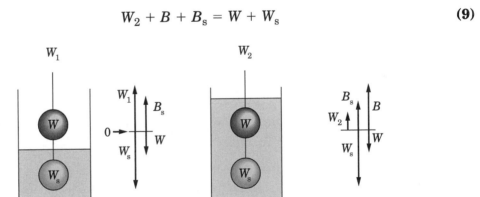

Figure 18.2 Forces acting on an object held submerged by a high density sinker.

Using equations 8 and 9 and the fact that $V_w = V_0$ because the object is submerged, leads to the following equations:

$$W_1 - W_2 = W + W_s - B_s - (W + W_s - B - B_s) = B \tag{10}$$

$$W_1 - W_2 = B = \rho_w V_w g = \rho_w V_0 g \tag{11}$$

$$\frac{W}{W_1 - W_2} = \frac{\rho_0 V_0 g}{\rho_w V_0 g} = \frac{\rho_0}{\rho_w} = SG \tag{12}$$

$$SG = \frac{W}{W_1 - W_2} = \frac{mg}{m_1 g - m_2 g} = \frac{m}{m_1 - m_2} \qquad (13)$$

Equation 13 implies that the specific gravity of an object whose density is less than that of water can be determined by a measurement of the mass of the object m, the apparent mass with the object in air and a sinker submerged in water m_1, and the apparent mass with the object and the sinker submerged in water m_2.

The specific gravity of a liquid can be determined by measuring the mass of some object m, the apparent mass of the object submerged in water m_w, and the apparent mass of the object submerged in the liquid m_L. It is assumed that the object sinks in both liquids. By similar arguments as those used to derive equations 7 and 13, it can be shown that the specific gravity of the liquid is given by

$$SG \text{ (liquid)} = \frac{m - m_L}{m - m_w} \qquad (14)$$

EXPERIMENTAL PROCEDURE—OBJECT WITH DENSITY GREATER THAN WATER

1. Using the laboratory balance, determine the mass of each of the two unknown metal cylinders. Record these values as m in Data Table 1. The unknowns should have their composition stamped on them, or else ask your instructor for the composition of the unknowns. Record the composition of the unknowns in Data Table 1.

2. Place the clamp on the laboratory table and screw a threaded rod in the clamp. The laboratory balance has a hole in its base designed to support the balance above the table. Slip the rod into the hole in the base of the laboratory balance. Adjust the height of the balance far enough above the table to allow room for the 1000-ml beaker under the balance. Fill the beaker about three-fourths full of water.

3. Tie a piece of light string or thread around one of the unknowns and suspend it from one of the slots in the left arm underneath the balance. Determine the apparent mass with the unknown suspended completely below the surface of the water. Be sure that the unknown is neither touching the side of the beaker nor in any way supported other than by the string. Record the mass reading of the balance as m_1 in Data Table 1.

4. Repeat steps 1 through 3 for the second unknown.

5. From Appendix II, obtain the known SG of the metals of which the unknowns are made and record those values in Data Table 1.

CALCULATIONS—DENSITY GREATER THAN WATER

1. Using equation 7 calculate the SG for each of the unknowns. Record the experimental values of the specific gravity in Calculations Table 1.

2. Calculate the percentage error in your value of the SG for each of the unknowns compared to the known values. Record them in Calculations Table 1.

EXPERIMENTAL PROCEDURE—OBJECT WITH DENSITY LESS THAN WATER

1. Using the laboratory balance determine the mass of the cork and record it as m in Data Table 2.

2. Determine the apparent mass with the cork suspended in air and a sinker tied below submerged in water as shown in Figure 18.2(a). Record that value as m_1 in Data Table 2.

3. Determine the apparent mass with both the cork and the same sinker submerged in water as shown in Figure 18.2(b). Record that value as m_2 in Data Table 2.

CALCULATIONS—DENSITY LESS THAN WATER

Using equation 13, calculate the specific gravity SG for the cork and record it in Calculations Table 2.

EXPERIMENTAL PROCEDURE—LIQUID OF UNKNOWN SG

1. Use one of the unknown metal cylinders from the original procedure. Its mass m and its apparent mass in water have already been determined. Record those values in Data Table 3.

2. Using a very clean beaker, determine the apparent mass of the metal cylinder in alcohol. Record this as m_L in Data Table 3. Return the alcohol to its container when you have finished with it. Be very careful not to have any spark or flame near the alcohol.

3. Repeat steps 1 and 2 for the other unknown metal.

4. From Appendix II determine the known value of the specific gravity of alcohol and record it in Data Table 3.

CALCULATIONS—LIQUID OF UNKNOWN SG

1. Using equation 14, determine the specific gravity SG of the alcohol for the two trials with the different metals. Record those values in Calculations Table 3.

2. Calculate the percentage error in each of the two measurements of the specific gravity of alcohol. Record the results in Calculations Table 3.

LABORATORY REPORT

Data Table 1

Metal	Known SG	m (kg)	m_1 (kg)

Calculations Table 1

$SG = \dfrac{m}{m - m_1}$	% Error

Data Table 2

m (kg)	m_1 (kg)	m_2 (kg)

Calculations Table 2

$SG = \dfrac{m}{m_1 - m_2}$

Data Table 3

Metal	m (kg)	m_W (kg)	m_L (kg)

Known SG of alcohol =

Calculations Table 3

$SG = \dfrac{m - m_L}{m - m_W}$	% Error

SAMPLE CALCULATIONS

QUESTIONS

1. Do your data indicate that Archimedes' principle is valid? State clearly the evidence for your answer.

2. An object whose specific gravity is 0.900 is placed in a liquid whose specific gravity is 0.900. Describe how the object will behave. In other words, will it sink or float, or what will be its behavior?

3. An object whose specific gravity is 0.850 is placed in water. What fraction of the object is below the surface of the water?

4. Equation 14 can be derived in a manner similar to the derivation given for equations 7 and 13. Using free-body diagrams of the forces involved, write the equations describing the problem and derive equation 14. Note that the specific gravity SG referred to in that equation is ρ_L / ρ_w.

Laboratory 19

The Pendulum—Approximate Simple Harmonic Motion

PRELABORATORY ASSIGNMENT

Read carefully the entire description of the laboratory and answer the following questions based on the material contained in the reading assignment. Turn in the completed prelaboratory assignment at the beginning of the laboratory period prior to the performance of the laboratory.

1. What requirement must be satisfied by a force acting on a particle in order for the particle to undergo simple harmonic motion?

2. A particle of mass $M = 1.35$ kg has a force exerted on it such that $F = -0.850x$, where x is the displacement of the particle from equilibrium. The force F is in Newtons, and x is in meters. What is the period T of its motion? Show work.

3. A simple pendulum of length $L = 0.800$ m has a mass $M = 0.250$ kg. What is the tension in the string when it is at an angle of $\theta = 12.5°$? Show work.

4. In question 3, what is the component of the weight of M directed along the arc of the motion of M? Show work.

5. What is the period T of the motion of the pendulum in question 3? Assume that the period is independent of θ.

6. What would be the period T of the pendulum in question 3 if L was unchanged but $M = 0.500$ kg? Assume that the period is independent of θ.

7. Determine the period T of the pendulum in question 3 if everything else stays the same, but $\theta = 45.0°$. Do not assume that the period is independent of θ.

The Pendulum—Approximate Simple Harmonic Motion

OBJECTIVES

A bob of mass M suspended on a thin, light string of length L is a good approximation of a simple pendulum. The pendulum is set into motion by displacing the bob until the string makes an angle θ with the vertical and then releasing the bob. The period T of the pendulum is the time for one complete oscillation of the system. Measurements for several such pendulums of varying length L, mass M, and angular amplitude θ will be used to achieve the following objectives:

1. Verification that the period of a pendulum is independent of the mass M of the bob

2. Determination of how the period T of a pendulum depends on the length L of the pendulum

3. Verification that the period T of a pendulum depends very slightly on the angular amplitude of the oscillation for large angles, but that the dependence is negligible for small angular amplitude of oscillation

4. Determination of an experimental value of the acceleration due to gravity g by comparing the measured period of a pendulum with theoretical predictions

EQUIPMENT LIST

1. Pendulum clamp, string, and calibrated hooked masses
2. Laboratory timer
3. Protractor and meter stick

THEORY

A particle of mass M that moves in one dimension is said to exhibit simple harmonic motion if its displacement x from some equilibrium position is described by a single sine or cosine function. This will be the case when the particle is subjected to a force F that is directly proportional to the magnitude of the displacement and directed toward the equilibrium position. In equation form

$$F = -kx \tag{1}$$

describes a force that produces simple harmonic motion. The period T of the motion is the time for one complete oscillation, and it is determined by the mass M and the constant k. The equation that describes the dependence of the period T on M and k is

$$T = 2\pi \sqrt{\frac{M}{k}} \tag{2}$$

A pendulum is a system that does not exactly satisfy the above conditions for simple harmonic motion, but it approximates them under certain conditions. An ideal pendulum is a point mass M suspended at one end of a massless string with the other end of the string fixed as shown in Figure 19.1. The motion of the system takes place in a vertical plane when the mass M is released from an initial angle θ. The angular amplitude θ is defined by the angle that the string makes with the vertical.

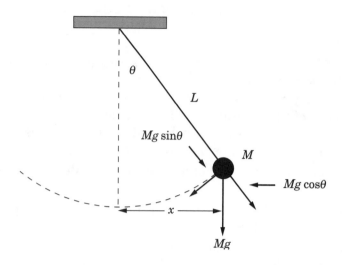

Figure 19.1 Force components acting on the mass bob of a simple pendulum.

The weight of the pendulum acts downward, and it can be resolved into two components. The component $Mg \cos\theta$ is equal in magnitude to the tension N in the string. The other component $Mg \sin\theta$ acts tangent to the arc along which the mass M moves. It is this component that provides the force that drives the system. In equation form, the force F along the direction of motion is

$$F = -Mg \sin\theta \tag{3}$$

For small values of the initial angle θ, the approximation $\sin\theta \approx \tan\theta \approx x/L$ can be used in equation 3, and it then becomes

$$F = -\frac{Mg}{L}x \tag{4}$$

Although equation 4 is an approximation, it is now of the same form as equation 1 with the constant $k = Mg/L$. Using that value of k in equation 2 gives

$$T = 2\pi \sqrt{\frac{M}{Mg/L}} = 2\pi \sqrt{\frac{L}{g}} \tag{5}$$

Equation 5 makes the theoretical prediction that the period T of a simple pendulum is independent of the mass M and the angular amplitude θ and depends only on the length L of the pendulum.

The exact solution to the period of a simple pendulum without making the small-angle approximation leads to an infinite series of terms with each successive term becoming smaller and smaller. Equation 6 gives the first three terms in the series. This is sufficient to determine the very slight dependence that does exist of the period T on the angular amplitude of the motion.

$$T = 2\pi \sqrt{\frac{L}{g}} \left[1 + \tfrac{1}{4} \sin^2\left(\tfrac{\theta}{2}\right) + \tfrac{9}{64} \sin^4\left(\tfrac{\theta}{2}\right) + \ldots\right] \tag{6}$$

For an ideal pendulum with no friction, the motion repeats indefinitely with no reduction in the amplitude as time goes on. For a real pendulum, there will always be some friction, and the amplitude of the motion decreases slowly with time. However, for small initial amplitudes, the change in the period as the amplitude decreases is negligible. This fact is the basis for the pendulum clock. Pendulum clocks, in one form or another, have been used for over 300 years. For over 100 years, devices to compensate for small changes in the length of the pendulum caused by temperature variations have been successfully used to build extremely accurate clocks.

EXPERIMENTAL PROCEDURE—LENGTH

1. The dependence of the period on the length of the pendulum will be determined with a constant mass and constant angular amplitude. Place a 0.2000-kg hooked calibrated mass on a string with a loop in one end. Adjust the position at which the other end of the string is clamped in the pendulum clamp until the distance from the point of support to the center of mass of the hooked mass is 1.0000 m. Note carefully that the length L of each pendulum will be from the point of support to the center of mass of the bob. The center of mass of the hooked masses will usually not be in the center because the hooked masses are not solid at the bottom. Estimate how much this tends to raise the position of the center of mass and mark the estimated center of mass on the hooked mass.

2. Displace the pendulum 5.0° from the vertical and release it. Measure the time Δt for 10 complete periods of motion and record that value in Data Table 1. It is best to set the pendulum in motion, and then begin the timer as it reaches the maximum displacement, counting 10 round trips back to that position. Repeat this process two more times, for a total of three trials with this same length. The pendulum should move in a plane as it swings. A tendency for the mass to move in an elliptical path will lead to error.

3. Repeat the procedure of step 2, using the same mass and an angle of 5.0° for pendulum lengths of 0.8000, 0.6000, 0.5000, 0.3000, 0.2000, and 0.1000 m. Do three trials at each length. Remember that the length of the pendulum is from the point of support to the center of mass of the hooked mass.

CALCULATIONS—LENGTH

1. Calculate the mean $\overline{\Delta t}$ and standard error α_t of the three trials for each of the lengths. Record those results in Calculations Table 1.

2. Calculate the period T from $T = \overline{\Delta t} / 10$ and record in Calculations Table 1.

3. According to equation 5, the period T should be proportional to $\sqrt{L}$. For each of the values of the length L calculate the $\sqrt{L}$ and record the results in Calculations Table 1. Perform a linear least squares fit to the data with T as the ordinate and $\sqrt{L}$ as the abscissa. According to equation 5, the slope of this fit should be equal to $2\pi/\sqrt{g}$. Equate the value of the slope determined from the least squares fit to $2\pi/\sqrt{g}$ treating g as unknown. Solve this equation for g and record that value as g_{exp} in Calculations Table 1. Also record the value of the correlation coefficient for the least squares fit.

EXPERIMENTAL PROCEDURE—MASS

1. The dependence of the period T on the mass M of the pendulum will be determined with the length L and amplitude θ held constant. Place a 0.0500-kg hooked mass on the end of the string and adjust the point of support of the string until the pendulum length is 1.0000 m. Displace the pendulum 5.0° and release it. Measure the time Δt for 10 complete periods of the motion and record it in Data Table 2. Repeat this measurement two more times for a total of three trials.

2. Keeping the length constant at $L = 1.0000$ m, repeat the procedure above for pendulum masses of 0.1000, 0.2000, and 0.5000 kg. Because the length of the pendulum is from the point of support to the center of mass, slight adjustments in the string length will be needed to keep the pendulum length constant for the different hooked masses.

CALCULATIONS—MASS

1. Calculate the mean $\overline{\Delta t}$ and standard error α_t of the three trials for each of the masses. Record those results in Calculations Table 2.

2. Calculate the period T from $T = \overline{\Delta t}/10$ and record in Calculations Table 2.

EXPERIMENTAL PROCEDURE—AMPLITUDE

1. The dependence of the period T on the amplitude of the motion will be determined with the length L and mass M held constant. Construct a pendulum 1.0000 m long with a mass of 0.2000 kg. Measure the time Δt for 10 complete periods of the motion with an amplitude of 5.0°. Repeat the measurement two more times for a total of three trials at this amplitude. Record all results in Data Table 3.

2. Repeat the procedure above for amplitudes of 10.0°, 20.0°, 30.0°, and 45.0°. Do three trials for each amplitude and record the results in Data Table 3.

CALCULATIONS—AMPLITUDE

1. Calculate the mean $\overline{\Delta t}$ and standard error α_t of the three trials for each of the amplitudes. Record those results in Calculations Table 3.

2. Calculate the period T from $T = \overline{\Delta t}/10$ and record the results in Calculations Table 3 as T_{exp}.

3. Equation 6 is the theoretical prediction for how the period T should depend on the amplitude. Using the values of $L = 1.000$ m and $M = 0.2000$ kg in equation 6, calculate the value of T predicted for each of the values of the amplitude θ. Record those values in Calculations Table 3 as T_{theo}.

4. For the experimental values of the period T_{exp}, calculate the ratio of the period at the other angles to the period at $\theta = 5.0°$. Call this ratio $T_{\text{exp}}(\theta)/T_{\text{exp}}(5.0°)$. Record these values in Calculations Table 3.

5. For the theoretical values of the period T_{theo}, calculate the ratio of the period at the other angles to the period at $\theta = 5.0°$. Call this ratio $T_{\text{theo}}(\theta)/T_{\text{theo}}(5.0°)$. Record these values in Calculations Table 3.

GRAPHS

1. Consider the data for the dependence of the period T on the length L. Graph the period T as the ordinate and $\sqrt{L}$ as the abscissa. Also show on the graph the straight line obtained by the linear least squares fit to the data.

2. Consider the data for the dependence of the period T on the mass M. Graph the period T as the ordinate and the mass M as the abscissa.

Laboratory 19

The Pendulum-Approximate Simple Harmonic Motion

LABORATORY REPORT

Data Table 1

Length (m)	Δt_1 (s)	Δt_2 (s)	Δt_3 (s)
1.0000			
0.8000			
0.6000			
0.5000			
0.3000			
0.2000			
0.1000			

Calculations Table 1

L (m)	$\overline{\Delta_t}$ (s)	α_t (s)	T (s)	$\sqrt{L}$ ($\sqrt{m}$)
1.0000				
0.8000				
0.6000				
0.5000				
0.3000				
0.2000				
0.1000				
Slope =		$g_{\exp}$ =		r =

Data Table 2

Mass (kg)	Δt_1 (s)	Δt_2 (s)	Δt_3 (s)
0.0500			
0.2000			
0.2000			
0.5000			

Calculations Table 2

Mass (kg)	$\overline{\Delta t}$ (s)	α_t (s)	T (s)
0.0500			
0.1000			
0.2000			
0.5000			

Data Table 3

Angle	Δt_1 (s)	Δt_2 (s)	Δt_3 (s)
5.0°			
10.0°			
20.0°			
30.0°			
45.0°			

Calculations Table 3

Angle	$\overline{\Delta t}$ (s)	α_t (s)	T_{exp} (s)	T_{theo} (s)	$\dfrac{T_{exp}(\theta)}{T_{exp}(5°)}$	$\dfrac{T_{theo}(\theta)}{T_{theo}(5°)}$
5.0°						
10.0°						
20.0°						
30.0°						
45.0°						

SAMPLE CALCULATIONS

QUESTIONS

1. In general, what is the precision of the measurements of T? Answer this question by considering what percentage is α_t of $\overline{\Delta t}$ for the measurements as a whole.

2. Do your data confirm the expected dependence of the period T on the length L of a pendulum? Consider the correlation coefficient for the least squares fit in your answer.

3. Comment on the accuracy of your experimental value for the acceleration due to gravity g.

4. What does the theory predict for the shape of the graph of period T versus M? Do your data confirm this expectation? Also calculate the mean and standard error of the periods for the four masses and comment on how this relates to mass independence of T.

5. Do your measured values for the period T as a function of the amplitude θ confirm the theoretical predictions? State clearly what is expected and what your data show.

6. The measurements of the period T were done by measuring the time for 10 periods. Why is the time for more than one period measured? If there is an advantage to measuring for 10 periods, why not measure for 1000 periods? Other than the fact that it would take too long, is there a valid reason why measuring for 1000 periods is not a good idea?

Simple Harmonic Motion—Mass on a Spring

OBJECTIVES

Simple harmonic motion is oscillatory motion that can be described by a single sine or cosine function. An object undergoes simple harmonic motion when it is subject to a force proportional to its displacement from an equilibrium position. In equation form $F = -kx$ describes such a force. A mass on the end of a spring is subject to a force that can be expressed by the above equation. In this laboratory, measurements of the motion of a mass on the end of a spring will be used to accomplish the following objectives:

1. Direct determination of the spring constant k of a spring by measuring the elongation of the spring for specific applied forces

2. Indirect determination of the spring constant k from measurements of the variation of the period T of oscillation for different values of mass on the end of the spring

3. Comparison of the two values of the spring constant k

4. Demonstration that the period T of oscillation of a mass on a spring is independent of the amplitude of the motion

EQUIPMENT LIST

1. Spring and masking tape
2. Table clamps, right angle clamps, and rods
3. Laboratory balance and calibrated hooked masses
4. Laboratory timer
5. Meter stick

THEORY

An object that experiences a restoring force proportional to its displacement from an equilibrium position is said to obey Hooke's law. In equation form this relationship can be expressed as

$$F = -kx \tag{1}$$

where k is a constant whose dimensions are N/m. The minus sign indicates that the force is in the opposite direction of the displacement. If a spring is the object exerting such a force, the constant k is called the "spring constant."

An object subject to a force described by equation 1 will undergo a type of oscillatory motion that is called "simple harmonic motion." The name comes from the fact that such motion can be described by a single sine or cosine function of time. If the object is displaced from its equilibrium position by some value A and then released, the object will oscillate back and forth about the equilibrium position. The values of its displacement x from the equilibrium position will range between $x = A$ and $x = -A$. The quantity A is called the "amplitude of the motion." For the initial conditions described above, the displacement x as a function of time t is given by

$$x = A \cos\left(\frac{2\pi t}{T}\right) \tag{2}$$

where T is the period of the motion. The period T is equal to the time for one complete oscillation from the maximum displacement on one side of equilibrium $(+A)$, to the maximum displacement on the other side of equilibrium $(-A)$, and back to the original position $(+A)$.

Consider a mass m placed on the end of a spring hanging vertically as shown in Figure 20.1. The original equilibrium position of the lower end of the spring is shown in Figure 20.1(a), and the position of the lower end of the spring when the mass is applied is shown in Figure 20.1(b). For purposes of determining the oscillatory motion, the position shown in Figure 20.1(b) can be considered as the new equilibrium position, and displacements can be measured from that point. In Figure 20.1(c) the mass is shown pulled down to a displacement A from this equilibrium position. When released, the mass will oscillate with amplitude A.

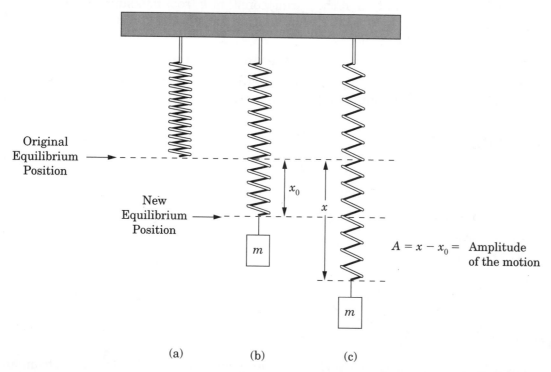

Figure 20.1 New equilibrium position with mass m placed on a spring.

The period of oscillation of the spring is independent of the amplitude A. It depends only on the spring constant k and the mass m. The period T is given by

$$T = 2\pi\sqrt{\frac{m}{k}} \tag{3}$$

Equation 3 is strictly true only if the spring is massless. For real springs with finite mass, a portion of the spring mass must be included along with the mass m. If the mass per unit length of the spring is constant, it can be shown that one third of the spring mass m_s must be included in equation 3 along with m. This gives as the equation for a spring of finite mass m_s

$$T = 2\pi \sqrt{\frac{m + (m_s/3)}{k}} \qquad (4)$$

EXPERIMENTAL PROCEDURE—SPRING CONSTANT

1. Attach the table clamp to the edge of the laboratory table and screw a threaded rod into the clamp vertically as shown in Figure 20.2. Place a right-angle clamp on the vertical rod and extend a horizontal rod from the right-angle clamp. Hang the spring on the horizontal rod and attach it to the horizontal rod with a piece of tape. Screw a threaded vertical rod into a support stand that rests on the floor. Place a right-angle clamp on the vertical rod and place a meter stick in the clamp so that the meter stick stands vertically. Adjust the height of the clamp on the vertical rod until the zero mark of the meter stick is aligned with the bottom of the hanging spring as shown in Figure 20.2.

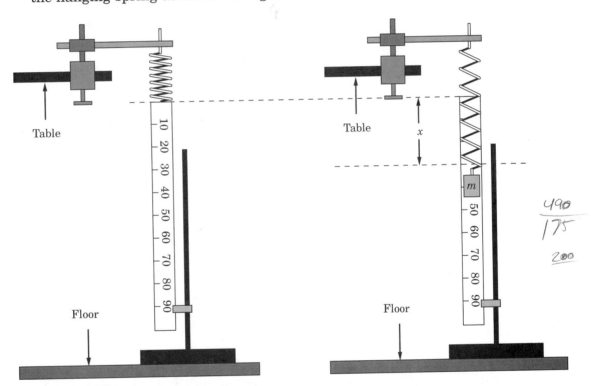

Figure 20.2 Arrangement to measure displacement of spring caused by mass m.

2. Place a hooked mass m of 0.1000 kg on the end of the spring. Slowly lower the mass m until it hangs at rest in equilibrium when released. Carefully read the position of the lower end of the spring on the meter-stick scale. Record the value of the mass m and the value of the displacement x in Data Table 1.

3. Repeat step 2, placing in succession 0.2000, 0.3000, 0.4000, and 0.5000 kg on the spring and measuring the displacement x of the spring. Record all values of m and x in Data Table 1.

CALCULATIONS—SPRING CONSTANT

1. Calculate the force mg for each mass and record the values in Calculations Table 1. Use the value of 9.800 m/s^2 for g. (It makes sense here to assume another significant figure in g in order not to be limited by that quantity for this data. Note that this is not the correct value to this number of significant figures but it makes no difference for our purposes.)

2. Perform a linear least squares fit to the data with mg as the ordinate and x as the abscissa. Calculate the slope of the fit and record it in Calculations Table 1 as the spring constant k.

3. Calculate the correlation coefficient for the linear least squares fit and record it in Calculations Table 1.

EXPERIMENTAL PROCEDURE—AMPLITUDE VARIATION

1. The dependence of the period T on the amplitude A will be done for a fixed mass. Place a hooked mass m of 0.2000 kg on the end of the spring. Slowly lower the mass until it hangs at rest when released. Note this position of the lower end of the spring.

2. Displace the mass downward 0.0200 m (i.e., so that $x - x_0 = A = 0.0200$ m as shown in Figure 20.1), release the mass, and let it oscillate. Measure the time for 10 complete periods and record it in Data Table 2 as Δt. Repeat the procedure two more times, for a total of three trials at this amplitude ($A = 0.0200$ m).

3. Repeat step 2 above for values of A equal to 0.0400, 0.0600, 0.0800, and 0.1000 m. Make three trials for each amplitude and measure the time for 10 periods for each trial. Record all results in Data Table 2.

CALCULATIONS—AMPLITUDE VARIATION

1. Calculate the mean $\overline{\Delta t}$ and standard error α_t for the three trials for each amplitude. Record the results in Calculations Table 2.

2. Calculate the period T from $T = \overline{\Delta t}/10$. Record the results in Calculations Table 2.

EXPERIMENTAL PROCEDURE—MASS VARIATION

1. The data just taken for the period as a function of amplitude should have shown that the period is independent of the amplitude. Therefore, the dependence of the period T on the mass m can be done for any amplitude. Place a hooked mass of 0.0500 kg on the spring and let it hang at rest. Displace the mass slightly below the equilibrium, release it, and let the system oscillate. Measure the time for 10 periods of the motion and record that time in Data Table 3 as Δt. Repeat the procedure two more times, for a total of three trials with this mass.

2. Repeat the procedure of step 1 for values of the mass m equal to 0.1000, 0.2000, 0.3000, 0.4000, and 0.5000 kg. Perform three trials of the time for 10 periods for each mass and record the results in Data Table 3.

3. Using a laboratory balance, determine the mass of the spring m_s and record it in Data Table 3.

CALCULATIONS—MASS VARIATION

1. Calculate the mean $\overline{\Delta t}$ and standard error α_t for the three trials for each mass. Record the results in Calculations Table 3.

2. Calculate the period T from $T = \overline{\Delta t}/10$. Record the results in Calculations Table 3.

3. Calculate the quantity $\sqrt{m + (m_s/3)}$ for each of the values of m. Record the results in Calculations Table 3.

4. According to equation 4, the period T should be proportional to $\sqrt{m + (m_s/3)}$ with $2\pi/\sqrt{k}$ with T as the ordinate and $\sqrt{m + (m_s/3)}$ as the abscissa. Determine the slope of this fit and equate it to $2\pi/\sqrt{k}$, treating k as unknown. Solve the resulting equation for k and record it in Calculations Table 3. Also record the value of the correlation coefficient of the least squares fit in Calculations Table 3.

5. Calculate the percentage difference between the value of k determined indirectly by this procedure and the value of k determined earlier by measuring the elongation per unit force.

GRAPHS

1. Graph the data from Calculations Table 1 for force mg versus displacement x with mg as the ordinate and x as the abscissa. Also show on the graph the straight line obtained from the linear least squares fit to the data.

2. Graph the data from Calculations Table 2 for the period T versus the mass m with T as the ordinate and m as the abscissa.

3. Graph the data from Calculations Table 3 for the period T versus $\sqrt{m + (m_s/3)}$ with T as the ordinate and $\sqrt{m + (m_s/3)}$ as the abscissa. Also show on the graph the straight line obtained from the linear least squares fit to the data.

PRELABORATORY ASSIGNMENT

Read carefully the entire description of the laboratory and answer the following questions based on the material contained in the reading assignment. Turn in the completed prelaboratory assignment at the beginning of the laboratory period prior to the performance of the laboratory.

1. What is the name given to a point on a vibrating string at which the displacement is always zero? What is the name given to a point at which the displacement is always a maximum?

2. What are the conditions (with respect to the points of zero amplitude and maximum amplitude) that must hold in order to produce a standing wave on a vibrating string?

3. How is the length of the string L related to the wavelength λ for standing waves?

4. What is the longest possible wavelength λ for a standing wave in terms of the string length L?

5. An arrangement like that shown in Figure 21.1 has a vibrator frequency of $f = 120$ Hz. The length of the string from the vibrator to the point where the string touches the top of the pulley is 1.20 m. What is the wavelength of the standing wave corresponding to the third resonant mode of the system?

6. What is the velocity of the wave for the same system described in question 5, but for the case of the fifth resonant mode?

7. Suppose that the system described in question 5 has string with a mass density ρ equal to 2.95×10^{-4} kg/m. What is the tension T in the string for the second resonant mode?

OBJECTIVES

If a string is fixed at one end and attached to a sinusoidally driven vibrator at the other end, waves travel down the string from the vibrator and are reflected back from the fixed end. Under the proper conditions, standing waves are formed from the interference between the two waves traveling in opposite directions. The tension T in the string of length L, the mass per unit length of the string ρ, the frequency of the vibrator f, and the velocity of propagation of the waves V are the relevant variables of the problem. For a string of length L, measurements of the values of the tension T for which resonances occur will be used to accomplish the following objectives:

1. Demonstration that resonances occur only for certain discrete values of the tension T in the string

2. Determination of the value of the tension T required to produce resonances with a given number of nodes

3. Determination of the wavelength λ of the wave associated with a given resonance with a given number of nodes

4. Demonstration that the wavelengths λ associated with resonances are proportional to $\sqrt{T}$, where T stands for the values of the tension in the string that produce resonances

5. Determination of an experimental value for the frequency f from a linear least squares fit to the data for λ versus $\sqrt{T}$

6. Comparison of the experimental value for the frequency with the known value of 120 Hz

EQUIPMENT LIST

1. String vibrator (60 Hz AC), string
2. Clamps, pulley, support rods
3. Mass holder and slotted masses
4. Meter stick (preferably 2-m stick)
5. Laboratory balance capable of measuring to 0.00001 kg (one for the class)

THEORY

Waves are one means by which energy can be transported. Waves in a string are an example of a type of wave known as a transverse wave. The motion of the individual particles of the medium (in this case the string) move perpendicular or transverse to the motion of the energy that moves along the length of the string. Consider the situation in Figure 21.1, which shows a string tied to a vibrator at one end that then passes over a pulley at the other end with masses on the end of the string to provide tension in the string. As the vibrator moves up and down at a fixed frequency f, it causes a wave of frequency f to propagate down the string. The vibrator that will be used in this laboratory is driven by an electromagnet. The alternating current frequency is 60 Hz, but since the electromagnet attracts the steel blade twice in each cycle, the vibrator frequency is 120 Hz.

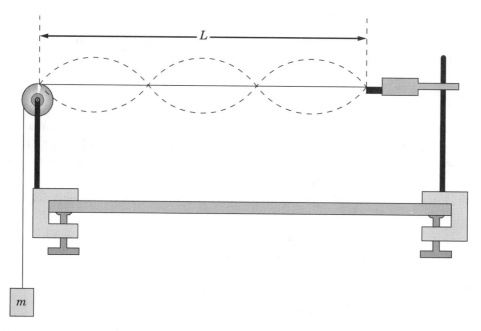

Figure 21.1 Experimental arrangement of vibrator and pulley.

The point at which the string passes over the pulley is a fixed point, and the wave is reflected from that point. Thus, the string is a medium in which two waves of the same speed, frequency, and wavelength are traveling in opposite directions. When these two waves interfere with each other, a standing wave is produced, provided that the proper relationship exists between the string length L and the wavelength λ of the wave.

When a standing wave is produced its characteristic features are the existence of nodes and antinodes at points along the string (Figure 21.2). A node N is a point for which there is no motion of the string—i.e., no displacement of the string from its equilibrium position. An antinode A is a point on the string for which the amplitude of vibration is a maximum at all times.

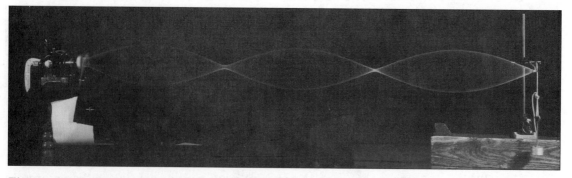

Figure 21.2 Standing wave with $n = 3$. (Photo courtesy Sargent-Welch Scientific Co.)

The conditions for the formation of a standing wave are that a node N must occur at each end of the string, and that an antinode must occur between each pair of nodes. The distance between two nodes is $\lambda/2$ or half of a wavelength of the wave. Therefore, in terms of the string length L, a standing wave is possible when the following equation holds:

$$L = n\left(\frac{\lambda}{2}\right) \tag{1}$$

where $n = 1, 2, 3, 4$.... Figure 21.3 shows the first four standing waves that are possible (i.e., $n = 1, 2, 3,$ and 4). From the figure it is clear that n can be thought of as the number of segments of half wavelengths that are in each standing wave. Equation 1 can be solved to obtain the wavelengths for which standing waves can occur. They are

$$\lambda = \frac{2L}{n} \tag{2}$$

where $n = 1, 2, 3, 4$.... Each value of n determines what is known as a resonant mode of the system. In principle, there is no limit to the number of resonant modes of the system. In practice, it is very difficult to achieve better than about $n = 10$ with the system used in the laboratory.

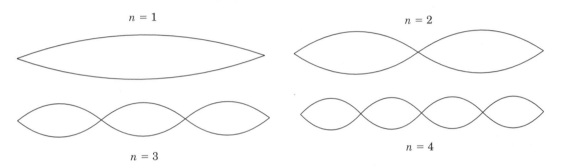

Figure 21.3 First four standing waves for waves on a string.

The frequency of the waves is fixed because the vibrator frequency is fixed to be 120 Hz. Thus, each different standing wave corresponding to each different n will have a different wave speed V. The wave speed is determined by two other variables of the experimental arrangement. They are the string tension T and the string mass per unit length ρ. The relationship between these quantities is given by

$$V = \sqrt{\frac{T}{\rho}} \tag{3}$$

Using the relationship $V = f\lambda$ and the expression for the speed V from equation 3, leads to the following relationship for the the wavelength λ in terms of the tension T:

$$\lambda = \frac{1}{f}\sqrt{\frac{T}{\rho}} \tag{4}$$

The experimental arrangement differs slightly from the ideal situation because the node at the vibrator end of the string is moving up and down rather than being fixed in space. This effect means that the wavelength is somewhat difficult to define. For example, in the case $n = 2$, the node will not be exactly in the center of the string; thus, each half of the string is of slightly different length. The effect becomes smaller and smaller as the number of segments increases. Usually it is suggested that the wavelength be determined by measuring only those segments that do not include the vibrator. Because it is somewhat difficult to locate the nodes precisely, there is usually less error involved if the simple assumption is made that the wavelength is the total length of the string divided by the number of segments. That is the assumption that will be made for purposes of determining the wavelength in this laboratory.

EXPERIMENTAL PROCEDURE

1. The mass per unit length of the string should be determined by the class as a whole. Carefully measure a 1.0000-m length of the type string to be used for the laboratory exercise. Using a laboratory balance capable of measuring to the nearest 0.00001 kg, determine the mass of the 1.0000 m length of string. Determine the value of ρ from this data and record it in the Data Table.

2. Clamp the vibrator and the pulley at opposite ends of the laboratory table, approximately 1.5 m apart. Cut a piece of string long enough to reach from the vibrator to the pulley and hang over the pulley as shown in Figure 21.1. Tie one end of the string to the vibrator and make a loop in the end of the string that hangs over the pulley. Place a 0.0500-kg mass holder on the loop.

3. Carefully measure the length of the string from the tip of the vibrator to the point at the very top of the pulley where the string touches the pulley. Record this distance as the string length L in the Data Table.

4. Plug in the vibrator and place on the mass holder whatever mass is required to produce a standing wave with two nodes in addition to the two nodes at the ends of the string. This corresponds to the case shown as $n = 3$ in Figure 21.3. (The modes with $n = 1$ and $n = 2$ will not be attempted because they require such very large string tension. In fact the tension required for the $n = 1$ case is so large that the string will quite likely break before it can be achieved.) In the process of determining the mass needed to produce this particular standing wave, it is helpful to pull slightly or lift slightly on the mass holder in order to determine if mass needs to be added or subtracted. Determine the amount of mass (to the nearest 0.002 kg if possible) needed to produce the largest amplitude of vibration with $n = 3$. Record the value of the mass m in the Data Table for the standing wave with $n = 3$. Be sure to include the mass of the mass holder in the total.

5. Remove mass from the mass holder until a standing wave with $n = 4$ is achieved as shown in Figure 21.3. Again determine the mass (to the nearest 0.002 kg if possible) needed to produce the largest amplitude of vibration with $n = 4$. Record this value of the mass m in the Data Table.

6. Continue this process, removing mass each time to produce standing waves for the cases $n = 5, 6, 7, 8,$ and 9. Note that in each case, n also refers to the number of segments into which the string length is divided. Each time, determine the mass needed to produce the largest possible amplitude for that standing wave. For all of these measurements determine the mass m to the nearest 0.001 kg if possible. Record the values of m for $n = 5, 6, 7, 8,$ and 9 in the Data Table.

CALCULATIONS

1. Using the measured values of L in equation 2, calculate the wavelengths λ for each of the standing waves $n = 3$ through $n = 9$ and record the results in the Calculations Table.

2. From the measured values of m, calculate the tension T from $T = mg$. For the value of g, use 9.80 m/s^2. Record the values of T in the Calculations Table.

3. Calculate the values of $\sqrt{T}$ for each of the values of T and record them in the Calculations Table. Note that when taking the square root, generally an extra significant figure is allowed in the value of the square root. For example, if a measured tension T is 7.83 N the recorded value of $\sqrt{T}$ should be 2.798.

4. According to equation 4, λ should be proportional to $\sqrt{T}$ with $1/f\sqrt{\rho}$ as the constant of proportionality. Perform a linear least squares fit to the data with λ as the ordinate and $\sqrt{T}$ as the abscissa. Calculate the slope of the fit and record it in the Calculations Table. Equate the slope to $1/f\sqrt{\rho}$, treating f as an unknown. Using the known value of ρ, solve the resulting equation for f. Record that value of the frequency as f_{exp} in the Calculations Table. Also calculate the correlation coefficient of the fit and record it in the Calculations Table.

5. Calculate and record the percentage error of f_{exp} compared to the known value of the frequency $f = 120$ Hz.

GRAPHS

Make a graph of the wavelength λ as the ordinate and $\sqrt{T}$ as the abscissa. Also show on the graph the straight line obtained from the linear least squares fit to the data.

LABORATORY REPORT

Data Table

n	m (kg)
$\rho =$ kg/m	$L =$ m

Calculations Table

$T = mg$ (N)	λ (m)	$\sqrt{T}\ (\sqrt{\text{N}})$
Slope =	Corr. coeff. =	
$f_{\text{exp}} =$	% error =	

SAMPLE CALCULATIONS

QUESTIONS

1. What is the accuracy of your experimental value for the frequency? State clearly the basis for your answer.

2. For the way in which this data was analyzed, the precision of the measurement would be related to the uncertainty in the measured slope. If instead, each measurement of the tension and wavelength were used to calculate an independent value for the frequency using the equation $f = (1\backslash\lambda)\sqrt{T/\rho}$, the standard error of those values would be a measure of the precision of the frequency. Calculate the seven values of the frequency for the data, the mean, and standard error of those values, and then comment on the precision of the data.

3. Calculate the velocity for the $n = 2$ and the $n = 8$ standing waves.

4. Calculate the tension T that would be required to produce the $n = 1$ standing wave for your string.

5. What would be the effect if the string stretched significantly as the tension increased? How would that have affected the data?

Laboratory 22

Speed of Sound—Resonance Tube

PRELABORATORY ASSIGNMENT

Read carefully the entire description of the laboratory and answer the following questions based on the material contained in the reading assignment. Turn in the completed prelaboratory assignment at the beginning of the laboratory period prior to the performance of the laboratory.

1. What is the equation that relates the speed V, the frequency f, and the wavelength λ of a wave?

2. How are standing waves produced?

3. What name is given to a point in space where the wave amplitude is zero at all times?

4. What name is given to a point in space where the wave amplitude is a maximum at all times?

5. What are the conditions that must be satisfied in order to produce a standing wave in a tube open at one end and closed at the other end?

6. For an ideal resonance tube, an antinode occurs at the open end of the tube. What property of real resonance tubes slightly alters the position of this antinode?

7. A student using a tuning fork of frequency 512 Hz observes that the speed of sound is 340 m/s. What is the wavelength of this sound wave?

8. A student using a resonance tube determines that three resonances occur at distances of $L_1 = 0.172$ m, $L_2 = 0.529$ m, and $L_3 = 0.884$ m below the open end of the tube. The frequency of the tuning fork used is 480 Hz. What is the average speed of sound from these data?

Speed of Sound—Resonance Tube

OBJECTIVES

A traveling wave is characterized by a speed V, a frequency f, and a wavelength λ. The relationship between these three quantities is given by $V = f\lambda$. When two waves of the same speed and frequency travel in opposite directions in some region of space, they can produce standing waves. When standing waves are produced in a tube, the amplitude of vibration becomes very large, and the system is said to be in resonance. A tube partially filled with water acts as a resonance tube for producing standing waves. In this laboratory, a tuning fork will be used to produce sound waves in a resonance tube to accomplish the following objectives:

1. Determination of several effective lengths of the closed tube at which resonance occurs for each tuning fork

2. Determination of the wavelength of the wave for each tuning fork from the effective length of the resonance tube

3. Determination of the speed of sound from the measured wavelengths and known tuning fork frequencies

4. Comparison of the measured speed of sound with the accepted value

EQUIPMENT LIST

1. Resonance tubes (with length scale marked on the tube)
2. Tuning forks (range, 500 to 1040 Hz), and rubber hammer
3. Thermometer (one for the class)

THEORY

For traveling waves of speed V, frequency f, and wavelength λ, the relationship between these three quantities is given by

$$V = f\lambda \tag{1}$$

If the frequency and wavelength of a wave are known, the speed of a traveling wave can be determined using equation 1. In practice, it is difficult to measure the properties of a traveling wave directly. Instead, experimentally it is easier to arrange for the traveling waves to interfere in such a way that standing waves result. Standing waves are produced by the interference between two waves of exactly the same speed, frequency, and wavelength traveling in the same region in opposite directions.

This laboratory uses a resonance tube to produce standing waves from the sound waves emitted from a tuning fork. The can shown in Figure 22.1 contains water, and as the can is moved up and down its support rod, the level of the water in the tube can be varied. The water acts as the closed end of the tube, and changing the water level changes the effective length of the resonance tube.

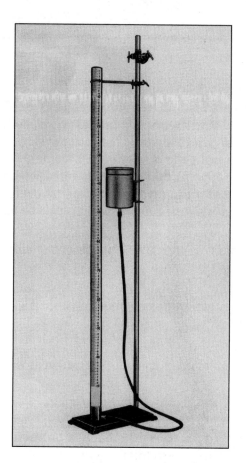

Figure 22-1 Resonance tube apparatus (Photo courtesy of Sargent Welch Scientific Co.).

A tuning fork is held by hand just above the open end of the tube and struck with a rubber hammer to make it vibrate. Sound waves travel down the tube, and they are reflected when they strike the water. Because of these reflected waves, there are traveling waves going in both directions inside the tube, which means that standing waves can be produced. The sound waves reflected from the closed end of the tube undergo a phase change of 180°, which places them completely out of phase with the incident sound waves. This means that the amplitude of the combination of incident and reflected waves must be zero at the closed end of the tube. Such a point in space where the wave amplitude must be zero at all times is called a "node," or N. From similar considerations of the relative phase between the incident and reflected waves, the open end of the tube is a point where the combined wave amplitude must be a maximum at all times. Such a point is called an "antinode," or A.

The standing waves that are produced in the tube are said to be in resonance with the tube, and this can occur only when there is a node at the closed end of the tube and an antinode at the open end of the tube. The speed of sound is fixed, and for a given tuning fork, the frequency is fixed. Therefore, the resonance conditions can be satisfied only for certain specific lengths of the tube that bear the proper relationship to the wavelength of the wave.

The necessary relationship between the length of the tube and the wavelength of the wave is illustrated in Figure 22.2 for the first four resonances of the tube. Sound waves are an example of a type of wave known as longitudinal. The amplitude of a sound wave is determined by pressure variations in the air along the direction of wave motion. The sound waves in the figure are pictured as if they were transverse waves merely for ease of representation.

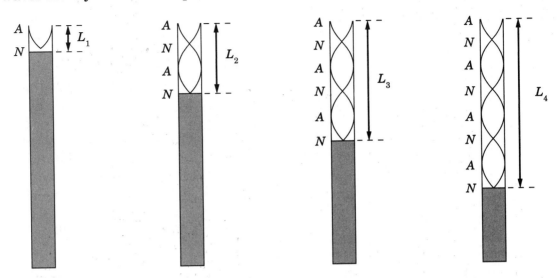

Figure 22.2 Nodes and antinodes for the first four resonances of a tube closed at one end.

The resonances are pictured from left to right as they are encountered when the level of the water in the tube is lowered, thus increasing the effective length of the tube closed at one end. The distances L_1, L_2, L_3, and L_4 refer to the distance from the top of the tube to the water level for the first four resonances that occur in the tube. The locations of all of the nodes N and antinodes A that occur for each of these resonances are also shown. In the first resonance there is a single node and antinode. The second resonance then introduces an additional node and antinode. Each successive resonance adds an additional node and antinode. In every case the distance between a node and the next antinode is one fourth of a wavelength ($1/4\ \lambda$). The distance between nodes is half of a wavelength ($1/2\ \lambda$).

In this laboratory, the location of several of the resonances that occur for each tuning fork will be experimentally determined. If the situation were ideal, the following relationships would be implied by Figure 22.2 for the first four resonances shown:

$$L_1 = {}^1/_4\ \lambda \qquad L_2 = {}^3/_4\ \lambda \qquad L_3 = {}^5/_4\ \lambda \qquad L_4 = {}^7/_4\ \lambda \qquad \textbf{(2)}$$

Examine Figure 22.2 carefully to be sure that you understand how the relationships given in equations 2 are implied by the figure.

In fact, the relationships given in equations 2 are not valid for a real resonance tube because there is a small effect due to the diameter of the tube that is not taken into account in those equations. For a real tube, the point at which the upper antinode actually occurs is just outside the end of the tube. The exact location depends on the diameter of the tube. Thus, equations 2 are not directly useful to determine the wavelength λ of the wave.

The end effect is the same for each of the resonances. Therefore, it will have no effect if differences between the locations of the individual resonances are considered. Considering the differences between adjacent resonances gives the following:

$$L_2 - L_1 = L_3 - L_2 = L_4 - L_3 = \lambda/2 \qquad (3)$$

Equations 3 imply that if several resonances for a given tuning fork are located, each of the differences between the resonances provides an accurate determination of the wavelength of the wave. The frequency of the tuning fork is known. Therefore, when the wavelength is known, equation 1 allows a determination of the speed of sound.

In fact, if equations 3 are used directly and the results are then averaged, there is a loss in some information contained in the data. Equations 3 would produce three values of the wavelength from the first four resonances. If these three values for the wavelength were then averaged, that would amount to taking the sum of twice the three differences and then dividing by three. In that process, all but the first and last resonance positions cancel from the calculation. In effect, one might as well not have measured them. Note that there is nothing incorrect about such a procedure, but it does lose some of the information contained in the data. This is a classic example of the fact that there is often more than one way to analyze data, but usually some techniques give more information than others.

The problem described above is solved if each wavelength is computed not from the adjacent differences but from the differences between each resonance and the first resonance. The resulting equations for the wavelength are given below. A subscript has been placed on the wavelength, but it is still understood that each of the wavelengths λ_1, λ_2, and λ_3 refer to the same wavelength calculated from three different sets of resonances. The equations are

$$\lambda_1 = 2(L_2 - L_1) \qquad \lambda_2 = (L_3 - L_1) \qquad \lambda_3 = {}^2\!/_3(L_4 - L_1) \qquad (4)$$

The speed of sound in air depends slightly on the temperature of the air. For a limited range of temperatures, the dependence is approximately linear. If V_T stands for the speed of sound at a temperature of $T°C$, to an excellent approximation it is given by

$$V_T = (331.5 + 0.607\, T)\ \text{m/s} \qquad (5)$$

This equation will be used to determine the accepted value for the speed of sound in air.

EXPERIMENTAL PROCEDURE

Note carefully that tuning forks should be struck only with the rubber hammer. Care must be taken to ensure that neither the hammer nor a vibrating tuning fork comes in contact with the tube.

1. Measure the room temperature of the air and record it in Data Table 1.

2. Adjust the water level until the can is essentially empty when the the tube is almost full. The water level in the tube should come at least to within 0.05 m of the open end of the tube. It may be necessary to remove some water from the can when the water level is lowered near the bottom of the tube.

3. One partner should hold a tuning fork over the top of the tube while the fork is struck repeatedly with the rubber hammer. It is important to keep the fork vibrating continuously with a large amplitude. With the tuning fork vibrating, another partner should slowly lower the water level from the top while listening for a resonance. The sound should become very loud when a resonance is reached. Attempt to measure the position of each resonance to the nearest millimeter. Raise and lower the water level several times to produce three trials for the measured position of the first resonance. Record the values of the three trials in Data Table 2. Record the frequency of the tuning fork in Data Table 2.

4. Repeat the procedure in step 3 to locate as many other resonances as possible. Depending on the frequency of the tuning fork used, either three or four resonances should be attainable. Record in Data Table 2 the location of the number of resonances that are attainable.

5. Using a second tuning fork of different frequency, repeat steps 1 through 3. Record in Data Table 3 the frequency of the tuning fork and the position of as many resonances as are attainable.

CALCULATIONS

1. Using equation 5, calculate the accepted value of the speed of sound from the measured room temperature. Record it in Data Table 1.

2. Calculate the mean and standard error of the three trials for the location of each of the resonances. Record each of the means and standard errors in the appropriate place in Calculations Tables 2 and 3.

3. Using equations 4, calculate the wavelengths that are appropriate. If four resonances were found, then all three values of λ can be determined. If only the first three resonances were measured, then only two values of λ can be determined. If this is the case, just leave the Calculations Table blank at the appropriate position. Be sure to use the mean values of the lengths to calculate the wavelengths.

4. Calculate the mean and standard error for the number of independent wavelengths measured for each tuning fork. Record those values in the Calculations Tables as $\bar{\lambda}$ and α_λ.

5. From the values of $\bar{\lambda}$ and the known values of the tuning fork frequencies, calculate the experimental value for V, the speed of sound.

6. Calculate the percentage error of the experimental values of V compared to the accepted value of the speed of sound in Data Table 1.

LABORATORY REPORT

Data Table 1

Room temperature =	°C
Accepted speed of sound =	m/s

Data Table 2

Tuning fork frequency =			Hz	
Trial	L_1 (m)	L_2 (m)	L_3 (m)	L_4 (m)
1				
2				
3				

Calculations Table 2

$\overline{L_1}$ =	m	$\overline{L_2}$ =	m	$\overline{L_3}$ =	m	$\overline{L_4}$ =	m
α_{L1} =	m	α_{L2} =	m	α_{L3} =	m	α_{L4} =	m
$\lambda_1 = 2(L_2 - L_1)$		m	$\lambda_2 = (L_3 - L_1)$ =		m	$\lambda_3 = {}^2/_3(L_4 - L_1)$ =	m
$\overline{\lambda}$ =	m	α_λ =	m	$V = f\overline{\lambda}$ =	m	% error =	

SAMPLE CALCULATIONS

Data Table 3

Trial	L_1 (m)	L_2 (m)	L_3 (m)	L_4 (m)
Tuning fork frequency =			Hz	
1				
2				
3				

Calculations Table 3

$\overline{L_1} =$ m	$\overline{L_2} =$ m	$\overline{L_3} =$ m	$\overline{L_4} =$ m
$\alpha_{L1} =$ m	$\alpha_{L2} =$ m	$\alpha_{L3} =$ m	$\alpha_{L4} =$ m
$\lambda_1 = 2(L_2 - L_1)$ m	$\lambda_2 = (L_3 - L_1) =$ m	$\lambda_3 = {}^2\!/_3(L_4 - L_1) =$ m	
$\bar{\lambda} =$ m	$\alpha_\lambda =$ m	$V = f\bar{\lambda} =$ m	% error =

SAMPLE CALCULATIONS

QUESTIONS

1. What is the accuracy of each of your measurements of the speed of sound? State clearly the evidence for your answer.

2. What is the precision of each of your measurements of the speed of sound? State clearly the evidence for your answer.

3. Equations 2 provide a means to determine the end correction for the tube. Using the value of $\overline{\lambda}$ for the first tuning fork, calculate values for L_1 and L_2 from those equations. They should be larger than the measured values of L_1 and L_2 by an amount equal to the end correction. Repeat the calculation for the second tuning fork. Compare these values for the end correction and comment on the consistency of the results.

4. Suppose that the temperature had been 10 C° higher than the value measured for the room temperature. How much would that have changed the measured value of $L_2 - L_1$ for each tuning fork? Would $L_2 - L_1$ be larger or smaller at this higher temperature?

5. Draw a figure showing the fifth resonance in a tube closed at one end. Show also how the length of the tube L_5 is related to the wavelength λ.

Laboratory 23

Specific Heat of Metals

PRELABORATORY ASSIGNMENT

Read carefully the entire description of the laboratory and answer the following questions based on the material contained in the reading assignment. Turn in the completed prelaboratory assignment at the beginning of the laboratory period prior to the performance of the laboratory.

1. What is the definition of specific heat?

2. What is the name for a device that provides a thermally isolated environment in which substances exchange heat?

A heated piece of metal at a temperature T_1 is placed into a calorimeter containing water and a stirrer. The temperature of the calorimeter, water, and the stirrer is initially T_2, where $T_1 > T_2$. The system is stirred continuously until it comes to equilibrium at a temperature of T_3. Answer questions 3 to 5 concerning what happens.

3. The final equilibrium temperature of the system is such that

   (a) $T_3 > T_1 > T_2$, (b) $T_1 < T_2 < T_3$, (c) $T_1 < T_3 < T_2$, or (d) $T_1 > T_3 > T_2$

4. The heat lost by the metal ΔQ_m and the heat gained by the calorimeter system ΔQ_c obey which of the following relationships? (a) $\Delta Q_m > \Delta Q_c$, (b) $\Delta Q_m < \Delta Q_c$, or (c) $\Delta Q_m = \Delta Q_c$

5. What is the purpose of stirring the system continuously?

6. A 350-g piece of metal is at an initial temperature of 22.0°C. It absorbs 1000 cal of thermal energy, and its final temperature is 45.0°C. What is the specific heat of the metal? Show your work.

7. A 250.0-g sample of metal shot is heated to a temperature of 98.0°C. It is placed in 100.0 g of water in a brass calorimeter cup with a brass stirrer. The total mass of the cup and the stirrer is 50.0 g. The initial temperature of the water, stirrer, and calorimeter cup is 20.0°C. The final equilibrium temperature of the system is 30.0°C. What is the specific heat of the metal sample? (The specific heat of brass is 0.092 cal/g-C°.) Show your work.

OBJECTIVES

The amount of heat required to change the temperature of a given mass of material by some fixed amount is different for each substance. The specific heat c is the quantity that is a measure of this property of matter. The units for c are cal/g-C°. Note that this is an exception to the usual use of SI units in this manual. For example, in these units the specific heat of water c_w has a value of 1.00 cal/g-C°, which means that one calorie of heat will change one gram of water by one Celsius degree. In this laboratory, metal shot will be heated to a known temperature, placed in a calorimeter containing water, and allowed to exchange heat with the water and its container. Measurements on this system will be used to accomplish the following objectives:

1. Demonstration that the specific heat of different materials is different
2. Determination of the specific heat of two samples of different metals

EQUIPMENT LIST

1. Calorimeter and stirrer
2. Steam generator
3. Metal shot (two different kinds of metal)
4. Two thermometers (preferably one 0–100°C and the other 0–50°C) and glycerin
5. Bunsen burner or electric heating plate

THEORY

When two objects of different temperatures are placed in thermal contact with each other, they exchange thermal energy until they are in thermal equilibrium at the same temperature. This exchanged thermal energy is known as heat. The factors affecting the amount of temperature change that each object experiences are the mass of each object and the specific heat of each substance. The specific heat c is defined as the amount of heat per unit mass required to change the temperature by one degree. The units used in this laboratory for specific heat are cal/g-C°. In such an exchange the thermal energy or heat for any one object in the system can be written in the form

$$\text{Heat} = \text{Thermal energy change} = mc\Delta T \tag{1}$$

where m is the mass in grams, c is the specific heat in cal/g-C°, and ΔT is the temperature change in C°. The units of the thermal energy change or heat will be calories. In any experimental arrangement designed to investigate this concept, it is crucial that a minimum of heat be either lost from the system or gained from outside the system. When there are no losses or gains from the system, all exchanges of heat are among the objects of the system. A device used to produce a thermally isolated environment to ensure that all heat exchanges are inside the system is called a "calorimeter." The calorimeter used in this laboratory consists of two metal cups held apart by a plastic ring that produces an insulating air space between the cups. The cups also have an insulating plastic top containing a hole into which the stirrer is placed and a rubber stopper with a hole into which a thermometer is placed. A picture of a calorimeter is shown in Figure 23.1.

Figure 23.1 Calorimeter. (Photo courtesy of Sargent Welch Scientific Company)

In this laboratory, metal shot will be heated to the temperature of steam in a steam generator and then placed into a calorimeter containing some water near room temperature. In this process, the metal shot will lose heat, and that heat will be gained by the water, the calorimeter cup, and the stirrer in the calorimeter. The use of the stirrer to mix the system ensures that all parts of the system come to thermal equilibrium as quickly as possible. In equation form, the statement that the heat lost by the metal shot equals the heat gained by the other parts of the system is

$$m_m\, c_m(T_m - T_e) = m_w\, c_w(T_e - T_o) + m_{cal}\, c_{cal}(T_e - T_o) + m_s c_s(T_e - T_o) \qquad \textbf{(2)}$$

where the ms are masses and the cs are specific heats. The subscripts are m for metal, w for water, cal for calorimeter, and s for stirrer. The initial temperature of the metal shot is T_m, the original temperature of the water, cup, and stirrer is T_0, and the final equilibrium temperature of the system is T_e.

EXPERIMENTAL PROCEDURE

1. Place about 250 g of one kind of metal shot into the cup designed to fit into the steam generator.

2. Fill the steam generator about half full of water. Keep the water level below the bottom of the cup that is placed in the generator. The cup should be heated by steam from the water. Use whatever means of heating is provided (Bunsen burner or electric heating plate) to heat the generator.

3. Place a thermometer that reads 0–100°C into the metal shot. If the thermometer is a mercury thermometer, be very careful not to break the bulb as you work it down into the shot. Place paper or some other insulation around the top of the cup to keep the outside air from cooling the top layer of shot.

4. While the shot is heating, determine the mass of the inner calorimeter cup and the stirrer separately and record their mass in the Data Table. Obtain from your instructor the values for c_{cal} and c_s (the specific heats of the calorimeter cup and the stirrer) and record them in the Data Table.

5. Place about 100 g of water in the inner calorimeter cup and determine the combined mass of the cup, stirrer, and water. Determine the mass of the water by subtraction and record it in the Data Table. The initial temperature of the water ideally should be a few degrees below the room temperature. This will tend to compensate for interaction with the air.

6. Assemble the calorimeter, placing the second thermometer in the rubber stopper. Use glycerin on the thermometer so that it may be easily slipped into the hole in the stopper. *If using a glass thermometer be extremely careful. If the thermometer breaks, you can cut yourself severely.* Stir the water slowly to be sure that all parts of the system are in equilibrium. Record the temperature of the water, calorimeter, and stirrer as T_0 in the Data Table. Take this reading for T_0 just prior to performing step 8 below.

7. Note the temperature of the metal shot every few minutes. Move the bulb around carefully in the shot to be sure all parts are in equilibrium. The temperature should rise and then level off at essentially the boiling point of water. This will probably be a few degrees below 100°C depending on the local elevation above sea level. Record the value (to the nearest 0.1°) of the maximum temperature as T_m in the Data Table.

8. Remove the thermometer from the shot and quickly transfer the shot to the water in the inner calorimeter cup. It is very important not to splash water out of the cup in this process. This can easily happen when removing the insulating plastic cover or when placing the shot in the water.

9. Place the insulating cover back on the calorimeter and slowly stir the water while watching the thermometer in the calorimeter. When the maximum temperature is reached, record its value (to the nearest 0.1°) as the equilibrium temperature T_e.

10. Remove the inner calorimeter cup and determine the mass of the cup, stirrer, water, and metal shot. Record that value in the Data Table. Determine the mass of the shot used by subtraction and record that value in the Data Table.

11. Discard the water but do not lose any of the shot. Place the wet shot on paper towels, spread it out, and allow the shot to dry. Before leaving the laboratory and when the shot is completely dry, return it to its proper container. Be careful not to mix the two kinds of metal shot.

12. From Appendix II, determine the known value of the specific heat for the type of metal of the shot and record it in the Data Table.

13. Repeat all of the above procedure for a second type of metal shot.

CALCULATIONS

1. In equation 2, all of the variables are known except for the specific heat of the metal shot. Treating it as the unknown, solve equation 2 for the specific heat of the metal shot and record that value as the experimental value in the Calculations Table.

2. Calculate the percentage error between your experimental value and the known value of the specific heat for each type of metal.

Laboratory 23
Specific Heat of Metals

LABORATORY REPORT

Data Table

Mass cup = _____ g   Mass stirrer = _____ g

Metal	Mass in grams				Temperatures (°C)		
	Cup + stirrer + water	Cup + stirrer + water + shot	Water	Shot	T_o	T_m	T_e

Calculations Table

Metal	Known specific heat (cal/g − °C)	Experimental specific heat (cal/g − °C)	% error

SAMPLE CALCULATIONS

QUESTIONS

1. What is the accuracy of your results for the specific heats of the metals? Suggest any change in the procedure that would improve the results. (*Hint*: The larger the change in temperature of the water, the better precision with which it can be measured.)

2. Can you make any quantitative statement about the precision of your results? If you cannot make such a statement, suggest what measurements would allow you to do so.

3. Suppose the shot were wet and thus included some water at the same temperature as the shot when it was placed in the calorimeter. How would this affect the results?

4. Water is often used as a heat storage medium. Does it perform that function extremely well, or is it used just because it is easily obtained and works fairly well? What property of water determines the answer to this question? How unique is water with respect to this property?

5. What would the change in temperature of the water have been if 200 g of water had been used with each of the samples? Would this have tended to improve the results, make them worse, or have no effect?

Laboratory **24**

Linear Thermal Expansion

PRELABORATORY ASSIGNMENT

Read carefully the entire description of the laboratory and answer the following questions based on the material contained in the reading assignment. Turn in the completed prelaboratory assignment at the beginning of the laboratory period prior to the performance of the laboratory.

1. An object changes in length because of a change in its temperature. What are the three factors on which this change in length depends?

2. What are the units for the linear thermal coefficient of expansion α?

3. Most materials expand when the temperature is raised. What can you conclude about α for a material that contracts when its temperature is raised?

4. A copper rod has a length of 1.117 m when its temperature is 22.0°C. If the temperature of the rod is raised to 275.0°C, what is the new length of the rod? [The value of α for copper is 16.8×10^{-6} (C°)$^{-1}$.]

5. Iron rails are used to build a railroad track. If each rail is 10.000 m long when it is placed in the track at a temperature of 20.0°C, how much space must be left between the rails so that they just touch each other when the temperature is 40.0°C? [The value of α for iron is 11.4×10^{-6} (C°)$^{-1}$.]

6. A rod in a steam jacket is measured to have a length of 0.600 m at a temperature of 22.0°C. Steam is then passed through the jacket for several minutes until the rod is at a temperature of 98.0°C. The increase in the length of the rod is measured by a micrometer screw arrangement to be 1.19 mm. What is the linear thermal coefficient of expansion α for the rod?

7. Considering the value of α found for the rod in question 6, from what kind of metal is it probably made?

Linear Thermal Expansion

OBJECTIVES

Most materials expand as their temperature increases. Although this effect takes place in all three dimensions, this laboratory will consider the expansion in only one dimension. The increase in length of a long thin rod will be determined. Empirically it is found that the change in length of an object is proportional to its initial length and to the change in temperature of the object. The constant of proportionality is called the "linear coefficient of thermal expansion," or α. In this laboratory, measurements of the length of several metal rods as a function of temperature will be used to accomplish the following objectives:

1. Demonstration that for the same temperature change the amount of thermal expansion depends on the type of metal

2. Determination of the value of the linear coefficient of thermal expansion α for several metals

3. Comparison of the experimental values for α with the known values

EQUIPMENT LIST

1. Linear expansion apparatus consisting of steam jacket containing a metal rod

2. Steam generator, rubber tubing, and beaker to catch steam condensation

3. Bunsen burner and stand, or electric heating plate

4. Meter stick

5. Thermometer (0–100°C)

6. Ohmmeter if using micrometer screw–type indicator

THEORY

When most materials are heated they expand as their temperature increases. However, note carefully that the expansion depends not on the heat input, but rather the temperature change. The expansion of materials with increased temperature takes place in three dimensions, but this laboratory will consider only one-dimensional changes in the length of a rod.

Consider a metal rod of some initial length L_0 at some initial temperature T_0. The rod is heated, and its temperature is increased to some temperature T_1. The length of the rod will increase to a new length L_1. The change in length of the rod ΔL is given by

$$\Delta L = L_1 - L_0 \tag{1}$$

This change in length ΔL is found to be proportional to the original length of the rod L_0 and to the change in the temperature ΔT, where $\Delta T = T_1 - T_0$. In equation form, the result is

$$\Delta L = \alpha \, L_0 \, \Delta T \qquad (2)$$

where α is a constant called the "linear coefficient of thermal expansion." Solving equation 2 for the constant α gives

$$\alpha = \frac{\Delta L}{L_0 \, \Delta T} \qquad (3)$$

From the form of equation 3, it is clear that α is the fractional change in length per unit change in the temperature. Since the fractional change in length $\Delta L/L_0$ has no dimensions, the units of α are $(C°)^{-1}$. Strictly speaking, α varies slightly with temperature. However, over the range of temperature used in this laboratory α can be assumed to be approximately constant.

The apparatus to be used in this laboratory is shown in Figure 24.1. It consists of a steam jacket containing a metal rod about 0.60 m long. The jacket is held by supports at either end. One of the end supports has a thumbscrew whose purpose is to keep that end of the rod fixed. The other end support contains an indicator to measure the change in length of the rod.

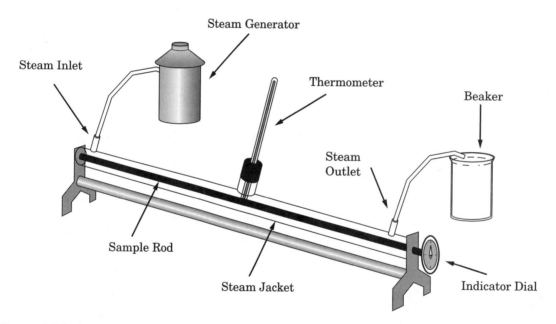

Figure 24.1 Experimental arrangement for thermal expansion apparatus.

There are two types of indicators. One has a micrometer screw with a rotary dial of 100 divisions. Each division is 0.01 mm; therefore, one complete turn of the dial is equivalent to a linear translation of 1 mm. Some micrometer screw indicators have a set of binding posts on each end support. These can be used to construct an electrical circuit to indicate when the micrometer screw makes contact with the rod. When using a micrometer screw–type indicator, it is important to remember to back the screw away from the rod before the rod is allowed to expand in order to prevent damage to the micrometer screw.

The second type of indicator contains a plunger-activated dial that reads directly in 0.01-mm increments. One complete revolution of the dial corresponds to 1 mm of linear displacement, with a total of 3.5 mm of displacement possible. When using this type of indicator, contact is made with the rod, and the rod is allowed to expand against the plunger, which always remains in contact with the rod.

EXPERIMENTAL PROCEDURE

1. Remove the rod from the steam jacket and measure the length of the rod with a meter stick. Measure to the nearest 0.001 m and record this length as L_0 in the Data Table.

2. Replace the rod in the steam jacket and secure the jacket in the support ends. If using a device that has binding posts, connect the leads from an ohmmeter to each of the binding posts. Actually, even if the apparatus does not have binding posts, the ohmmeter can be used by making contact with the end supports when a measurement is to be made.

3. Using a one-holed rubber stopper, place a thermometer in the opening provided for that purpose. The opening is located in the center of the steam jacket. The thermometer bulb should just barely touch the rod. If the apparatus has been standing unused for several hours or more, record the temperature after the thermometer is in contact with the rod. If the steam jacket has been recently heated, run cool water through the jacket until the entire system is at equilibrium at a temperature near room temperature. Record the temperature (to the nearest 0.1C°) in the Data Table as T_0. Adjust the indicator dial until contact is made with the rod. If using the micrometer-type device, contact is indicated by the ohmmeter. Record the indicator dial setting as D_0 in the Data Table.

4. If using the micrometer-type indicator, back the screw out several turns at this time. If using the plunger-type indicator, leave it in contact with the rod. *It is extremely critical that there be no disturbance of the rod between this reading and the final reading after the rod has been heated.*

5. Connect the steam supply to the steam jacket with a rubber hose. At the other end of the jacket, connect a hose from the steam outlet to a beaker to catch the steam condensation. Pass steam through the jacket for several minutes and monitor the temperature of the rod. When the temperature has reached its maximum value, record that value of the temperature (to the nearest 0.1C°) as T_1.

6. If using the plunger-type indicator, simply read the value on the indicator dial. If using the micrometer screw–type device, turn the screw in until it touches the rod as indicated by the ohmmeter. Record the reading as D_1 in the Data Table.

7. Repeat steps 1 through 6 for other rods of different metals. Be extremely careful not to burn yourself on the heated steam jacket. Before beginning the procedure, run cool water through the apparatus until the new rod and jacket are in equilibrium near room temperature.

8. From Appendix II, determine the known value of α for each of the rods measured and record them in the Data Table.

CALCULATIONS

1. Calculate the increase in length ΔL for each rod from $\Delta L = D_1 - D_0$ and record each of them in the Calculations Table.

2. Calculate the increase in temperature ΔT for each rod from $\Delta T = T_1 - T_0$ and record each of them in the Calculations Table.

3. Using equation 3, calculate the linear coefficient of thermal expansion α for each rod and record each of them in the Calculations Table.

4. Calculate the percentage error for each experimental value of α compared to the known value. Record them in the Calculations Table.

LABORATORY REPORT

Data Table

Metal	Known α (°C)$^{-1}$	L_0 (m)	T_0 (°C)	T_1 (°C)	D_0 (m)	D_1 (m)

Calculations Table

Metal	ΔL (m)	ΔT (°C)	α (°C)$^{-1}$	% error

SAMPLE CALCULATIONS

QUESTIONS

1. What is the accuracy of your measurements of α? State clearly the basis for your answer.

2. The change in length ΔL could be measured with more accuracy if it were larger. This could be accomplished by heating the rod directly with a Bunsen burner to a temperature considerably higher than 100°C. Would that be a reasonable alternative? In what way is the steam heat a more workable technique?

3. The original length L_0 of the rod was measured only to the nearest 1 mm. Does this cause a significant error in the final result? If so, why does it? If not, why does it not?

4. Would the measured value of α have been the same or different if lengths were determined in centimeters instead of meters? State clearly the basis for your answer.

5. A washer is made of brass and has an inside diameter of 2.000 cm and an outside diameter of 3.000 cm at 20.0°C. A solid rod of aluminum has a diameter of 2.000 cm at 20.0°C and just fits inside the washer as shown in Figure 24.2. Both the washer and aluminum rod are raised to a temperature of 150.0°C. Will the rod still fit inside the washer? If so, how much smaller is the rod than the opening in the washer? If not, how much larger is the rod than the opening in the washer? Show your work.

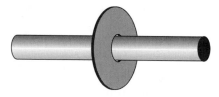

Fig 24.2 Brass washer on an aluminum rod.

Laboratory 25
The Ideal Gas Law

PRELABORATORY ASSIGNMENT

Read carefully the entire description of the laboratory and answer the following questions based on the material contained in the reading assignment. Turn in the completed prelaboratory assignment at the beginning of the laboratory period prior to the performance of the laboratory.

1. What is the ideal gas law?

2. What are the conditions under which a real gas approximately obeys the ideal gas law?

3. What are the SI units for pressure and volume? What is the value of Boltzmann's constant k_B?

4. Which temperature scale must always be used in the ideal gas law?

5. The temperature in a room is measured to be $T_C = 24.5°C$. What is the absolute temperature of the room in kelvins?

6. If the volume of the room in question 5 is 50.0 m^3, and the pressure is 1.013×10^5 N/m^2, what is the value of N, the number of molecules in the room?

7. A gas at constant temperature has a volume of 25.0 m^3, and the pressure of the gas is 1.50×10^5 N/m^2. What is the volume of the gas if the pressure of the gas is increased to 2.50×10^5 N/m^2 while the temperature remains fixed? Is this an example of Boyle's law or Charles' law?

8. A gas at constant pressure has a volume of 35.0 m^3, and its temperature is 20.0°C. What is its volume if its temperature is raised to 100.0°C at the same pressure? Is this an example of Boyle's law or Charles' law?

OBJECTIVES

A gas containing N molecules is confined to a container of volume V at pressure P and temperature T. The four variables of the problem are not independent but are related. For an ideal gas the relationship between them is simple, and it is given by $PV = Nk_BT$. This relationship is known as the ideal gas law, and k_B is a universal constant known as Boltzmann's constant. The temperature T must be the absolute temperature in kelvins. This laboratory will use measurement of the temperature T, the pressure P, and the volume V of a sample of air in a large plastic syringe to accomplish the following objectives:

1. Demonstration that the pressure P of a gas at a fixed temperature is proportional to the quantity $(1/V)$, where V is the gas volume

2. Demonstration that the volume V of a gas at a fixed pressure is proportional to the temperature T of the gas

3. Determination of an experimental value for the constant that relates Celsius temperature T_C to the absolute temperature T.

4. Comparison of the experimental value for this constant with its known value of 273.15.

EQUIPMENT LIST

1. Homemade gas law apparatus,* petroleum jelly or stopcock grease, and string

2. Thermometer (0–100°C) and vernier calipers

3. 600-ml beaker and tongs to fit the beaker

4. Masses to allow application of up to 6 kg in 1-kg increments

5. Electric heating plate or Bunsen burner (If a Bunsen burner is used, a stand to hold the beaker while it is heated is also needed.)

6. One large container of ice for the class

THEORY

Consider a gas containing N molecules confined to a volume V at a pressure P and temperature T. The relationship between these quantities is of fundamental importance. In the most general case, all of these quantities vary over a wide range, and

---

* The apparatus can be constructed from a large plastic syringe (either 35 ml or 60 ml volume) marked in 1-ml increments. The needle is removed, and the end of the syringe is melted closed (airtight). The outer cylinder of the syringe is mounted on a small wooden support. A small flat wooden platform is attached to the top of the plunger. These devices can be easily and cheaply constructed. See Figure 25.1 for a picture.

the equation of state that relates them is inherently very complex. However, if only cases in which the density of the gas is low are considered, the equation of state is greatly simplified. A gas that satisfies these conditions is referred to as an ideal gas. Although there are no true ideal gases, most real gases behave to a good approximation as an ideal gas near room temperature and atmospheric pressure. The equation of state of an ideal gas is given by

$$PV = Nk_BT \qquad (1)$$

The SI units of pressure P are N/m^2, and the SI units of volume V are m^3. The constant k_B is called Boltzmann's constant, and it has the value 1.38×10^{-23} J/K. In the ideal gas law the temperature must always be expressed as the absolute temperature in kelvins. The relationship between temperature in Celsius T_C and the absolute temperature in kelvins T is given by

$$T = T_C + 273.15 \qquad (2)$$

In this laboratory, equation 1 will be verified in two different ways. Consider first the case of a fixed amount of gas (thus N is constant) at some given constant temperature. Under these conditions the quantity Nk_BT is a constant that shall be designated as C_1. In terms of C_1, equation 1 can be rewritten in the form

$$P = C_1 \, (1/V) \qquad (3)$$

Consider the apparatus shown in Figure 25.1. The sliding rubber seal of the syringe allows a fixed amount of gas that is trapped in the cylinder to vary in volume V as the pressure P is changed. The total pressure P consists of two parts, the atmospheric pressure P_a and any additional pressure, which is called gauge pressure, P_g. As shown in Figure 25.1 the gauge pressure is applied by placing mass m on the platform of the plunger. The gauge pressure is caused by the weight mg of the mass per unit area A of the cylinder giving

$$P_g = F/A = mg/A \qquad (4)$$

Substituting $P_g + P_a$ for P in equation 3 leads to the following:

$$P_g = C_1(1/V) - P_a \qquad (5)$$

Equation 5 states that there is a linear relationship between the applied or gauge pressure P_g and the quantity $1/V$ with C_1 as the slope and $-P_a$ as the intercept. This aspect of the ideal gas law is known as Boyle's law. In the first part of the laboratory, the volume V of a gas sample will be measured as a function of the gauge pressure P_g. Experimental values for C_1 and P_a will then be obtained from a fit of the data to the form of equation 5.

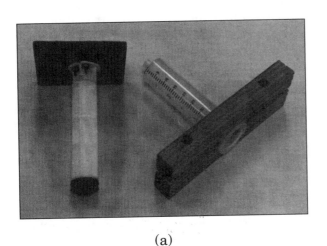

(a)   (b)

Figure 25.1 Syringe with needle removed and end closed so that it is airtight. The sliding seal allows a fixed amount of air to change volume as pressure or temperature is changed.

Consider now the case of a fixed amount of gas at some constant pressure P. Under these conditions the quantity Nk_B/P is a constant that shall be designated as C_2. In terms of the constant C_2 and the Celsius temperature T_C, equation 1 can be rewritten as

$$V = C_2\, T_C + C_2\,(273.15) \tag{6}$$

Equation 6 states that there is a linear relationship between the volume V of a gas and the Celsius temperature T_C with C_2 as the slope and $C_2(273.15)$ as the intercept. This aspect of the ideal gas law is essentially the law of Charles and Gay-Lussac. In the second part of the laboratory, the volume of a sample of gas will be measured as a function of the Celsius temperature T_C. Experimental values for C_2 and $C_2(273.15)$ will be obtained by a fit of the data to the form of equation 6. Examination of equation 6 also shows that if the ratio of the two constants in the equation is designated as K, that ratio is given by the following:

$$K = \frac{\text{Intercept}}{\text{Slope}} = 273.15 \tag{7}$$

Using equation 7, an experimental value for K can be obtained from the intercept and slope of the fit to the volume versus temperature data. The negative of the value for the constant $(-K)$ is an experimental value for temperature on the Celsius scale at which absolute zero occurs.

EXPERIMENTAL PROCEDURE—CONSTANT TEMPERATURE

1. Pull the plunger from the syringe and place a short length of string into the cylinder with a loose end hanging out. Put a thin coat of petroleum jelly or stopcock grease on the rubber tip of the plunger. Replace the plunger in the syringe. The string allows a small air leak in order to adjust the volume of air trapped by the plunger. Adjust the volume of air to 25 ml if using the 35-ml syringe and 45 ml if using the 60-ml syringe, and then withdraw the string. (1 ml = 1 cm^3 = 1 x 10^{-6} m^3)

2. Place the support block on the syringe across the opening of the beaker as shown in Figure 25.1. Place mass m on the plunger platform as shown in the figure. Use values for m from 1 kg to 6 kg in increments of 1 kg. Determine the volume of the gas trapped in the syringe for each value of m. In order to determine the volume more accurately, push down slightly on the plunger and allow the plunger to spring back before each measurement of the volume is taken. Record the values of V for each m in Data Table 1.

3. Using the vernier calipers determine the inside diameter D of the syringe cylinder and record it in Data Table 1.

4. Determine the temperature of the room air in Celsius T_C and record it in Data Table 1.

5. Standard atmospheric pressure at sea level is 1.013×10^5 N/m^2. Determine the present local barometric pressure in these units and record that value in Data Table 1.

CALCULATIONS—CONSTANT TEMPERATURE

1. Calculate the cross-sectional area A of the syringe from the measured diameter D of the cylinder, where $A = \pi D^2/4$. Record the value of A in Calculations Table 1.

2. Calculate the force F corresponding to each mass m and record the values of F in Calculations Table 1. ($F = mg$ with $g = 9.80$ m/s^2)

3. Using equation 4, calculate the gauge pressure P_g for each of the values of F. Record the values of P_g in Calculations Table 1.

4. For each of the measured values of V, calculate the quantity $1/V$ and record the results in Calculations Table 1.

5. Perform a linear least squares fit with P_g as the ordinate and $1/V$ as the abscissa. Determine the slope, intercept, and correlation coefficient of the fit. Record the slope as C_1 and the negative of the intercept as $(P_a)_{exp}$ in Calculations Table 1. Record the correlation coefficient in Calculations Table 1.

6. Calculate the percentage error for the value of $(P_a)_{exp}$ compared to the current atmospheric pressure and record the results in Calculations Table 1.

EXPERIMENTAL PROCEDURE—CONSTANT PRESSURE

1. Using a string as before, adjust the volume V of the syringe to 25 ml if using the 35-ml syringe and 45 ml if using the 60-ml syringe.

2. Place tap water in the beaker and place the syringe support on the the beaker. The syringe cylinder should be almost completely immersed in the water. Place a 2-kg mass on the plunger platform. Place the beaker in a position to be heated and bring the water to a boil. *The beaker must be securely supported while being heated.* Allow at least 5 minutes after the water begins to boil for the air in the syringe to come to equilibrium with the boiling water. Measure the volume V and the temperature of the water T_C and record them in Data Table 2.

3. Remove the 2-kg mass from the plunger. Using the beaker tongs, carefully remove the beaker from the source of heat and set it on the laboratory table. Replace the 2-kg mass on the plunger. Add small pieces of ice to the water and stir the mixture after each addition of ice until all the ice melts. Monitor the temperature of the water at all times. Determine the volume V of the syringe for water temperatures of about 75°C, 50°C, and 25°C. For each temperature at which the volume is determined, have no ice in the water and allow a few minutes at that temperature to establish equilibrium between the water and the air in the syringe. Record the volume V and the temperature T_C (to the nearest 0.1°C) in Data Table 2.

4. For the last temperature add enough ice so that an ice and water equilibrium is established. Some water may have to be discarded. Allow at least 5 minutes for the air to come to equilibrium with the ice and water and record the volume V of the trapped air and the temperature T_C (to the nearest 0.1° C) in Data Table 2.

CALCULATIONS—CONSTANT PRESSURE

1. Perform a linear least squares fit with V as the ordinate and T_C as the abscissa. Determine the slope, intercept, and correlation coefficient of the fit and record them in Calculations Table 2.

2. Using equation 7, calculate an experimental value of the constant K from the values of the slope and intercept from the fit. Record this value as K_{exp} in Calculations Table 2.

3. Calculate the percentage error in the value of K_{exp} compared to the known value for K, which is 273.15. Record this percentage error in Calculations Table 2.

GRAPHS

1. Make a graph of the data in Calculations Table 1 with P_g as the ordinate and $1/V$ as the abscissa. Also show on the graph the straight line obtained by the linear least squares fit to the data.

2. Make a graph of the data in Calculations Table 2 with V as the ordinate and T as the abscissa. Choose a scale that allows the data to be extrapolated back to $V = 0$. In effect this will mean a temperature scale from approximately $-300°C$ to $100°C$. Also show on the graph the straight line obtained from the fit to the data extrapolated to $V = 0$.

LABORATORY REPORT

Data Table 1

m (kg)	V (m^3)

$D =$	m
$T_C =$	°C
$P_a =$	N/m^2

Calculations Table 1

F (N)	P_g (N/m^2)	$1/V$ (m^{-3})

$A =$	m^2	Corr. coeff. =	
$C_1 =$	N−m	$(P_a)_{exp} =$	N/m^2
% error =			

Data Table 2

T_C (°C)	V (m^3)

Calculations Table 2

Slope =	m^3/°C
Intercept =	m^3
Corr. coeff. =	
$K =$	
% error =	

SAMPLE CALCULATIONS

QUESTIONS

1. How well do your data confirm the ideal gas law? State your answer to this question as quantitatively as possible.

2. What is the accuracy of your value of $(P_a)_{exp}$ compared to the actual atmospheric pressure?

3. What is the accuracy of your value for K?

4. Consider the value of the constant C_1 in Calculations Table 1. It should be equal to Nk_BT. Equate the measured value of C_1 to Nk_BT and solve the resulting equation for N. In this case, N is the number of molecules in the syringe when the temperature was held constant.

5. Consider the value of the slope in Calculations Table 2. It should be equal to Nk_B/P. Equate the measured value of the slope to Nk_B/P and solve for N. In this case, N is the number of molecules in the syringe when the pressure was held constant. Remember that the value of P_g for this part of the laboratory was caused by the 2.00-kg mass. Use the known value of the atmospheric pressure P_a.

Laboratory 26

Equipotentials and Electric Fields

PRELABORATORY ASSIGNMENT

Read carefully the entire description of the laboratory and answer the following questions based on the material contained in the reading assignment. Turn in the completed prelaboratory assignment at the beginning of the laboratory period prior to the performance of the laboratory.

1. Electric field lines are drawn (a) from positive charges to negative charges, (b) from negative charges to positive charges, (c) from the largest charge to the smallest charge, or (d) from the smallest charge to the largest charge.

2. The points where the potential is the same (in three dimensional space) have the same voltage. (a) True (b) False

3. The points where the potential is the same (in three dimensional space) lie on a surface. (a) True (b) False

4. The relationship between the direction of the electric field lines and the equipotential surfaces is (a) field lines are everywhere parallel to surfaces, (b) field lines always intersect each other, (c) field lines are everywhere perpendicular to surfaces, or (d) field lines always make angles between 0° and 90° with surfaces.

5. For this laboratory, why are the measured equipotentials lines instead of surfaces?

6. If two electrodes have a source of potential difference of 100 V connected to them, how many equipotential surfaces exist in the space between them?

7. How much work is done in moving a charge of 10.0 μC 1.00 m along an equipotential of 10.0 V?

8. For this laboratory, why is it important to center the electrodes on the resistance paper?

9. In the performance of this laboratory, what is the recommended maximum allowed potential difference from one end of an electrode to the other end?

10. How are you supposed to decide how many points to measure for each equipotential for a given electrode configuration?

Equipotentials and Electric Fields

OBJECTIVES

In this experiment, the relationship between the equipotential surfaces and electric field lines in the region around several different electrode configurations will be investigated. Electrodes, drawn on carbon impregnated paper with conducting paint and connected to a direct current power supply, will simulate statically charged electrodes to accomplish the following objectives:

1. Determination of the location of equipotential surfaces in the region around oppositely charged electrodes by measurements made with a voltmeter

2. Investigation of the shape of the equipotential surfaces for several specific electrode arrangements

3. Construction of electric field lines from the measured equipotentials by drawing lines perpendicular to the equipotentials

4. Comparison of the experimentally determined shapes of electric field lines with several familiar electrode arrangements (line charge, two line charges of opposite sign, and parallel plates)

5. Determination of the dependence of the magnitude of E on the distance r from a line of charge

EQUIPMENT LIST

1. Corkboard and push pins to attach power supply and voltmeter to electrodes

2. Carbon-impregnated gridded resistance paper

3. Conducting paint or conducting pen (either silver or carbon-based)

4. Direct current power supply (20 V, low current)

5. High-impedance voltmeter (preferably digital)

THEORY

Consider two electrodes of arbitrary shape some distance apart carrying equal and opposite charges. There will then exist a fixed potential difference or voltage between the electrodes. Suppose that this potential difference is 20 V. If the electrode with the negative charge is arbitrarily assumed to be at zero potential, then the electrode with the positive charge is at a potential of +20 V. Given these assumptions, in the space surrounding these electrodes there will exist points that are at the same potential. For example, for the case described above, there will exist some points for which the potential is +10 V, other points for which the potential is +15 V, and still other points for which the potential is +5 V. In a three-dimensional space, all points at the same potential will form a surface, and there will be a different surface for each value of the potential between 0 V and 20 V. In fact, there will exist an infinite number of such surfaces because one could divide the 20-V total potential difference into an infinite number of steps. Each of these surfaces with the same value of potential is called an "equipotential surface." In this laboratory, the equipotentials for a few simple, but often used, electrode configurations will be determined.

In addition to the equipotential surfaces that exist in the region around charged electrodes, an electric field is also present. By definition, the electric field is a vector field, which can be represented by lines drawn from the positively charged electrode to the negatively charged electrode. The direction of the electric field lines at every point in space is the direction of the force that would be exerted on a positive test charge placed at that point in space. To ensure that the test charge does not disturb the other charges, the test charge must be small. In fact, in the exact definition, the limit must be taken as the test charge approaches zero. The magnitude of the electric field is the force per unit charge on the positive test charge as the magnitude of the test charge approaches zero. The units of electric field are N/C. The number of field lines per unit area at a given point is a measure of the magnitude of the electric field. Thus, a region where there are a large number of lines per unit area is a region of large electric field.

The electric field lines must always exist in a fixed geometrical relationship with the equipotential surfaces for any electrode configuration. Simply stated, the relationship is that the electric field lines are everywhere perpendicular to the equipotential surfaces. Since the electrodes themselves are equipotential surfaces, the electric field lines must also intersect the electrodes perpendicularly. This is usually a very helpful guide when attempting to determine the shape of electric fields around an electrode arrangement.

If the change in potential, ΔV, is measured between two points separated by a displacement, Δx, then it can be shown that the following equation is true:

$$E = - \frac{\Delta V}{\Delta x} \qquad \textbf{(1)}$$

where E stands for the electric field. To be exact, the limit of equation 1 must be taken as $\Delta x \rightarrow 0$, but equation 1 can be useful as an approximation if Δx is small. According to equation 1, if $\Delta V > 0$ for a Δx in a given direction, then E is in the opposite direction of the displacement Δx; but if $\Delta V < 0$, then the field is in the direction of Δx. Equation 1 also shows that another proper unit for the electric field is V/m. Another important fact that can be seen from equation 1 is that it takes no work to move an electric charge on an equipotential surface, because along the equipotential, $\Delta V = 0$, and thus $E = 0$ along the equipotential surface.

Since the electrodes that can be drawn on the carbon paper are limited to two dimensions, they represent a slice taken through a real three-dimensional electrode configuration. As a result, the equipotentials mapped by this laboratory will be equipotential lines rather than surfaces. Any two-dimensional electrode arrangement can be produced by drawing the desired electrode shape on the carbon paper with conducting paint or a conducting pen and then attaching a source of potential difference to the electrodes. A direct-current power supply will provide the source of potential difference and will serve to keep the voltage between the two electrodes fixed at whatever value is chosen from the power supply. The electrode to which the negative terminal of the power supply is attached will arbitrarily be chosen to be the zero of potential, and all measurements will be made relative to that electrode. The apparatus used in the laboratory is shown in Figure 26.1. A voltmeter will then be used to find the points on the paper in the region of the electrodes that are at some given value of potential. Once enough points at that potential have been located to establish the shape of the equipotential, the equipotential line can be constructed by joining the points with a smooth curve. The number of data points needed to establish the shape must be decided by the student.

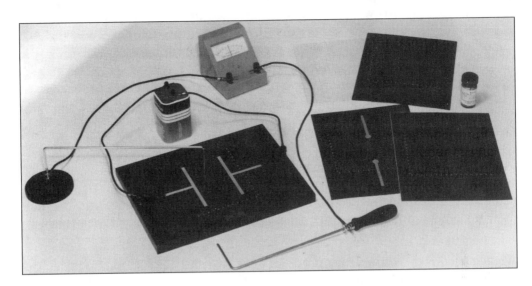

Figure 26.1 Equipotential mapping apparatus shown here using a battery rather than a power supply. (Photo courtesy of Sargent-Welch Scientific Company)

EXPERIMENTAL PROCEDURE

1. Use a paint brush or conductive ink pen to draw the three electrode configurations shown in Figure 26.2. Note carefully the following important cautions: (a) Place the conductive paper on a hard surface to draw the electrodes. Do not draw while the paper is on the corkboard. (b) Make sure that the conductive paint or ink flows smoothly and evenly when drawing electrodes and that a solid line is obtained.

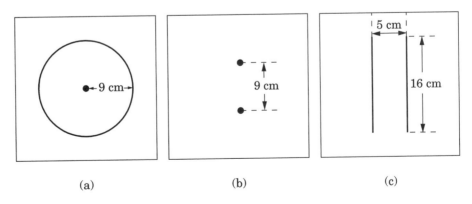

Figure 26.2 Electrode configurations to be mapped.

2. Draw the three electrode configurations pictured in Figure 26.2 and described below. For each configuration, use one clean sheet of carbon paper and arrange the electrodes as nearly centered on the paper as possible because edge effects can be important in some cases. (a) Line of charge perpendicular to the paper and guard cylinder of radius 9.0 cm—draw a small dot at the center of the paper and then draw a circle of radius 9.0 cm centered on the dot. Be sure that the dot is centered on one of the grid markings. (b) Two lines of opposite charge perpendicular to the paper—draw two small dots 9.0 cm apart symmetrically located on the paper. (c) Parallel plate capacitor—draw two straight lines 16.0 cm long and 5.0 cm apart symmetrically located on the paper.

3. For each electrode configuration in turn, place the metal push pins in the electrodes and connect the two leads from the power supply to the pins. For the parallel plate and two-line charge configuration there is symmetry, and the assignment of which electrode is negative and which is positive is arbitrary. For the line charge and guard ring, choose the line charge as positive and the guard ring as negative.

4. In each case, set the potential difference between the electrodes to be 20.0 V. To set this value, connect the voltmeter between the electrodes with the negative voltmeter lead connected to the negative power supply output, and the positive voltmeter lead connected to the positive power supply output. Once this value is set it should remain fixed.

5. For each electrode configuration, make sure that all connections are secure and that the push pins are pressed firmly into the corkboard, which will ensure good contact with the electrodes. In order to check that the electrodes themselves have the proper conductivity, connect one lead of the voltmeter to one of the electrode push pins, and then using the other lead of the voltmeter as a probe, touch it to various parts of the same electrode. For a properly conductive electrode, the maximum voltage between any two points on the same electrode should be less than 0.2 V. Repeat this test for the other electrode.

6. Determine the equipotentials by connecting the negative voltmeter lead to the electrode push pin that is connected to the negative output of the power supply (which is to be considered the zero of potential). As illustrated in Figure 26.3, the other voltmeter lead then serves as a probe, and it is used to measure the potential at any point on the paper by touching the probe to the paper at that point. Be sure that the probe used has a sharp point and that the probe is held perpendicularly to the paper so that only the point of the probe touches the paper. A given equipotential line (for example the 10.0-V equipotential) is mapped by moving the

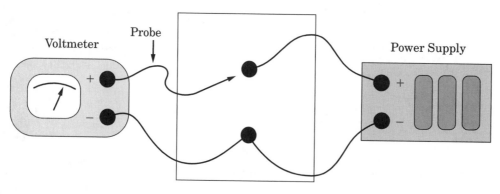

Figure 26.3 Power supply and voltmeter connections for mapping equipotentials.

probe around to find the points at which the voltmeter reading is 10.0 V and then connecting these points to produce a smooth curve or line. For each equipotential, obtain enough points to define the shape of that equipotential line clearly.

7. Using the procedure described in step 6, map in pencil the following equipotential lines for each electrode configuration:

    (a) Line charge—15.0, 10.0, 6.50, 4.50, 3.50, 2.50, 1.50, and 0.75 V.

    (b) Two-line charges—16.0, 13.0, 11.5, 10.0, 8.50, 7.00, and 4.00 V.

    (c) Parallel plates—4.00, 8.00, 12.0, and 16.0 V.

8. In general, electric field lines were stated to be everywhere perpendicular to equipotential surfaces. Because our electrodes are confined to the plane of the paper, the equipotentials are lines, but it is still true that the electric field lines are everywhere perpendicular to these equipotential lines. For each set of electrodes, draw a set of lines that are perpendicular to your measured equipotential lines. These are the electric field lines. Place arrows on them to indicate their direction from positive charge to negative charge. Distinguish them in some way from the equipotential lines, either by dotting the lines or drawing them in a different color.

9. Make the following measurements for the electrode configuration of the line of charge. Measure the value of the change in potential at the distances from the line of charge listed in the Data Table. Tape the two voltmeter probes together with a small piece of insulating material, holding the points of the probes apart at a fixed distance, Δx, of about 0.00300 m. Place the probes symmetrically about the grid positions on the paper at the values listed in the Data Table in such a manner that the gap between the probes is centered on each grid position in turn. It is critical that the gap be as precisely centered on each position as possible. Record the values of ΔV and the value of Δx in the Data Table.

CALCULATIONS

1. For the measurements made in step 9, calculate the approximate value of E at each point as $\Delta V/\Delta x$. Record these values of E in the Calculations Table.

2. Perform a linear least squares fit to this data of E versus $1/r$. Note that the ordinate is to be E and the abscissa is to be $1/r$. Determine the slope, intercept, and the correlation coefficient for the fit and record them in the Calculations Table.

GRAPHS

1. Construct accurate drawings on 1-cm-by-1-cm graph paper showing the electrodes and your measured equipotentials and electric field lines for each electrode configuration.

2. Make a graph of the data for E versus $1/r$ for the line charge data. Also show on the graph the straight line obtained in the least squares fit.

LABORATORY REPORT

Data Table

r (m)	ΔV (volt)
0.0150	
0.0200	
0.0300	
0.0400	
0.0500	
0.0600	
0.0700	
0.0800	

$\Delta x = $ _____ m

Calculations Table

$E = \frac{\Delta V}{\Delta x} \left(\frac{\text{volt}}{\text{m}} \right)$	$1/r \, (\text{m}^{-1})$

Slope = _____

Intercept = _____

Regress. coeff. = _____

SAMPLE CALCULATIONS

QUESTIONS

1. Although the magnitudes of the equipotentials and electric fields for electrode configurations (a) and (b) are consistent with interpretation of those arrangements as line charges, the shapes of the equipotentials and electric field lines are the same as those for a point charge and a dipole charge. Compare your graphs with those in your textbook for a point charge and a dipole. Comment on their similarities and differences, if any.

2. Point A in Figure 26.4 represents a point halfway between the (+) electrode and the 16.00 equipotential on the plot of your data. Calculate the electric field E_A at the point A using equation 1. Note that the value of ΔV is given by the difference in potential between the (+) electrode and 16.00 V. In other words, if the (+) electrode has a potential of 20.00 V then $\Delta V = 4.00$ V. The value of Δx to be used is the distance from the center of the (+) electrode to the 16.00-V equipotential measured in the direction from the (+) electrode toward the (−) electrode. In a similar manner, calculate the field E_C at point C, which is halfway between the (−) electrode and the 4.00-V equipotential. Point B is at the center of the electrodes. Calculate E_B using ΔV between the 8.50-V equipotential and the 11.50-V equipotential, and using for Δx the distance between those equipotentials measured along the line between the electrodes. Record the values of E_A, E_B, and E_C.

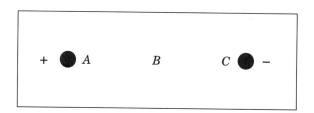

Figure 26.4 Point A near positive electrode, C near negative electrode, and B at center.

$E_A = \underline{\hspace{2cm}}$ V/m $E_B = \underline{\hspace{2cm}}$ V/m $E_C = \underline{\hspace{2cm}}$ V/m

3. Are the results for the E field at points A, B, and C in the preceding question consistent with what you would expect for the relative values at these points? State your reasoning.

4. According to theory, the E field in the region between the plates in electrode configuration (c) should be constant. Calculate the E field from equation 1 at points A, B, and C defined by Figure 26.5 below using the same process as in question 2.

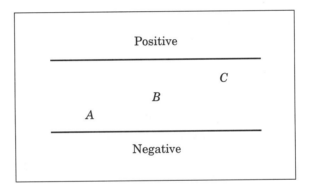

Figure 26.5 Point A near negative electrode displaced slightly from the center, C near the positive electrode displaced same amount in the other direction, and B at center in the middle.

$E_A =$ _____ V/m $E_B =$ _____ V/m $E_C =$ _____ V/m

5. Are the values for the E field at the points in question 4 approximately constant within the experimental uncertainty?

6. According to theory, the electric field E for a line of charge should be proportional to $1/r$, where r is the distance from the line of charge. Considering the graph of the data and the value of the correlation coefficient, comment on whether or not your data for the line of charge confirms this dependence.

Laboratory 27

Capacitance Measurement with a Ballistic Galvanometer

PRELABORATORY ASSIGNMENT

Read carefully the entire description of the laboratory and answer the following questions based on the material contained in the reading assignment. Turn in the completed prelaboratory assignment at the beginning of the laboratory period prior to the performance of the laboratory.

1. What is the definition of capacitance?

2. What are the units of capacitance?

3. A capacitor is said to have a charge of 10 μC. What is the charge on the positively charged plate? What is the charge on the negatively charged plate?

4. A 1.50-μF capacitor has a voltage across the plates of 6.00 V. What is the charge on the capacitor?

5. Which of the following are true for three capacitors C_1, C_2, and C_3 in parallel with a battery of voltage V? More than one answer may be true.

(a) $V = V_1 = V_2 = V_3$; (b) $Q_e = Q_1 = Q_2 = Q_3$; (c) $Q_e = Q_1 + Q_2 + Q_3$; or
(d) $V = V_1 + V_2 + V_3$

6. Which of the following are true for three capacitors C_1, C_2, and C_3 in series with a battery of voltage V? More than one answer may be true.

(a) $V = V_1 = V_2 = V_3$; (b) $Q_e = Q_1 = Q_2 = Q_3$; (c) $Q_e = Q_1 + Q_2 + Q_3$; or
(d) $V = V_1 + V_2 + V_3$

7. Three capacitors of capacitance 5.00 μF, 8.00 μF, and 11.00 μF are connected in parallel. What is the equivalent capacitance of the combination?

8. Three capacitors of capacitance 5.00 μF, 8.00 μF, and 11.00 μF are connected in series. What is the equivalent capacitance of the combination?

Capacitance Measurement with a Ballistic Galvanometer

OBJECTIVES

A capacitor is a circuit element consisting of two conducting surfaces separated by an insulating material called a "dielectric." When the terminals of a battery or other source of electromotive force are connected to the conducting plates of a capacitor, the plates will acquire equal and opposite charges. The magnitude of the charge acquired by either plate Q depends on the voltage V of the source and on the capacitance C of the capacitor. In this laboratory, capacitors will be charged to a known voltage and then discharged through a ballistic galvanometer. Measurements of the deflection of the galvanometer for a series of capacitors with known and unknown values of capacitance will be used to accomplish the following objectives:

1. Determination of the deflection of the galvanometer for several capacitors of known capacitance charged to a known voltage

2. Demonstration that for a fixed charging voltage of the capacitors the galvanometer deflection is proportional to the capacitance of the capacitors

3. Determination of the capacitance of several unknown capacitors from calibration of capacitance versus the galvanometer deflection

4. Experimental determination of the capacitance of series and parallel combinations of capacitors and comparison of the results with the theoretical predictions

EQUIPMENT LIST

1. Ballistic galvanometer, support stand, telescope, and scale
2. DC power supply (0–20 V)
3. DC voltmeter (0–20 V)
4. Five or six known capacitors in range 0.5–3.0 μF (can be in form of a decade box)
5. Three unknown capacitors in range 1–2 μF
6. Switch (single pole, double throw)

THEORY

Part I: Capacitors

Consider a capacitor of capacitance C with a voltage V across the plates. The charges on each of the plates are equal in magnitude and opposite in sign. If Q stands for the magnitude of the charge on either plate, the relationship between these quantities is given by

$$C = \frac{Q}{V} \tag{1}$$

Equation 1 can be taken as the definition of capacitance, which is charge per unit voltage. The units of capacitance are C/V, which has been given the name farad with the symbol F. A 1-F capacitor has a very large capacitance, and the capacitors used in this laboratory will be on the order of 10^{-6} F or one μF. Figure 27.1 shows the symbol used for a capacitor in circuit diagrams. The figure illustrates the fact that a capacitor whose charge is Q has a charge of $+Q$ on one plate and $-Q$ on the other plate. Therefore, the net charge on a capacitor is always zero.

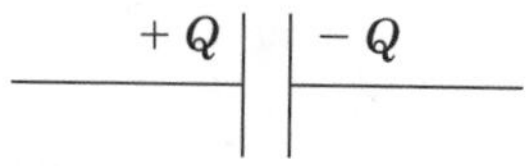

$$+Q \;\lvert\;\lvert\; -Q$$

Figure 27.1 Capacitor with charge Q has $+Q$ on one plate and $-Q$ on the other.

Part 2: Parallel Combinations

In Figure 27.2, three capacitors C_1, C_2, and C_3 are shown connected in parallel to a battery of voltage V. Let Q_1, Q_2, and Q_3 stand for the charges on the capacitors and let V_1, V_2, and V_3 stand for the voltage across the capacitors. There exists some capacitance C_e that is equivalent to the three capacitors in parallel. An expression for C_e in terms of C_1, C_2, and C_3 is desired.

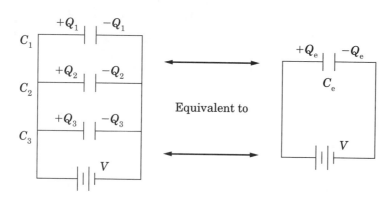

Figure 27.2 Circuit diagram for three capacitors in parallel.

For capacitors connected in parallel, the voltages are all the same, and they are equal to the battery voltage V. The charge on the equivalent capacitor is Q_e. The charges on the individual capacitors Q_1, Q_2, and Q_3 are all different, but the sum of the individual charges is equal to the equivalent charge. In equation form, these statements are

$$V = V_e = V_1 = V_2 = V_3 \tag{2}$$

$$Q_e = Q_1 + Q_2 + Q_3 \tag{3}$$

Each of the capacitors, including the equivalent capacitor, obeys equation 1. That means that the following relationships are true.

$$Q_1 = C_1 V_1 \quad Q_2 = C_2 V_2 \quad Q_3 = C_3 V_3 \quad Q_e = C_e V_e \tag{4}$$

Substituting the expressions for Q from equation 4 into equation 3 and then making use of equation 2 leads to the following expression:

$$C_e = C_1 + C_2 + C_3 \tag{5}$$

Equation 5 is the desired expression for the equivalent capacitance in terms of the individual capacitances. It states that for capacitors in parallel, the equivalent capacitance is simply the sum of the individual capacitances. The extension of equation 5 to the case of any number of capacitors in parallel should be clear.

Part 3: Series Combinations

Three capacitors C_1, C_2, and C_3 are shown connected in series in Figure 27.3. Again, let Q_1, Q_2, and Q_3 stand for the charges on the three capacitors and let V_1, V_2, and V_3 stand for the voltage across the three capacitors. What is desired is an expression for the equivalent capacitance C_e in terms of the individual capacitances C_1, C_2, and C_3.

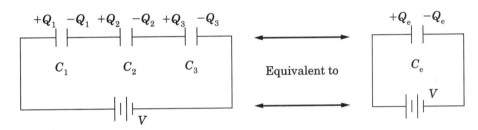

Figure 27.3 Circuit diagram for three capacitors in series.

When connected in series, the charges are the same for each capacitor and for the equivalent capacitor. This can be seen from an examination of Figure 27.3 by noting, for example, that the plate of C_1 that is connected to C_2 has no other connection to the circuit. That means that the net charge on that part of the circuit must be zero. This can be accomplished only if the charges on all the capacitors are the same. In series, the sum of the voltages on the individual capacitors is equal in magnitude to the voltage of the battery. That is true because the sum of all the changes in voltage around a closed loop must be zero. In equation form, these statements are

$$Q_e = Q_1 = Q_2 = Q_3 \tag{6}$$

$$V = V_e = V_1 + V_2 + V_3 \tag{7}$$

Again, equations 4 hold for all the capacitors. Solving each of the expressions in equations 4 for the voltage and using them in equation 7 leads to

$$\frac{Q_e}{C_e} = \frac{Q_1}{C_1} + \frac{Q_2}{C_2} + \frac{Q_3}{C_3} \tag{8}$$

From equation 6, it is clear that one can divide through equation 8 by the common value of the charge, which leads to the desired expression for the equivalent capacitance C_e in terms of the individual capacitances:

$$\frac{1}{C_e} = \frac{1}{C_1} + \frac{1}{C_2} + \frac{1}{C_3} \tag{9}$$

The extension of equation 9 to the case of any number of capacitors in series should be clear.

Part 4: Galvanometer

A galvanometer is a device for detecting the magnitude of a current. A current is a flow of charge Q in some time interval Δt. In most cases, galvanometers are designed to respond to the current itself $Q/\Delta t$. A ballistic galvanometer, on the other hand, is designed to detect the total charge Q that flows through the galvanometer during some very short time interval Δt. There is no interest in the value of the time interval, and the total charge Q is the only quantity of interest. A galvanometer acts as a ballistic galvanometer when the time constant of its motion is large compared to the time Δt during which the charge Q flows.

Measurements will be performed with a ballistic galvanometer whose total deflection when a charge Q is passed through the galvanometer is proportional to the magnitude of the charge Q. Equation 1 implies that if a series of capacitors are charged to the same voltage V, then the charge Q that each capacitor acquires is a measure of the capacitance of that capacitor. This statement along with the statement about the characteristics of a ballistic galvanometer implies that when a series of capacitors are charged to the same voltage V, the deflection that each produces on a ballistic galvanometer will be proportional to the capacitance of the capacitors. In this laboratory, several capacitors of known capacitance will be charged to the same voltage. When they are discharged through the ballistic galvanometer, the deflection produced essentially provides a calibration of the galvanometer deflection in terms of the capacitance.

The experimental arrangement for a ballistic pendulum is shown in Figure 27.4. The deflection of the coil in the galvanometer is read on a scale in front of the device. A telescope is used to view the scale as reflected from a plane mirror on the coil. As the coil rotates, the position on the scale that can be seen rotates, and this is a measure of the deflection of the coil.

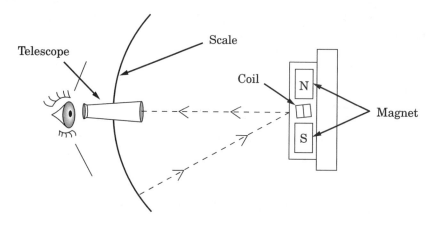

Figure 27.4 Experimental arrangement for the ballistic galvanometer.

EXPERIMENTAL PROCEDURE—CALIBRATION

1. Do not touch the galvanometer until given explicit instructions to do so by the instructor. The gold suspensions in the galvanometers are delicate and expensive. The galvanometer will probably have already been set up and adjusted by the instructor. When given permission, proceed to the next step.

2. Construct the circuit shown in Figure 27.5, using the known capacitor with the largest value.

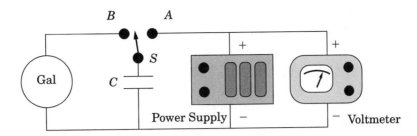

Figure 27.5 Circuit diagram to charge C and then discharge it through galvanometer.

3. Adjust the telescope to focus on the scale. The angle of the telescope can be adjusted to align with the scale. Note that the scale is labeled red on one side of zero and black on the other side.

4. With the largest known capacitor in the circuit, charge the capacitor to some known voltage by throwing switch S to position A. Then discharge the capacitor through the galvanometer by throwing the switch to position B. Note the deflection of the galvanometer. Adjust the voltage applied to the capacitor until you find the voltage that will produce approximately 90% deflection for this capacitor. Record that value of voltage as V in Data Tables 1 and 2. Make all measurements with this voltage.

5. Measure the deflection for each of the known capacitors when they have been charged with voltage V and then discharged through the galvanometer. Perform three trials for each capacitor. Record the values of the deflection D and the known value of each capacitor in Data Table 1.

CALCULATIONS-CALIBRATION

1. Calculate the mean $\overline{D}$ and standard error α_D for the three trials for each capacitor and record them in Calculations Table 1.

2. Perform a linear least square fit to the data with the capacitance C as the ordinate and the deflection $\overline{D}$ as the abscissa. Determine the slope K, the intercept I, and the correlation coefficient r for the fit and record them in Calculations Table 1.

EXPERIMENTAL PROCEDURE—UNKNOWN CAPACITANCE

1. Label the three unknown capacitors C_1, C_2, and C_3. For each of the unknown capacitors, determine the deflection of the galvanometer when they have been charged to voltage V and then discharged through the galvanometer. Perform three trials for each capacitor. Record the value of the deflection D for each trial.

2. Connect capacitor C_1 and C_2 in parallel and place them in the circuit in the position of C in Figure 27.5. Determine the deflection when the combination is charged with voltage V and then discharged through the galvanometer. Record the deflections in Data Table 2.

3. Repeat step 2 with capacitors C_2 and C_3 in parallel.

4. Connect capacitors C_1 and C_2 in series and place them in the circuit in the position of capacitor C in Figure 27.5. Determine the deflection when the combination is charged with voltage V and then discharged through the galvanometer. Record the deflections in Data Table 2.

5. Repeat step 4 with capacitors C_2 and C_3 in series.

CALCULATIONS—UNKNOWN CAPACITANCE

1. Calculate the mean $\overline{D}$ and standard error α_D for each of the three trials. Record them in Calculations Table 2.

2. For the individual capacitors, the series combinations, and the parallel combinations calculate the experimental values for the capacitance from $C_{\text{exp}} = K\overline{D} + I$, where K is the slope and I is the intercept of the least squares fit. The slope K has units of μF/deflection and the intercept has units of μF. Record these values of the experimental capacitance in Calculations Table 2.

3. Calculate a theoretical value for the series and parallel combinations of the capacitors using equations 5 and 9. In the calculations, use the measured values for the individual capacitors. Record these theoretical values for the series and parallel combinations in Calculations Table 2 in the appropriate places.

4. Calculate the percentage difference between the experimental and theoretical values for the series and parallel combinations and record them in Calculations Table 2.

GRAPHS

Make a graph with the capacitance C as the ordinate and the deflection $\overline{D}$ as the abscissa. Also show on the graph the straight line obtained by the linear least squares fit to the data.

Laboratory 27

Capacitance Measurement with a Ballistic Galvanometer

LABORATORY REPORT

Data Table 1

$C\ (\mu F)$	D_1	D_2	D_3

Voltage = _____ V

Calculations Table 1

$\overline{D}$	α_D

$K =$	$\mu F/$defl
$I =$	μF
$r =$	

Data Table 2

Capacitor	D_1	D_2	D_3
C_1			
C_2			
C_3			
C_1 & C_2 Parallel			
C_2 & C_3 Parallel			
C_1 & C_2 Series			
C_2 & C_3 Series			

Voltage = _____ V

Calculations Table 2

$\overline{D}$	α_D	$C_{ex}(\mu F)$	$C_{th}(\mu F)$	% diff

QUESTIONS

1. What is the precision of your measurements of the unknown capacitors? State clearly the basis for your answer.

2. If possible, obtain values for the unknown capacitors. Determine the accuracy of your measurements of the capacitance.

3. Based on the precision and accuracy of your results, is there any evidence for a systematic error in the measurements? State clearly the basis for your answer.

4. How well do your data confirm the theoretical equations for the parallel combination of capacitors? Answer the question as quantitatively as possible.

5. How well do your data confirm the theoretical equation for the series combination of capacitors? Answer the question as quantitatively as possible.

6. Suppose that you were given an unknown capacitor whose value is about twice as large as the largest known capacitor that you measured. What simple change in the procedure would allow you to determine its capacitance using the value for K that you have already determined?

Laboratory 28
___

Measurement of Electrical Resistance and Ohm's Law

PRELABORATORY ASSIGNMENT

Read carefully the entire description of the laboratory and answer the following questions based on the material contained in the reading assignment. Turn in the completed prelaboratory assignment at the beginning of the laboratory period prior to the performance of the laboratory.

1. If a circuit element carries a current of 3.71 A, and the voltage drop across the element is 8.69 V, what is the resistance of the circuit element? Show your work.

    $R =$ _____ Ω

2. A resistor is known to obey Ohm's law. When there is a current of 1.72 A in the resistor it has a voltage drop across its terminals of 7.35 V. If a voltage of 12.0 V is applied across the resistor, what is the current in the resistor? Show your work.

    $I =$ _____ A

3. The resistivity of copper is 1.72×10^{-8} $\Omega-$m. A copper wire is 15.0 m long, and the wire diameter is 0.0500 cm. What is the resistance of the wire? Show your work.

    $R =$ _____ Ω

4. A wire of length 10.0 m has a resistance of 2.75 Ω. What is the resistance of a piece of that same wire whose length is 25.0 m? Show your work.

    $R =$ _____ Ω

5. A wire of cross-sectional area 5.00×10^{-6} m² has a resistance of 1.75 Ω. What is the resistance of a wire of the same material and length as the first wire, but whose cross-sectional area is 8.75×10^{-6} m²? Show your work.

R = _____ Ω

6. A wire of length 10.0 m and cross-sectional area 6.00×10^{-6} m² has a resistance of 3.75 Ω. What is the resistance of a wire of the same material as the first wire, but whose length is 30.0 m, and whose area is 9.00×10^{-6} m²? Show your work.

R = _____ Ω

7. Three resistors of resistance 20.0 Ω, 30.0 Ω, and 40.0 Ω are connected in series. What is their equivalent resistance? Show your work.

R = _____ Ω

8. Three resistors of resistance 15.0 Ω, 25.0 Ω, and 35.0 Ω are connected in parallel. What is their equivalent resistance? Show your work.

R = _____ Ω

Measurement of Electrical Resistance and Ohm's Law

OBJECTIVES

In this experiment, measurements of the voltage across a wire coil and the current in the wire coil will be used to accomplish the following objectives:

1. Definition of the concept of electrical resistance of matter using coils of wire as an example

2. Demonstration of the dependence of the resistance on the length, cross-sectional area, and resistivity of the wire

3. Demonstration of the equivalent resistance of resistors in series and in parallel arrangements

EQUIPMENT LIST

1. Resistance coils (standard set available from Sargent-Welch or Central Scientific consisting of 10-m and 20-m length of copper and German silver wire)

2. Ammeter (range, 0–2 A; direct current)

3. Voltmeter (range, 0–30 V; direct current, preferably digital readout)

4. Direct-current power supply (0–20 V at 2 A will provide the necessary 2 A for all but the German silver coil)

THEORY

If a potential difference V is applied across some element in an electrical circuit, the current I in the element is determined by a quantity known as the resistance R. The relationship between these three quantities serves as a definition of the quantity resistance. This relationship, and thus the definition of R is

$$R = \frac{V}{I} \qquad (1)$$

An object that is a pure resistor has its total electrical characteristics determined by equation 1. Other circuit elements may have other important electrical characteristics in addition to resistance such as capacitance or inductance. The resistance of any circuit element, whether it has other significant electrical properties or not, is given by the ratio of voltage to current as described in equation 1. For any given circuit element, the value of this ratio may change as the voltage and current changes. Nevertheless, the ratio of V to I defines the resistance of the circuit element at that particular voltage and current. The units of resistance are thus volt/ampere, which is given the name ohm. The symbol for ohm is Ω.

Certain circuit elements obey a relationship that is known as Ohm's law. For these elements, the quantity R (equal to V/I) is a constant for different values of V and thus different values of I. Therefore, in order to show that a circuit element obeys Ohm's law, it is necessary to vary the voltage V (the current I will then also vary) and observe that the ratio V/I is in fact constant. In this experiment such measurements will be performed on five different coils of wire to show that they do obey Ohm's law and to determine the resistance of the coils.

The resistance of any object to electrical current is a function of the material from which it is constructed, as well as the length, cross-sectional area, and temperature of the object. At constant temperature the resistance R is given by

$$R = \rho \frac{L}{A} \tag{2}$$

where R is the resistance (Ω), L is the length (m), A is the cross sectional area (m^2), and ρ is a constant dependent on the material called the resistivity (Ω-m). Actually ρ is a function of temperature, and if the coils of wire that are used in this experiment heat up as a result of the current in them, this may be a source of error.

Circuit elements in an electrical circuit can be connected in series or parallel. Consider the case of three resistors, R_1, R_2, and R_3, connected in series as shown in Figure 28.1. For resistors in series, the current is the same for all the resistors, but the voltage drop across each resistor depends on the value of the resistors. For resistors in series the equivalent resistance R_e of the three resistors is given by the equation

$$R_e = R_1 + R_2 + R_3 \tag{3}$$

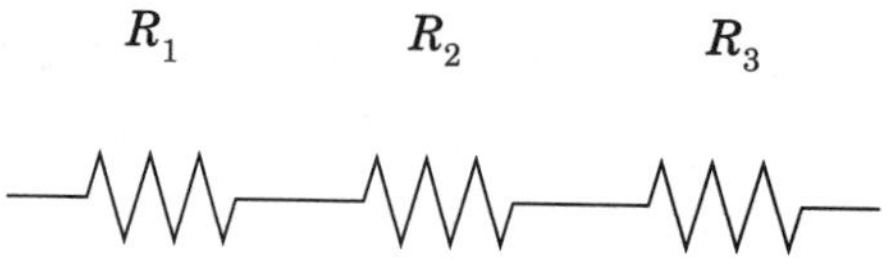

$$R_1 \qquad\qquad R_2 \qquad\qquad R_3$$

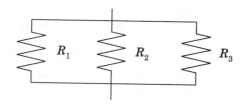

Figure 28.1 Resistors in series.

Consider the case of three resistors in parallel as shown in Figure 28.2. For resistors in parallel the current is different in each resistor, but the voltage across each resistor is the same. In this case the equivalent resistance R_e of the three resistors in terms of the individual resistors is given by

$$\frac{1}{R_e} = \frac{1}{R_1} + \frac{1}{R_2} + \frac{1}{R_3} \tag{4}$$

One of the objectives of this experiment will be to confirm the behavior of resistors in series and parallel which has been described above.

$$R_1 \qquad\qquad R_2 \qquad\qquad R_3$$

Figure 28.2 Resistors in parallel.

EXPERIMENTAL PROCEDURE

1. Connect the ammeter A, the voltmeter V, and the power supply PS to the first resistor as shown in Figure 28.3. Note that the basic circuit is the power supply in series with a resistor. In order to measure the current in the resistor, the ammeter is placed in series with it. In order to measure the voltage drop across the resistor, the voltmeter is placed in parallel with it.

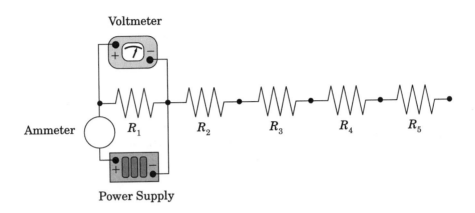

Figure 28.3 Measurement of current and voltage for resistor R_1

2. Vary the current through resistor R_1 in steps of 0.250 A up to 1.000 A. For each specified value of the current, measure the voltage across the resistor and record the values in Data Table 1. The resistors will heat up and may be damaged by allowing the current to pass through them for long periods. For this reason, measurements should be made quickly at each value of the current, and the power supply should be turned up only while measurements are actually being taken. *DO NOT LEAVE THE VOLTAGE TURNED UP WHILE DECIDING WHAT TO DO NEXT!*

3. Repeat step 2 for each of the five resistors. Be sure that for each resistor the ammeter is in series with that resistor and the power supply and the voltmeter is in parallel with the resistor. If the ammeter and voltmeter being used have multiple scales, be sure that the scales agree at the crossover point. Record all values in Data Table 1.

4. Connect the first four resistors in series and measure the equivalent resistance of the combination. To accomplish this, use two values of current, 0.500 A and 1.000 A, and measure the value of the voltage drop for these two values of current. Record these values of voltage in Data Table 2, and from these values of voltage and current, a value for the resistance will be later calculated.

5. Measure the voltage drop across the combination of the second, third, and fourth resistors in series for the current values of 0.500 A and 1.000 A and record in Data Table 2.

6. Connect R_1 and R_2 in parallel as shown in Figure 28.4 and measure the resistance of this parallel combination. Measure the voltage drop across the combination for current values of 0.500 A and 1.000 A and record in Data Table 2 for later calculation.

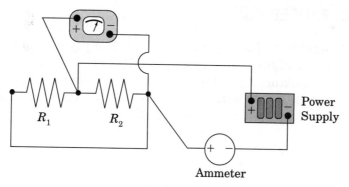

Figure 28.4 Resistors R_1 and R_2 in parallel.

7. Connect R_1 and R_3 in parallel as shown in Figure 28.5 and again measure the voltage drop for current values of 0.500 A and 1.000 A and record in Data Table 2.

8. Connect R_2 and R_3 in parallel and perform the same measurements as described in steps 6 and 7. Record the results in Data Table 2.

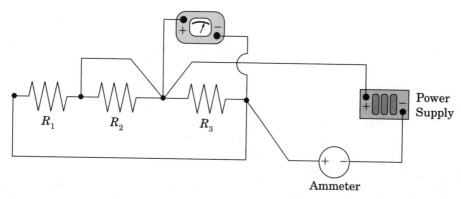

Figure 28.5 Resistors R_1 and R_3 in parallel.

CALCULATIONS

1. The first four coils are made of copper with a value of resistivity of $\rho = 1.72 \times 10^{-8}$ Ω-m. The fifth coil is made of an alloy known as German silver, which has a resistivity of $\rho = 28.0 \times 10^{-8}$ Ω-m. The first, second, and fifth coils are 10.0 m long, and the third and fourth coils are 20.0 m long. The diameters of the first, third, and fifth coils are 0.0006439 m, and the diameters of the second and fourth coils are 0.0003211 m. Using these values in equation 2, calculate the value of the resistance for each of the five coils and record the results in Calculations Table 1 as the theoretical values for the resistance R_{theo}.

2. If equation 1, which defines R is solved for V, the result is $V = IR$. This equation states that there is a linear relationship between the voltage and the current, and that the slope of the straight line in a graph of V versus I will be the resistance R. Perform a linear least squares fit to the data in Data Table 1 for the voltage as a function of the current for each of the five resistors. Make V the ordinate and I the abscissa in the fit. Determine the slope of the fit for the data on each resistor

and record it in Calculations Table 1 as the experimental value for the resistance R_{exp}. Also calculate and record the value of the correlation coefficient r for each of the least squares fits.

3. Calculate the percentage error in the values of R_{exp} compared to the values of R_{theo} for the five resistors and record the results in Calculations Table 1.

4. For the data of Data Table 2, calculate the values of the equivalent resistance for the various series and parallel combinations listed in the table. Record in Calculations Table 2 these experimental values of the equivalent resistance for each of the two trials as the value of the measured voltage divided by the appropriate current (V/0.500 and V/1.000). Calculate and record the mean of the two trials as $\overline{(R_e)}_{\text{exp}}$ in both Calculations Table 2 and Calculations Table 3.

5. Equations 3 and 4 give the theoretical expressions for equivalent resistance for series and parallel combinations of resistance. Calculate a theoretical value for the equivalent resistance for each series and parallel combination measured in Data Table 2. However, for the values of the individual resistances R_1, R_2, and R_3 in equations 3 and 4, use the experimental values determined from the least squares fit to the data on the individual resistors. These values are the appropriate ones to use because often these resistor coils have been abused over the years by being overheated, and their present resistance may no longer be the same as that calculated by equation 2. Record this theoretical value for the equivalent resistance in each case as $(R_e)_{\text{theo}}$ in Calculations Table 3.

6. Calculate the percentage difference between the values of $\overline{(R_e)}_{\text{exp}}$ and $(R_e)_{\text{theo}}$ for each of the series and parallel combinations measured and record the results in Calculations Table 3.

GRAPHS

Construct graphs of the data in Data Table 1 with V as the ordinate and I as the abscissa. Use only one piece of graph paper for all five resistors, making five small graphs on that one sheet. Choose different scales for each graph if needed, but make the five graphs as large as possible while still fitting on one page. Also, show on each small graph the straight line for the least squares fit.

Measurement of Electrical Resistance and Ohm's Law

LABORATORY REPORT

Data Table 1

I (A)	V_{R1} (V)	V_{R2} (V)	V_{R3} (V)	V_{R4} (V)	V_{R5} (V)
0.250					
0.500					
0.750					
1.000					

Data Table 2

Combination	I (A)	V (V)
$R_1 R_2 R_3 R_4$ Series	0.500	
	1.000	
$R_2 R_3 R_4$ Series	0.500	
	1.000	
$R_1 R_2$ Parallel	0.500	
	1.000	
$R_1 R_3$ Parallel	0.500	
	1.000	
$R_2 R_3$ Parallel	0.500	
	1.000	

SAMPLE CALCULATIONS

Calculations Table 1

	$R_1\,(\Omega)$	$R_2\,(\Omega)$	$R_3\,(\Omega)$	$R_4\,(\Omega)$	$R_5\,(\Omega)$
R_{theo}					
R_{exp}					
r					
% error R_{exp}					

Calculations Table 2

Combination	$(R_e)_{\text{exp}}^{\,1}\,(\Omega)$	$(R_e)_{\text{exp}}^{\,2}\,(\Omega)$	$\overline{(R_e)_{\text{exp}}}\,(\Omega)$
$R_1\,R_2\,R_3\,R_4$ Series			
$R_2\,R_3\,R_4$ Series			
$R_1\,R_2$ Parallel			
$R_1\,R_3$ Parallel			
$R_2\,R_3$ Parallel			

Calculations Table 3

Combination	$(R_e)_{\text{theo}}\,(\Omega)$	$\overline{(R_e)_{\text{exp}}}\,(\Omega)$	% diff
$R_1\,R_2\,R_3\,R_4$ Series			
$R_2\,R_3\,R_4$ Series			
$R_1\,R_2$ Parallel			
$R_1\,R_3$ Parallel			
$R_2\,R_3$ Parallel			

QUESTIONS

1. Do the individual resistors you have measured obey Ohm's law? In answering this question, consider the least squares fits and the graphs you have made for each resistor. Remember linear behavior of V versus I is the proof of ohmic behavior.

2. Evaluate the agreement between the theoretical values for the individual resistances and the experimental values. As noted in the Calculations section, there may be significant disagreement between these values if the coils have been overheated in the past. Do any of your experimental values suggest that any of your coils may have been abused in the past?

3. If a coil became heated during your measurements, its resistance would tend to increase with temperature. State if any of the graphs give evidence of heating during your measurements, which would show up at higher current as an increase in the voltage above that expected from extrapolating the data at lower current.

4. Evaluate the agreement between the experimental and theoretical values of the series combinations of resistors. Do the results support equation 3 as the model for series combination of resistors? The agreement is not expected to be perfect, but you should determine if the agreement is reasonable within the expected experimental uncertainty.

5. Evaluate the agreement between the experimental and theoretical values of the parallel combinations of resistors. Do the results support equation 4 as the model for the parallel combination of resistors within the expected experimental uncertainty?

6. According to the data given in step 1 of the Calculations section, the first and second coils have the same length and the third and fourth coils have the same length. They differ only in cross-sectional area. According to theory, what should be the ratio of the resistance of the second coil to the first and the fourth coil to the third? Calculate these ratios for your experimental results and compare the agreement with the expected ratio.

7. According to the data given in step 1 of the Calculations section, the first and third coils have the same cross-sectional area and the second and fourth coils have the same cross-sectional area. They differ only in length. According to theory, what should be the ratio of the resistance of the third coil to the first and the fourth coil to the second? Calculate these ratios for your experimental results and compare the agreement with the expected ratio.

8. Evaluate the extent to which you have accomplished the objectives of this laboratory.

PRELABORATORY ASSIGNMENT

Read carefully the entire description of the laboratory and answer the following questions based on the material contained in the reading assignment. Turn in the completed prelaboratory assignment at the beginning of the laboratory period prior to the performance of the laboratory.

1. When the Wheatstone bridge in Figure 29.1 is balanced, which of the following statements are true? (*More than one may be true.*) (a) There is no current in the unknown resistor. (b) There is no current in the galvanometer. (c) There is no voltage drop across the galvanometer. (d) The currents in R_1 and R_3 are the same. (e) The currents in R_1 and R_2 are the same. (f) The voltage across R_1 and R_3 is the same. (g) The voltage across R_1 and R_2 is the same.

2. The slide-wire form of the Wheatstone bridge makes use of the fact that (a) the resistance of a wire is equal to its length, (b) the resistance of a wire is proportional to its cross-sectional area, (c) the resistance of a wire is proportional to its length, or (d) the resistance of a wire is inversely proportional to its length.

3. A slide-wire Wheatstone bridge is used in the configuration shown in Figure 29.2 with the known resistor positioned as shown and equal to 10.00 Ω. The balance point is found to be at the position of 35.7 cm for the scale as shown in the figure. What is the value of the unknown resistance R_U? Show your work.

   $R_U =$ _____ Ω

4. A student makes three measurements of an unknown resistor. The values are 150.6, 151.2, and 150.8 Ω. What are the mean, standard deviation, and standard error of those values? Show work.

5. Does the measurement of resistance using the Wheatstone bridge depend on the value of the power supply voltage used? If it does, explain why; if it does not, explain why not.

6. What is the purpose of the 10-kΩ resistor in parallel with the switch that is in series with the galvanometer in Figure 29.2?

7. A resistor is coded as shown below. What is the nominal value of the resistor, and what is the precision of this value?

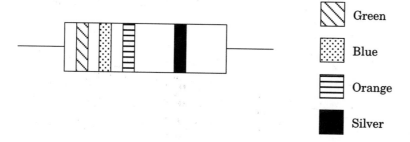

	Green
	Blue
	Orange
	Silver

OBJECTIVES

The Wheatstone bridge is a circuit designed to measure an unknown resistance by comparison with other known resistances. A slide-wire form of the Wheatstone bridge will be used to accomplish the following objectives:

1. Demonstration of the standard color code used to specify the value of commercially available resistors

2. Explanation of the principles on which the operation of the Wheatstone bridge is based

3. Determination of the value of several unknown resistances

EQUIPMENT LIST

1. Slide-wire Wheatstone bridge
2. Standard decade resistance box (1000 Ω)
3. Galvanometer
4. Switch with 10-kΩ resistor in parallel
5. Direct-current power supply or 1.5-V no. 6 dry-cell battery
6. Five color-coded resistors (range, 100 to 1000 Ω) to serve as unknowns
7. Assortment of spade, banana plug, and alligator clip leads
8. Multimeter with resistance scale

THEORY

Consider a circuit containing three resistors whose values are both known and adjustable, an unknown resistance, a power supply, and a galvanometer connected as shown in Figure 29.1. Current I from the power supply arrives at junction J, where it divides. The current in R_1 is I_1, and the current in R_3 is I_2, where $I = I_1 + I_2$. By experimentally varying the values of the resistances, a condition can be achieved in which there is no current in the galvanometer G. This condition is called the "balance condition." When there is no current in the galvanometer the current in R_1 has no place to go except through R_2; thus the current in R_2 is also equal to I_1. Similarly, when the balance condition holds, the current through R_4 must be the same as the current through R_3, and is thus equal to I_2.

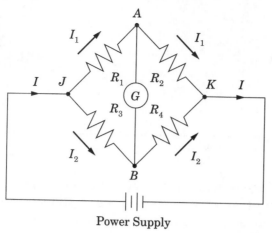

Power Supply

Figure 29.1 Wheatstone bridge circuit.

When there is no current in the galvanometer there is no potential difference between points A and B; in other words, points A and B are at the same potential. Thus, the change in potential from point J to point B (V_{JB}) is equal to the potential change from point J to point A (V_{JA}), and it follows that

$$V_{JA} = I_1R_1 = V_{JB} = I_2R_3 \quad \text{and thus} \quad I_1R_1 = I_2R_3 \qquad (1)$$

Similarly, the potential change across R_2 (V_{AK}) is the same as the potential change across R_4 (V_{BK}), so that

$$V_{AK} = I_1R_2 = V_{BK} = I_2R_4 \quad \text{and thus} \quad I_1R_2 = I_2R_4 \qquad (2)$$

If equation 1 is divided by equation 2, the currents cancel, and it follows that

$$\frac{R_1}{R_2} = \frac{R_3}{R_4} \qquad (3)$$

Therefore, when the balance condition has been experimentally achieved, equation 3 can be used to determine the value of an unknown resistance if three of the four values of resistance are known.

In this experiment, a slide-wire form of the Wheatstone bridge, which is shown in Figures 29.2 and 29.3 will be used. In it, resistances R_3 and R_4 are replaced by a uniform wire between the points J and K with a sliding contact key at point B. Since the wire has a uniform cross section, the resistance of the two portions of wire JB and BK are proportional to their lengths. The ratio of their lengths, JB/BK, is equal to the ratio of their resistances R_3/R_4. If R_1 is an unknown resistance R_U, and R_2 is a known variable resistor R_K, equation 3 becomes

$$\frac{R_U}{R_K} = \frac{JB}{BK} \quad \text{or solving for } R_U, \quad R_U = R_K\frac{JB}{BK} \qquad (4)$$

The 10-kΩ resistor and switch S in series with the galvanometer are designed to protect the galvanometer. Be sure that you are aware of their proper function as described in the procedure before completing the connection of the power supply to the circuit.

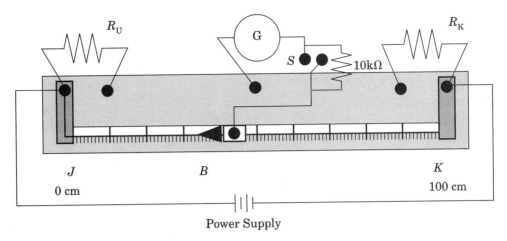

R_U G S $10k\Omega$ R_K

J B K

0 cm 100 cm

Power Supply

Figure 29.2 Slidewire form of the Wheatstone bridge.

The value of resistors routinely used in electronic instrumentation are coded by a series of colored bands on the resistor. The key to the resistor color-coding system is given in Table 29.1. The four bands are placed with three equally spaced bands close to one end of the resistor followed by a space and then a fourth band. The first two bands are the first two digits in the value of the resistor, and the third band gives the exponent of the power of 10 to be multiplied by the first two digits. Thus, a resistor whose first three bands are labeled Yellow-Violet-Red has a value of $47 \times 10^2 \ \Omega$, or $4700 \ \Omega$.

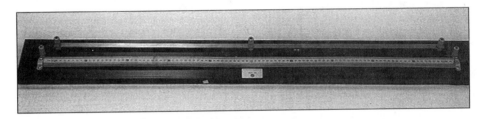

Figure 29.3 Slidewire form of the Wheatstone bridge.

EXPERIMENTAL PROCEDURE

1. Using the resistor code table, read the nominal values of the five unknown resistors and record them in Data Table 1. Record the smallest value as Unknown #1, and then the remaining ones in increasing order.

2. Adjust the power supply voltage to 1.50 V. Leave the power supply fixed at this value for all the measurements. All measurements should be made with this same voltage, which has been chosen so that the currents in all resistors of the circuit will be small. This ensures that there is no heating of the resistors. Any significant heating of the resistors could cause differential increases in resistance, which would lead to errors.

3. Place the first unknown resistor in the Wheatstone bridge circuit in the position of R_U in the circuit shown in Figure 29.2. Place the resistance box in the position of R_K in Figure 29.2 and choose a value for R_K approximately equal to the nominal value that you read from the resistor code for this unknown resistor. Record the value of R_K in Data Table 1. (Note that the value of R_K should be given to one

First Two Bands		Third Band		Fourth Band	
Color	Digits	Color	Exponent	Color	Precision
Black	0	Black	0	Colorless	20%
Brown	1	Brown	1	Silver	10%
Red	2	Red	2	Gold	5%
Orange	3	Orange	3		
Yellow	4	Yellow	4		
Green	5	Green	5		
Blue	6	Blue	6		
Violet	7	Violet	7		
Gray	8	Silver	−2		
White	9	Gold	−1		

First Digit ——
Second Digit ——
—— Precision
—— Exponent

Table 29.1 Resistor Color Code.

place beyond the last digit of the resistance box under the assumption that the uncertainty is not in the last digit set on the box, but in the digit beyond that. For example, if a resistance is set on the resistance box as 153 Ω, it should be recorded as 153.0 Ω because there is clearly not one unit of uncertainty in the 3.)

4. The 10-kΩ resistor and switch S in series with the galvanometer are designed to protect the galvanometer from excessive current. Be sure that each attempt to find a balance condition starts with switch S open. This places the resistor in series with the galvanometer and limits the current.

5. With switch S open, move the sliding contact at B until a balance is achieved— i.e., zero current in the galvanometer. This a rough balance.

6. With the system at rough balance, close switch S to achieve maximum sensitivity and make a fine adjustment to achieve the balance condition. Because the galvanometer may have a small zero offset, determine the point where there is no deflection of the meter. This may not be at the zero of the meter. Record in Data Table 1 the values of JB and BK, the length of the two sections of wire at balance. Note that the Wheatstone bridge has a scale of 1 mm as the smallest marked division. Therefore, measurements of JB and BK should be made to the nearest 0.1 mm.

7. Using the same unknown resistor, repeat steps 3 through 6 above with two other values of R_K, one value approximately 10% greater than the original R_K, and one value approximately 10% less than the original R_K.

8. Repeat steps 3 through 7 above for each of the other four unknown resistors.

9. Using the resistance scale on a multimeter, measure the value of each of the five unknown resistors and record those values in Data Table 2.

CALCULATIONS

1. Using equation 4, calculate and record the three measured values for each of the five unknown resistors in the Calculations Table.

2. Calculate and record the mean and standard error for the three measurements of each of the five resistors.

Laboratory 29
Wheatstone Bridge

LABORATORY REPORT

Data Table 1

	R_K (Ω)	JB (cm)	BK (cm)
Unknown #1			
Coded Value			
(Ω)			
Unknown #2			
Coded Value			
(Ω)			
Unknown #3			
Coded Value			
(Ω)			
Unknown #4			
Coded Value			
(Ω)			
Unknown #5			
Coded Value			
(Ω)			

Calculations Table

R_U (Ω)	$\overline{R_U}$ (Ω)	$^\alpha R_U$

Data Table 2

Unknown #	1	2	3	4	5
Resistance	Ω	Ω	Ω	Ω	Ω

SAMPLE CALCULATIONS

QUESTIONS

1. Assume there is an uncertainty of ± 0.03 cm in locating the position B of the contact on the slide wire. What percentage error does this introduce in the determination of JB/BK and thus the measured value of resistance when B is at 50 cm?

2. Assuming the same uncertainty of ± 0.03 cm in the position of B, what is the resulting percentage error in JB/BK and thus in resistance when B is at 10 cm?

3. Keeping in mind your answers to questions 1 and 2, can you suggest a reason why the values of the known resistor were chosen to be approximately equal to and ±10% of the coded value for the unknown?

4. Compare each of your Wheatstone bridge values of the five unknown resistors with its coded value by calculating the percentage error assuming the coded values as correct. From the resistor color coding state the precision of the resistors.

#1 Error = _____%   #2 Error = _____%   #3 Error = _____%

#4 Error = _____%   #5 Error = _____%   Precision = _____%

Are the percentage errors of the measurements within the precision of the resistors?

5. Compare your Wheatstone bridge values of the five unknown resistors with the values in Data Table 2 determined by using the resistance meter. Calculate the percentage errors for each resistor assuming the values in Data Table 2 as correct. In general, is the agreement between these values better or worse than the agreement with the coded values?

#1 Error = _____%   #2 Error = _____%   #3 Error = _____%

#4 Error = _____%   #5 Error = _____%

6. Express the values of the standard error for each of the five unknown resistors as a percentage of the mean value.

#1 % Standard Error = _____ %   #2 % Standard Error = _____ %

#3 % Standard Error = _____ %   #4 % Standard Error = _____ %

#5 % Standard Error = _____ %

7. Considering the stated precision of the coded values and the various comparisons that have been suggested above, state whether your Wheatstone bridge measurements of these resistors represent more reliable values for the actual values of these resistors than the coded values. Be very specific about the facts on which your opinion is based.

8. Suppose a higher voltage was used with the Wheatstone bridge. The currents in the circuit would be larger. Would larger currents result in increased sensitivity of the measurement? If so, state specifically why. If not, state specifically why not. What is the possible disadvantage of larger currents?

Laboratory 30

Bridge Measurement of Capacitance

PRELABORATORY ASSIGNMENT

Read carefully the entire description of the laboratory and answer the following questions based on the material contained in the reading assignment. Turn in the completed prelaboratory assignment at the beginning of the laboratory period prior to the performance of the laboratory.

1. Define "capacitive reactance."

2. Describe the role that capacitive reactance plays when there is an alternating current in the capacitor.

For Questions 3 to 7, consider the circuit diagram of Figure 30.1. Mark as true or false the statements concerning the conditions that hold when the bridge is balanced.

3. $V_{AB} = V_{BD}$ and $V_{AC} = V_{CD}$ (a) true (b) false

4. $V_{BC} = 0$ (a) true (b) false

5. $V_{AB} = V_{CD}$ and $V_{AC} = V_{BD}$ (a) true (b) false

6. $V_{AB} = V_{AC}$ and $V_{BD} = V_{CD}$ (a) true (b) false

7. The balance condition for the bridge is different for different frequencies. (a) true (b) false.

8. A bridge like the one shown in Figure 30.1 is balanced. The known capacitor C_K has a capacitance of 1.17 μF. The value of R_1 is 1000 Ω, and the value of R_2 is 525 Ω. What is the value of the unknown capacitor C_U?

9. A 3.75-μF capacitor is in series with a 6.85-μF capacitor. What is the equivalent capacitance C_e of the combination?

10. A 5.75-μF capacitor is in parallel with a 3.82-μF capacitor. What is the equivalent capacitance C_e of the combination?

OBJECTIVES

A capacitor is a circuit element consisting of two conducting surfaces separated by an insulating material called a "dielectric." When an alternating current of frequency f exists in a capacitor, the measure of opposition to that current is a quantity called the "capacitive reactance," or X_C. The potential difference across the capacitor is given by IX_C, where I is the current in the capacitor. Thus, X_C plays a role for alternating current that is analogous to the role played by resistors for direct current. This quantity depends on the angular frequency $\omega = 2\pi f$ and on the capacitance C and is given by $X_C = 1/\omega C$. In this laboratory, an alternating-current bridge circuit will be constructed with two capacitors and two resistors. When the bridge is balanced, the ratio of the value of the two capacitive reactances is equal to the ratio of the value of the two resistors. Assuming that the value of the ratio of the resistors and the value of one capacitor is known, the capacitance of the other capacitor can be determined. In this laboratory, measurements on an alternating-current bridge circuit will be used to accomplish the following objectives:

1. Determination of the capacitance of several unknown capacitors by establishing balance in a bridge circuit

2. Experimental determination of the capacitance of series and parallel combinations of capacitors and comparison of the results with the theoretical predictions

EQUIPMENT LIST

1. Capacitor of accurately known value (in the range 0.5–1 μF)
2. Three capacitors of unknown value (in the range 0.2–3 μF)
3. Two decade resistance boxes (1000 Ω and 10,000 Ω)
4. Sine-wave generator (variable frequency, 5-V peak-to-peak amplitude)
5. Alternating-current voltmeter (digital readout, capable of measuring high frequency)

THEORY

Consider the circuit shown in Figure 30.1, known as an alternating-current bridge circuit. A capacitor of unknown value C_U and a capacitor of known value C_K are in the circuit, along with resistors R_1 and R_2. A sine-wave generator is applied between points A and D in the circuit as shown in the figure. The bridge is said to be balanced when the potential difference V_{BC} between points B and C is zero. This potential difference is measured by placing an alternating-current voltmeter between points B and C. If potential difference V_{BC} is zero, then points B and C are at the same potential. That means that the potential difference V_{AB} between points A and B in the circuit is the same as the potential difference V_{AC} between points A and C in the circuit. Writing the potential differences in terms of the currents and the capacitive reactances gives

$$V_{AB} = I_1 \left(\frac{1}{\omega C_U} \right) = V_{AC} = I_2 \left(\frac{1}{\omega C_K} \right) \tag{1}$$

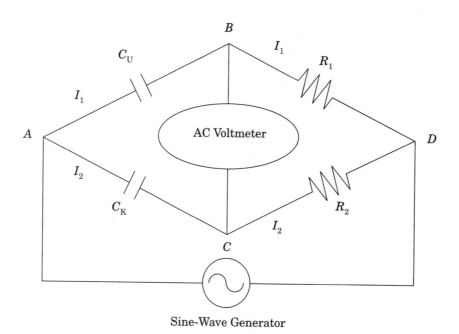

Figure 30.1 Alternating current bridge with two capacitors and two resistors.

When the balance condition is satisfied, the current in the resistor R_1 is I_1, the same as that in C_U. It is also true that the current in the resistor R_2 is I_2, the same as that in C_K when the balance condition is satisfied. It is also true that the potential difference V_{BD} and V_{CD} are equal. Writing these statements in terms of the currents and the resistances gives

$$V_{BD} = I_1 R_1 = V_{CD} = I_2 R_2 \tag{2}$$

If equations 1 and 2 are each solved for the ratio of the currents I_1/I_2, the result is

$$\frac{I_1}{I_2} = \frac{\omega C_U}{\omega C_K} = \frac{R_2}{R_1} \tag{3}$$

Solving equation 3 for the unknown capacitance C_U gives

$$C_U = C_K \frac{R_2}{R_1} \tag{4}$$

Equation 4 shows that if the value of C_K is known, a capacitor of unknown capacitance C_U can be determined by experimentally varying the ratio of the resistances until the ratio R_2/R_1 is found for which the bridge is in balance. Note that equation 4 is independent of the frequency f, and that the currents I_1 and I_2 do not need to be determined.

Figure 30.2 shows two capacitors of capacitance C_1 and C_2 in parallel. They are equivalent to a single capacitance C_e. The expression for this equivalent capacitance C_e in terms of the capacitances C_1 and C_2 is

$$C_e = C_1 + C_2 \qquad\qquad (5)$$

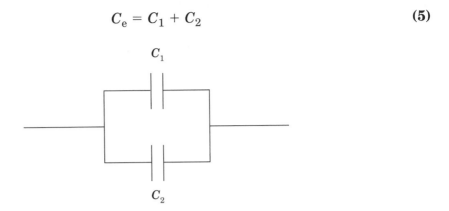

Figure 30.2 Two capacitors in parallel.

Figure 30.3 shows two capacitors of capacitance C_1 and C_2 in series. They are equivalent to a single capacitance C_e. The expression for this equivalent capacitance C_e in terms of the capacitances C_1 and C_2 is

$$\frac{1}{C_e} = \frac{1}{C_1} + \frac{1}{C_2} \qquad\qquad (6)$$

For a more complete description of capacitors in series and parallel, see the theory section of Laboratory 27.

Figure 30.3 Two capacitors in series.

In this laboratory the capacitance of several capacitors whose capacitance is unknown will be determined by placing them in a bridge of the type shown in Figure 30.1. In addition, series and parallel combinations of two capacitors at a time will be determined and compared with the theoretical predictions of equations 5 and 6.

EXPERIMENTAL PROCEDURE

1. Label the unknown capacitors as C_1, C_2, and C_3. Record the value of the known capacitor as C_K in the Data Table.

2. Construct the circuit shown in Figure 30.1 with capacitor C_1 in the position of C_U and the known capacitor in the position of C_K. Place the 1000-Ω resistance box set to value of 1000 Ω in the position of R_1. Place the 10,000-Ω resistance box in the position of R_2. Record the value of R_1 in the Data Table.

3. Plug in the sine-wave generator and place the output of the generator between the points A and D. Turn it up to maximum amplitude and set the frequency to 1000 Hz.

4. Place the voltmeter between points B and C in the circuit. Adjust the value of R_2 until the minimum voltage is read between points B and C. Record the value of R_2 that produces the minimum in the Data Table.

5. Repeat steps 2 through 4 for the other two unknown capacitors C_2 and C_3.

6. Repeat steps 2 through 4 for the following parallel combinations of capacitors placed in the position of C_U in the circuit in Figure 30.1: (a) C_1 and C_2 in parallel; (b) C_2 and C_3 in parallel.

7. Repeat steps 2 through 4 for the following series combinations of capacitors placed in the position of C_U in the circuit in Figure 30.1: (a) C_1 and C_2 in series; (b) C_2 and C_3 in series.

CALCULATIONS

1. Using equation 4, calculate the experimental values for the unknown capacitors and for the series and parallel combinations. Record the results as C_{exp} in the Calculations Table.

2. Using equations 5 and 6, calculate the theoretical equivalent capacitance for the series and parallel combinations corresponding to the measured series and parallel combinations. Use the experimentally determined values for the capacitance of capacitors C_1, C_2, and C_3 in the calculations. Record those results as C_{theo} in the Calculations Table.

3. Calculate the percentage error of the experimental values of the series and parallel combinations compared to the theoretical values for those quantities. Record the results in the Calculations Table.

LABORATORY REPORT

Data Table

Capacitor	$R_2\,(\Omega)$
$C1$	
$C2$	
$C3$	
$C1$ & $C2$ in parallel	
$C2$ & $C3$ in parallel	
$C1$ & $C2$ in series	
$C2$ & $C3$ in series	
$C_K =$ μF	$R_1 =$ Ω

Calculations Table

$C_{\text{exp}}\,(\mu\text{F})$	$C_{\text{theo}}\,(\mu\text{F})$	% error

---

SAMPLE CALCULATIONS

QUESTIONS

1. To what extent do your data confirm the theoretical equation for the parallel combination of capacitors? State your answer as quantitatively as possible.

2. To what extent do your data confirm the theoretical equation for the series combination of capacitors? State your answer as quantitatively as possible.

3. The laboratory procedure instructs you to use the maximum amplitude of the sine-wave generator. How would it affect the results if half the amplitude were used instead?

4. Suppose the bridge is rearranged into the form shown in Figure 30.4. What is the balance condition for this bridge?

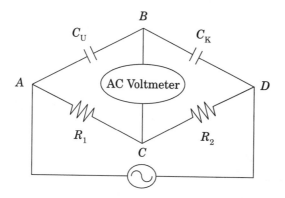

Figure 30.4 Different arrangement of the alternating current bridge.

PRELABORATORY ASSIGNMENT

Read carefully the entire description of the laboratory and answer the following questions based on the material contained in the reading assignment. Turn in the completed prelaboratory assignment at the beginning of the laboratory period prior to the performance of the laboratory.

1. Describe the principle on which the operation of a D'Arsonal-type galvanometer is based.

2. A galvanometer has (a) a meter deflection proportional to the current in the galvanometer, (b) a meter deflection proportional to the voltage across the galvanometer, (c) a fixed resistance, or (d) all of the above are true.

3. The galvanometer constant K is (a) the current for full scale deflection (units A), (b) the current for deflection of one scale division (units A/div), (c) the total current times the number of scale divisions (units A div), or (d) the reciprocal of the galvanometer resistance R_g (units 1/A).

4. In the procedure for determination of R_g and K, when the shunt resistor R_s is placed in parallel with the galvanometer, what happens to galvanometer deflection, and why does it happen?

5. To construct a voltmeter of a given full-scale deflection from a galvanometer, the appropriate resistance must be placed in (a) series or (b) parallel with the galvanometer.

6. To construct an ammeter of a given full-scale deflection from a galvanometer, the appropriate resistance must be placed in (a) series or (b) parallel with the galvanometer.

7. A galvanometer has $R_g = 150$ Ω and $K = 0.750 \times 10^{-4}$ A/div. The galvanometer has five divisions for a full-scale reading (i.e., $N = 5$). What value of resistance is needed, and how must it be connected to the galvanometer to form a voltmeter of 20.0 V full scale? Show your work.

8. What value of resistance is needed, and how must it be connected to form the galvanometer of question 7 into an ammeter of 2.50 A full scale? Show your work.

9. To measure voltage, a voltmeter is placed in a circuit in (a) series or (b) parallel. The resistance of an ideal voltmeter is _____.

10. To measure the current an ammeter is placed in a circuit in (a) series or (b) parallel. The resistance of an ideal ammeter is _____.

OBJECTIVES

A galvanometer is a device used to detect the presence of electrical current. In this laboratory, measurements made with several circuits containing a galvanometer will be used to accomplish the following objectives:

1. Determination of the internal resistance of the galvanometer R_g
2. Determination of the current sensitivity of the galvanometer K
3. Transformation of the galvanometer into a voltmeter of given full-scale deflection by placing the appropriate value of resistance in series with the galvanometer
4. Transformation of the galvanometer into an ammeter of given full-scale deflection by placing the appropriate value of resistance in parallel with the galvanometer
5. Comparison of the accuracy of the voltmeter and ammeter constructed from the galvanometer with a standard voltmeter and a standard ammeter

EQUIPMENT LIST

1. Direct-current power supply (capable of at least 3 V)
2. Galvanometer (D'Arsonal type zero centered, available from Central Scientific, Sargent-Welch and other supply companies)
3. Resistance box (variable in steps of 10 Ω between 2500 Ω and 3500 Ω)
4. A resistor of approximately 330 Ω (This should be either a 1% resistor or else a resistance meter should be provided for the class so that the value of this resistor can be accurately measured.)
5. Direct-current voltmeter (range of 0–3.00 V, preferably digital readout)
6. Direct current ammeter (range of 0–1.00 A, preferably digital readout)
7. Spool of #28 copper wire (one for the class)
8. Assorted leads

THEORY—GALVANOMETER CHARACTERISTICS

The D'Arsonal-type galvanometers used in this experiment are based on the fact that a wire coil located in the presence of a magnetic field will experience a torque when there is a current in the coil. This torque is exerted against a spring, and the deflection of a pointer attached to the coil is proportional to the current in the galvanometer. Since the coil has a fixed resistance R_g, the deflection of the pointer will also be proportional to the voltage across the terminals of the galvanometer. Therefore, a galvanometer can be calibrated to serve as either a voltmeter or an ammeter.

A galvanometer is characterized by its resistance R_g and a constant K called the "current sensitivity." This constant is simply the amount of current needed to deflect the galvanometer one scale division, and it is expressed in units of A/div. The values of R_g and K for the galvanometers used in the experiment are taken to be unknown. They will be determined experimentally by a series of measurements with known values of resistance in series and parallel with the galvanometer as described in the following procedure.

EXPERIMENTAL PROCEDURE TO DETERMINE R_g AND K

1. As shown in Figure 31.1 connect the galvanometer, power supply, and the decade resistance box in series, and then connect the voltmeter in parallel with the power supply. Set the resistance box R_1 to a value of 2500 Ω and adjust the power supply voltage carefully until the galvanometer deflects full scale. Record the voltmeter reading as V and the number of large divisions into which the scale is divided as N in Data Table 1.

2. A resistor connected in parallel with a device is referred to as a shunt resistor because it diverts part of the current that was originally going through the device. Use a composition resistor whose value is approximately 330 Ω as a shunt resistor. Using the ohmmeter, measure an accurate value for this shunt resistor R_s and record it in Data Table 1. Leaving the power supply voltage set exactly as above, connect the shunt resistor in parallel with the galvanometer as shown in Figure 31.2. The deflection of the galvanometer will now be less than full scale.

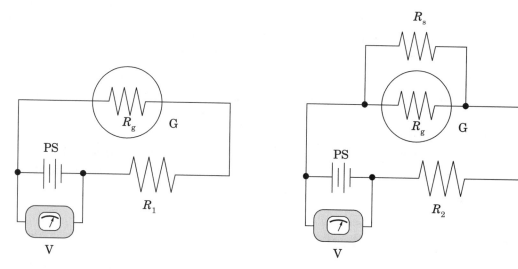

Figure 31.1 Original Circuit.   **Figure 31.2** Original Circuit plus shunt.

3. Still leaving the power supply voltage set, adjust the value of the resistance box from its present value to a somewhat lower value needed to cause the galvanometer to again deflect full scale. Make small adjustments and watch carefully so that the galvanometer does not deflect beyond full scale. A very large abrupt decrease in the value of the resistance box could divert enough current through the galvanometer to cause permanent damage to the galvanometer.

Record the value of the resistance box setting that gives a full scale deflection as R_2 in Data Table 1.

4. Turn the power supply to zero and remove the shunt resistor. The circuit is again like Figure 31.1; but now select a value of 3000 Ω for R_1, the resistance box, and repeat the procedure described in step 1. Record the value of V needed to produce a full-scale deflection for this resistance in Data Table 1.

5. Repeat steps 2 and 3 above, inserting the same shunt resistor R_s of 330 Ω. Determine the value of R_2 needed to produce a full-scale deflection with the shunt resistor in place and record it in Data Table 1.

6. Turn the power supply to zero and remove the shunt resistor. Set the resistance box to a value of $R_1 = 3500$ Ω and repeat steps 1, 2, and 3, recording the values of V and R_2 in Data Table 1.

CALCULATION OF R_g AND K

1. By applying Ohm's law to the circuits in Figure 31.1 and Figure 31.2, when the applied voltage V is the same and the galvanometer is at full-scale deflection in both cases, it can be shown that the resistance of the galvanometer R_g is given by

$$R_g = \frac{R_s}{R_2}(R_1 - R_2) \tag{1}$$

Using equation 1, calculate the three values of R_g determined by the three trials in Data Table 1. Also, calculate the mean $\overline{R_g}$ and standard error α_{R_g} for these measurements. Record all calculated values in Calculations Table 1.

2. The constant K is defined as the current needed to produce a deflection of one scale division, and the deflection in the above procedure was N scale divisions. Ohm's law applied to the circuit of Figure 31.1 leads to the following:

$$I = KN = \frac{V}{R_1 + R_g} \quad \text{or} \quad K = \frac{V}{N(R_1 + R_g)} \tag{2}$$

Using equation 2, determine K (the galvanometer current sensitivity) from the values of V and R_1 in Data Table 1 and the calculated values of R_g in Calculations Table 1. For each calculation of K, use the value of R_g determined from the V and R_1 being used to calculate K. Also, calculate the mean $\overline{K}$ and standard error α_K for the three values of K. Record all calculated quantities in Calculations Table 1.

THEORY—CONVERSION OF THE GALVANOMETER INTO A VOLTMETER

The galvanometer deflects full scale for a value of current given by $I_g = KN$. The voltage V_g across the galvanometer terminals that produces a full-scale deflection is given by $V_g = I_g R_g = KNR_g$. If it is desired to measure a larger voltage than V_g, it is necessary to place a resistor R_V in series with the galvanometer so that most of the voltage is across R_V and the rest across the galvanometer. Figure 31.3 illustrates this idea. If a voltage of V_{FS} between terminals 1 and 2 in Figure 31.3 results in a current in the galvanometer equal to KN, then the series combination of R_V and the galvanometer act as a voltmeter with a full-scale voltage V_{FS}. In other words, the deflection of the galvanometer will be proportional to the voltage between terminals

1 and 2, with the galvanometer showing full-scale deflection when the voltage between terminals 1 and 2 is equal to V_{FS}. In equation form this is

$$I = KN = \frac{V_{FS}}{R_V + R_g} \tag{3}$$

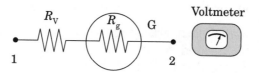

Voltmeter

Figure 31.3 Combination of Galvanometer and Series Resistor form a Voltmeter.

Solving equation 3 for R_V leads to the following expression:

$$R_V = \frac{V_{FS}}{KN} - R_g \tag{4}$$

Equation 4 can be used to solve for the value of R_V needed to turn a galvanometer of given K, N, and R_g into a voltmeter of full-scale voltage V_{FS}. For example, if a galvanometer has $K = 2.00 \times 10^{-4}$ A/div, $R_g = 100\ \Omega$, and $N = 5$, equation 4 can be used to find the value of R_V needed to make a voltmeter that measures a maximum voltage of 25.0 V ($V_{FS} = 25.0$ V). Placing those values in equation 4 and solving, gives $R_V = 24,900\ \Omega$. Therefore, a 24,900-Ω resistor in series with the galvanometer will produce between terminals 1 and 2 in Figure 31.3, a voltmeter that reads 25.0 V at full scale. Also note that since there are 5 divisions on the galvanometer scale, the scale marks 1, 2, 3, 4, and 5 stand for 5, 10, 15, 20, and 25 V.

As a final word on the theory of voltmeters, note that when a voltmeter is used to measure the voltage in a circuit, it must necessarily alter the original circuit because of the current in the voltmeter. If the current in the voltmeter is a minimum, then the alteration of the original circuit is held to a minimum. Since a voltmeter is always placed in a circuit in parallel it follows that the larger the resistance of the voltmeter, the less current it draws, and the more accurate is the voltmeter. In fact, the ideal voltmeter would have an infinite resistance.

EXPERIMENTAL PROCEDURE—GALVANOMETER INTO A VOLTMETER

1. Using equation 4, calculate the value of R_V needed to turn your galvanometer into a voltmeter that reads full-scale deflection for 5.00 V ($V_{FS} = 5.00$ V). In this calculation, use the mean values of K and R_g from Calculations Table 1. Record this value of R_V in Data Table 2.

2. Connect one side of the galvanometer to one side of the resistance box set to the value of R_V. Connect one end of a second lead to the other side of the galvanometer. Connect one end of a third lead to the other side of the resistance box. Between the loose ends of the second and third leads is a voltmeter that reads 5.00 V at full scale. This voltmeter will be referred to as the experimental voltmeter.

3. Compare the experimental voltmeter with the standard voltmeter by connecting them in parallel across the output of the power supply as shown in Figure 31.4.

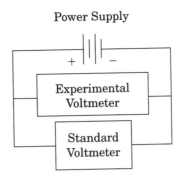

Power Supply

Experimental Voltmeter

Standard Voltmeter

Figure 31.4 Experimental and standard voltmeter in parallel with power supply.

Turn the power supply up slowly until the *experimental voltmeter* reads exactly 1.00 V and record the value read by the standard voltmeter at this point. Make this same comparison at 2.00, 3.00, 4.00, and 5.00 V as read on the *experimental voltmeter* and record all results in Data Table 2.

4. Following steps 1 through 3, calculate the value of R_V needed to make a voltmeter of 10.0-V full-scale deflection. Construct such a voltmeter and compare it to the standard voltmeter at 2.00, 4.00, 6.00, 8.00, and 10.00 V. Record all the results in Data Table 2.

5. Following the same procedure, construct a voltmeter that reads 15.0 V at full scale and compare it with the standard voltmeter at 3.00, 6.00, 9.00, 12.0, and 15.0 V. Record the results in Data Table 2.

6. Calculate the percentage error of the experimental voltmeter readings compared to the standard voltmeter and record the results in Calculations Table 2.

THEORY—CONVERSION OF THE GALVANOMETER INTO AN AMMETER

As previously described, a galvanometer deflects full scale when the current is $I_g = KN$. If it is desired to measure a current larger than I_g, it is necessary to place a small shunt resistance R_A in parallel with the galvanometer to divert part of the current away from the galvanometer as shown in Figure 31.5. The current I comes in at terminal 1 and divides at the junction. The current in the galvanometer is I_g, and I_A is the current in the shunt resistor, where $I = I_g + I_A$. Since R_g and R_A are in parallel, they have the same voltage across them, or in equation form, $I_g R_g = I_A R_A$. Combining the two previous equations leads to

$$IR_A = I_g(R_A + R_g) \tag{5}$$

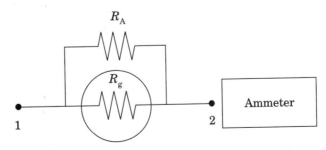

Figure 31.5 Galvanometer and shunt resistor in parallel form an Ammeter.

In order for the combination in Figure 31.5 to act as an ammeter of a given full-scale deflection I_{FS}, it is necessary that $I = I_{FS}$ when $I_g = KN$. Making these assumptions in the above equation gives

$$I_{FS}R_A = KN(R_A + R_g) \qquad (6)$$

Solving equation 6 for R_A leads to the following equation:

$$R_A = \frac{KNR_g}{I_{FS}} - KN \qquad (7)$$

Equation 7 can be used to calculate the value of the resistor R_A needed to cause the parallel combination shown in Figure 31.5 to be an ammeter whose full-scale current is I_{FS}.

Because an ammeter must be connected into a circuit in series, it will alter the original circuit as little as possible when it has as low a resistance as possible. The ideal ammeter therefore has zero resistance.

PROCEDURE—GALVANOMETER INTO AN AMMETER

1. Using equation 7, calculate the value of R_A needed to turn your galvanometer into an ammeter that reads 1.00 A at full scale. For values of R_g and K, use the mean value in Calculations Table 1. Record the value of R_A in Data Table 3.

2. Number 28 copper wire has a resistance of 0.00213 Ω/cm. Calculate the length of #28 copper wire needed to have a resistance equal to R_A. Record that value in Data Table 3.

3. Cut a piece of #28 copper wire a few centimeters longer than the length calculated in step 2. If the wire being used has an insulating coating, cut it away a few centimeters on each end of the wire. Attach the wire between the posts of the galvanometer in such a way that the length of wire between where one end touches one post and the other end touches the other post is equal to the length calculated in step 2. At the same time that the wire is attached between the posts, attach a short lead to each galvanometer post as shown in Figure 31.6. The two loose ends of the two leads are now an ammeter that reads 1.00 A at full scale. Refer to it as the experimental ammeter.

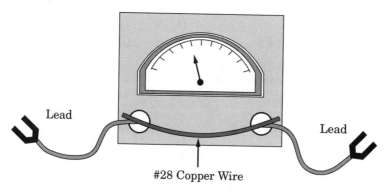

Lead   Lead

#28 Copper Wire

Figure 31.6 Galvanometer with #28 copper wire shunt resistor.

3. After making sure that the power supply is turned completely to zero, place the experimental ammeter in series with a standard ammeter and with the power supply. Very slowly turn the supply up until the experimental ammeter reads 0.200 A. Record the reading of the standard ammeter in Data Table 3. Continue this process, comparing the experimental ammeter to the standard ammeter at 0.400, 0.600, 0.800, and 1.000 A.

4. Calculate the percentage errors of the experimental ammeter readings compared to the standard ammeter and record the results in Calculations Table 3.

LABORATORY REPORT

Data Table 1

$R_1\,(\Omega)$	$V\,(\text{volts})$	$R_2\,(\Omega)$

$N =$ _____

$R_s =$ _____

Calculations Table 1

$R_g\,(\Omega)$	$\overline{R_g}\,(\Omega)$	α_{R_g}	$K\,(\text{A/div})$	$\overline{K}\,(\text{A/div})$	α_K

SAMPLE CALCULATIONS

Data Table 2

$V_{FS} = 5.00$ V $R_V = $ _____ Ω	Experimental V	1.00 V	2.00 V	3.00 V	4.00 V	5.00 V
	Standard V	V	V	V	V	V
$V_{FS} = 10.00$ V $R_V = $ _____ Ω	Experimental V	2.00 V	4.00 V	6.00 V	8.00 V	10.0 V
	Standard V	V	V	V	V	V
$V_{FS} = 15.00$ V $R_V = $ _____ Ω	Experimental V	3.00 V	6.00 V	9.00 V	12.0 V	15.0 V
	Standard V	V	V	V	V	V

Calculations Table 2
Percentage Error of Experimental Voltmeter Versus % Full Scale Reading

Type	20% FS	40% FS	60% FS	80% FS	100% FS
5-V Meter					
10-V Meter					
15-V Meter					

Data Table 3

$I_{FS} = 1.00$ A		$R_A = $ _____ Ω		Length $R_A = $ _____ cm	
Experimental I	0.200 A	0.400 A	0.600 A	0.800 A	1.000 A
Standard I					

Calculations Table 3
Percentage Error of Experimental Ammeter Versus % Full Scale Reading

Type	20% FS	40% FS	60% FS	80% FS	100% FS
1-A Meter					

QUESTIONS

1. Considering the standard error, comment on the precision of your measurements of R_g and K. Express the standard error as a percentage of the mean.

2. Are the absolute differences between each of the three experimental voltmeters and the standard voltmeter approximately of the same order of magnitude?

3. Are the percentage errors of the three experimental voltmeters approximately constant?

4. Consider the 5-V experimental voltmeter. Does it tend to read too high or too low as compared to the standard voltmeter?

5. Presumably a change in the value of R_V would cause better agreement of the 5-V experimental voltmeter with the standard voltmeter. Would R_V need to be a larger or smaller resistance and why?

6. Are the absolute differences between the experimental ammeter readings and the standard ammeter readings approximately of the same order of magnitude?

7. Are the percentage errors of the experimental ammeter approximately constant?

8. Does the experimental ammeter tend to read too high or too low?

9. Presumably, by a change in the value of R_A, the experimental voltmeter could be made to show better agreement with the standard ammeter. Does R_A need to be a larger or smaller resistance to accomplish this and why?

10. It is stated in the laboratory instructions that the same voltage V is applied to both circuits in Figures 31.1 and 31.2 and the galvanometer deflects full scale in both cases. Thus, the galvanometer current is $I_g = KN$ for both circuits. In Figure 31.2 let the current in R_S be called I_S. From the application of Ohm's law to these circuits, derive the expression given as equation 1 for R_g.

Laboratory 32

Potentiometer and Voltmeter Measurements of the Electromotive Force (emf) of a Dry Cell

PRELABORATORY ASSIGNMENT

Read carefully the entire description of the laboratory and answer the following questions based on the material contained in the reading assignment. Turn in the completed prelaboratory assignment at the beginning of the laboratory period prior to the performance of the laboratory.

1. What is the emf and the internal resistance of a typical new dry cell? What might be typical values for the emf and internal resistance of a very old dry cell?

2. A dry cell has an emf of 1.48 V and an internal resistance of 1.11 Ω. What is its terminal voltage when it is connected to a load resistance of 25.00 Ω?

3. A voltmeter can be used to measure the terminal voltage of a dry cell. What determines whether or not this terminal voltage is a good approximation to the emf of the cell?

4. On what experimental condition is the operation of a potentiometer based, and what advantage does this experimental condition produce for the measurement of the emf?

5. In a measurement using a potentiometer, a known emf and an unknown emf are compared to the voltage drop (a) across a voltmeter, (b) across a galvanometer, (c) along a wire of uniform cross section, or (d) across a power supply of fixed voltage.

6. A dry cell has an emf of 1.35 V and an internal resistance of 2000 Ω. What will be its terminal voltage when measured with a voltmeter whose input impedance is 10,000 Ω?

7. What will be the terminal voltage of the dry cell of question 6 when measured with a voltmeter of input impedance 10 MΩ?

8. When the potentiometer described in this laboratory has been properly standardized, to what emf does 1.0000 m of wire correspond?

9. An unknown emf is measured with a properly standardized potentiometer of the type described in this laboratory. It balances when the traveling plug is at position 1.3, and the slider is 95.7 cm. What is the emf of the dry cell?

Potentiometer and Voltmeter Measurements of the emf of a Dry Cell

OBJECTIVES

When a dry cell provides current to a circuit, its terminal voltage is the electromotive force (emf) of the cell minus the voltage drop across the internal resistance of the dry cell. A potentiometer is a device that measures the emf of a voltage source under the condition that no current is provided by the source. Measurements on a dry cell with a potentiometer and a voltmeter will be used to accomplish the following objectives:

1. Illustration of the principles of operation of a potentiometer
2. Comparison of the emf of several dry cells determined by a potentiometer and by a voltmeter
3. Determination of the internal resistance of a dry cell

EQUIPMENT LIST

1. Slide-wire potentiometer
2. Galvanometer
3. Single-pole, single-throw switch with a 10-kΩ resistor in parallel
4. Single-pole, double-throw switch
5. Standard cell
6. Several dry cells (range of emf, 1.5 to 1.3 V)
7. Direct-current power supply (voltage and current determined by whatever needed for the form of potentiometer used)
8. Voltmeter (input resistance at least 2 MΩ)
9. Student-type voltmeter (input resistance $\approx$ 2 kΩ)
10. Decade resistance box (100 kΩ maximum)
11. Assorted connecting leads and tap key

THEORY

Part I: Dry Cell Characteristics

A new dry cell produces an electromotive force (emf) between 1.50 and 1.60 V with an internal resistance in the range of a fraction to a few ohms. The symbol for emf is $\mathcal{E}$, and the symbol for internal resistance is R_i. The voltage that appears across the dry cell terminals when the dry cell is providing current I to a circuit is called the "terminal voltage," or V. In equation form, the terminal voltage is given by

$$V = \mathcal{E} - IR_i \tag{1}$$

Equation 1 states that when current is being provided, the terminal voltage is always less than the emf of the cell by the amount of the voltage drop across R_i. Equation 1 also shows that if no current is provided by the cell its terminal voltage equals its emf or, in other words, $V = \mathcal{E}$ if $I = 0$.

A resistor placed across the terminals of a source of emf is called a "load resistor," or R_L. The current drawn from the dry cell in this case is determined by the values of R_L and R_i. Note that the dry cell can be represented by a pure emf $\mathcal{E}$ in series with its internal resistance R_i as shown in Figure 32.1. The terminal voltage is the voltage between the dry cell terminals (shown as 1 and 2 in Figure 32.1). Applying Ohm's law to Figure 32.1 gives

$$\mathcal{E} = I(R_i + R_L) \quad \text{or} \quad I = \frac{\mathcal{E}}{(R_i + R_L)} \tag{2}$$

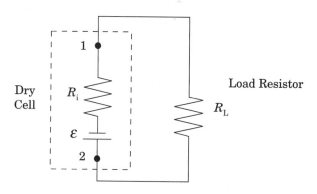

Figure 32.1 Dry cell with a load resistance.

Substituting equation 2 into equation 1 leads to the following:

$$V = \mathcal{E} - \mathcal{E}\left(\frac{R_i}{R_i + R_L}\right) = \mathcal{E}\left(1 - \frac{R_i}{R_i + R_L}\right) = \mathcal{E}\left(\frac{R_L}{R_i + R_L}\right) \tag{3}$$

Equation 3 shows that the terminal voltage V is strongly dependent on the relative values of R_L and R_i. If, for example, $R_L \gg R_i$, then $R_L/(R_i + R_L) \to 1$, and V is essentially equal to $\mathcal{E}$. On the other hand, if R_L and R_i are comparable, then V may differ considerably from $\mathcal{E}$. If $R_L = R_i$, for instance, then $V = \mathcal{E}/2$.

As a dry cell ages, its emf decreases, and its internal resistance increases. A very old dry cell might have, for example, an emf of only 1.30 V and an internal resistance of several thousand ohms. When R_i becomes this large, it means that for any significant current drawn from the cell most of the voltage drop occurs across the internal resistance. As a result, the terminal voltage of the cell becomes very small, and the cell is essentially nonfunctional.

Part 2: Voltmeter Measurements

When a voltmeter is placed across the terminals of a dry cell, the input impedance of the voltmeter is effectively a load resistor. Thus, the voltage measured by the voltmeter is essentially the same as that given by equation 3, with R_L replaced in that equation by R_V, the voltmeter resistance. The equation that results is

$$V = \varepsilon\left(\frac{R_V}{R_i + R_V}\right) \tag{4}$$

The value of R_V is strongly dependent on the quality of the voltmeter. A typical student voltmeter for use in this laboratory might have R_V as low as 3000 Ω. If such a meter is used to measure the terminal voltage of a "good" dry cell with R_i on the order of 1 Ω, the terminal voltage would be $(3000)/(3001) = 0.9997$ of the emf. For many purposes this would be an acceptable estimate of the emf. On the other hand, if this same voltmeter were used to measure the terminal voltage of a cell with R_i on the order of 5000 Ω, the terminal voltage would be $(3000)/(8000) = 0.375$ of the emf. Clearly this is not a useful estimate of the emf.

A typical value of R_V for a modern voltmeter of good quality might be in the range of 10 to 20 MΩ. For such a voltmeter, the terminal voltage measured for both of the dry cells described above would be an extremely good approximation of the true emf.

Part 3: Measurements with a Potentiometer

The potentiometer is a device whose operation is based on the principle of comparing a source of emf whose value is known to the unknown emf source being measured. The comparison is made in such a way that the unknown emf is placed in the circuit in the opposite direction to a known voltage difference. The experimental proof that the known emf is balanced against the known potential difference is determined by finding the point at which there is no current in the unknown emf. The advantage of this approach is that the true emf is being measured because there is no voltage drop across the internal resistance.

Figure 32.2 shows schematically the wiring of a simple slide-wire potentiometer. It consists of a wire (shown in the figure from A to B) of uniform cross-sectional area through which a constant current is maintained by the power supply. There is a progressive drop in the potential along the wire from A to B. Since the wire is uniform in cross-sectional area, the potential change along the length is proportional to the length of the wire. The switch labeled S_1 allows either cell ε_S or ε_U to be connected in parallel with a portion of the wire AB. The cell designated ε_S is a standard cell whose emf is accurately known, and ε_U stands for the cell of unknown emf that is to be determined.

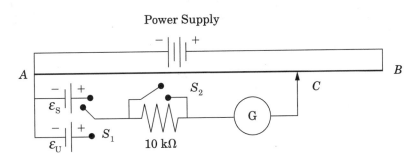

Figure 32.2 Simple slidewire potentiometer.

The current in the wire AB is chosen to be large enough so that the potential change from A to B is greater than either $\mathcal{E}_U$ or $\mathcal{E}_S$. With switch S_1 set so that $\mathcal{E}_S$ is in the circuit, the slider can be moved along the wire until the point C is located where the voltage drop from A to C is exactly equal to $\mathcal{E}_S$. This is accomplished experimentally by finding the point C for which there is no current in the galvanometer that is in series with $\mathcal{E}_S$. After throwing switch S_1 to the other position, which places $\mathcal{E}_U$ in the circuit, a new point D is found where the voltage drop from A to D is equal to $\mathcal{E}_U$. Again, this point is found experimentally by finding the point where there is no current in the galvanometer. Since the potential drop along the wire is proportional to the length of the wire, a ratio exists between the lengths of the wire, AD and AC, and the two emfs. This can be expressed as

$$\frac{\mathcal{E}_U}{\mathcal{E}_S} = \frac{AD}{AC} \quad \text{or} \quad \mathcal{E}_U = \mathcal{E}_S \left(\frac{AD}{AC}\right) \tag{5}$$

Using equation 5, the value of the unknown emf $\mathcal{E}_U$ can be determined from the measured ratio of the lengths of wire and the known value of $\mathcal{E}_S$.

In practice, the potentiometer can be made more accurate by extending its total length by including additional coils of wire each of the same length (Figure 32.3). Also, many potentiometers are built in such a way that they can be calibrated so that a given length is exactly proportional to some fixed convenient value of voltage. Specific instructions will be given below for making measurements with a potentiometer of this type, which has a slide wire 1 m in length and 15 other 1-m coils of wire in series with the slide wire. There is a banana-plug receptacle at the junction of each coil of wire so that any number of the 15 coils of wire can be incorporated in the circuit along with the chosen length of the slide wire. The potentiometer is designed to be standardized to ensure that each meter of wire has a voltage drop of 0.100 V. Thus, the maximum emf that can be measured with this standardization is 1.600 V. Note that when the standardization procedure has been accomplished, it allows one to read the emf of an unknown directly because a given length of the potentiometer corresponds to a fixed amount of voltage.

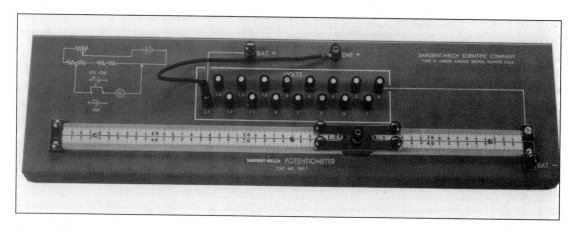

Figure 32.3 Slidewire form of the potentiometer with a total length of 16 m of wire. (Photo courtesy of Sargent-Welch Scientific Company)

EXPERIMENTAL PROCEDURE—DRY CELL EMF BY POTENTIOMETER

1. Before making any measurements with the potentiometer, note that special care is required in the use of the standard cell. The cell should remain upright at all times and never be allowed to turn on its side or upside down. Be certain that the cell is placed in the circuit with the correct polarity and that during initial adjustment, switch S_2 is open. This places the 10-kΩ resistor in series with the standard cell and galvanometer, which serves to limit the current drawn from the standard cell.

2. The circuit diagram for the potentiometer described in this experiment is shown in Figure 32.4. Connect the circuit as shown, but have the circuit approved by your instructor before the power supply is turned on. (*Note that the circuit diagram below and the detailed procedure given below apply to a 16-m-long potentiometer with provisions to calibrate it directly in volts. If another type of potentiometer is provided, consult your instructor for specific instructions in its use.*)

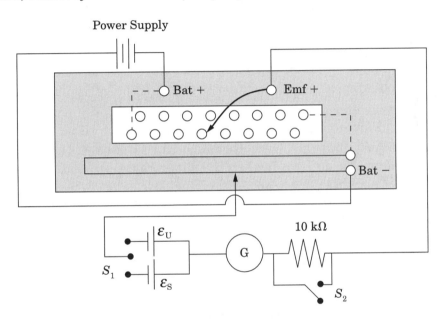

Figure 32.4 Slidewire potentiometer with 16 m total wire length.

3. Following the steps detailed below for standardizing and using the potentiometer, measure the emf of three unknown dry cells. The three unknown dry cells should include one that is fairly new and one that is fairly old. For each unknown cell, make three independent measurements by standardizing the potentiometer and then measuring the emf of the cell three separate times. The purpose of standardizing before each trial on a given cell is to include the process of standardizing in the determination of the precision of the measurements.

4. Procedure to standardize the potentiometer:

   (a) Throw switch S_1 to the standard cell position.

   (b) The potentiometer is designed to be calibrated in such a way that each meter of wire has a potential drop of 0.1 V across it. Therefore, the positions for the banana-plug lead coming from the terminal labeled emf+ are labeled 0.1, 0.2, 0.3, etc. Adjust the position of the banana plug and the slider until the total corresponds to the emf of the standard cell used. For example, if the standard cell has an emf of 1.0566 V, place the plug in position 1.0 and the slider at a length of 56.6 cm.

(c) With switch S_2 open, press and release the slider key, noting the galvanometer deflection.

(d) Adjust the power supply for minimum galvanometer deflection when the slider key is pressed.

(e) Close switch S_2 and press and release the slider key, noting the galvanometer deflection.

(f) Adjust the power supply for zero galvanometer deflection when the slider key is pressed.

(g) The potentiometer has now been standardized to 0.1 V per meter of wire. The power supply should remain fixed at its setting for any measurements taken with this standardization. Note that the maximum emf that could be measured is 1.6000 V because the total length of the potentiometer is 16 m.

5. Procedure to measure an unknown emf (to be followed after the potentiometer has been standardized by the steps described above):

(a) Throw switch S_1 to unknown emf position.

(b) With switch S_2 open, touch the slider key, noting the galvanometer deflection.

(c) Move the traveling plug and the slider position until the point is found that gives the minimum galvanometer deflection when the slider key is pressed.

(d) Close switch S_2, and then move the slider until the point is found that gives zero galvanometer deflection when the slider key is pressed.

(e) Read the value of the unknown emf as the position of the traveling plug plus the slider position. For example, if the plug is at 1.4 and the slider at 35.5 cm, the emf of the unknown cell is 1.4355 V.

EXPERIMENTAL PROCEDURE—DRY CELL VOLTAGE BY VOLTMETER

1. Using a voltmeter whose internal resistance is on the order of 10 MΩ, place the voltmeter on terminals of one of your unknown dry cells. The value of the voltmeter reading is the terminal voltage of the unknown dry cell. Repeat this procedure for each of the unknown emfs and record the values in Data Table 2 under the column labeled "High-Impedance Voltmeter."

2. Using a student-type low-impedance voltmeter, measure the terminal voltage of each of the unknown dry cells. Record the values in Data Table under the column labeled "Low-Impedance Voltmeter."

EXPERIMENTAL PROCEDURE—INTERNAL RESISTANCE OF A DRY CELL

1. Place the terminals of the high-impedance voltmeter across the terminals of the dry cell with the lowest emf as measured by the potentiometer. Note the value of the voltmeter reading, which should be essentially the same as determined earlier by the potentiometer.

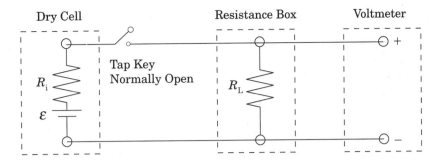

Figure 32.5 Circuit to measure internal resistance of the dry cell.

2. Set a decade resistance box with a 10 kΩ decade to the maximum value of 100 kΩ. Place the decade resistance box in parallel with the voltmeter and the dry cell with a tap key as shown in Figure 32.5. The resistance box is serving as a load resistor, and the voltmeter now reads the terminal voltage of the cell when the tap key is momentarily pressed. *Just touch the key momentarily and read the voltmeter at that instant. If the key is left pressed for any length of time, it will drain the cell as it provides current for the load resistor, thus making the results meaningless.*

3. Touch the tap key and note the value of the terminal voltage for the 100-kΩ load resistor. If there is at least a 5% decrease in the voltmeter reading below the original emf of the cell, record this value of the terminal voltage and the value of 100 kΩ for the load resistance in Data Table 3. If the voltmeter reading does not fall by at least 5% when the 100-kΩ resistor is placed across the cell, lower the decade resistance box value and touch the tap key again to read the terminal voltage. When the terminal voltage is approximately 5% less than the cell emf, record the value of the load resistance and the terminal voltage in Data Table 3.

4. Continue this process of lowering the resistance box while noting the terminal voltage for values of the terminal voltage that are about 10% and then about 15% less than the emf of the cell. Record the exact value of the terminal voltage and the load resistance that gives that terminal voltage in Data Table 3.

CALCULATIONS—DRY CELL EMF BY POTENTIOMETER

1. Calculate the mean and standard error of the three determinations of the emf of each unknown dry cell and record the values in Calculations Table 1.

CALCULATIONS—DRY CELL VOLTAGE BY VOLTMETER

1. Calculate the percentage difference between the values of the terminal voltage measured with the high-impedance voltmeter and the emf values determined with the potentiometer, assuming the potentiometer values to be correct. For the potentiometer measurements, use the mean values calculated above. Record the calculated percentage differences in Calculations Table 2.

2. Repeat the calculations described in step 1 for the terminal voltage values measured with the low impedance voltmeter. Record the results in Calculations Table 2.

CALCULATIONS—INTERNAL RESISTANCE OF A DRY CELL

1. Equation 3 relates the terminal voltage of a dry cell to its emf, internal resistance, and load resistance. Using the data in Data Table 3 for the terminal voltage versus the load resistance, solve equation 3 for the internal resistance and determine a value for the internal resistance from each of the three values of load resistance and terminal voltage. Record the values of R_i in Calculations Table 3.

Laboratory 32

Potentiometer and Voltmeter Measurements of the emf of a Dry Cell

LABORATORY REPORT

Data Table 1

Unknown emf	Trial 1	Trial 2	Trial 3
#			
#			
#			

Calculations Table 1

$\overline{emf}$	α_{emf}

Data Table 2

Unknown emf	High-Impedance Voltmeter	Low-Impedance Voltmeter
#		
#		
#		

Calculations Table 2

Unknown emf	% Diff High-Imp. Voltmeter	% Diff Low-Imp. Voltmeter

Data Table 3

ε	Load resistance R_L (Ω)	Terminal Voltage (Volts)

Calculations Table 3

R_i (Ω)

SAMPLE CALCULATIONS

QUESTIONS

1. Consider the mean values of the emf of each cell as determined by the potentiometer. Comment on the precision of the measurements.

2. What can you say about the accuracy of the measurements discussed in question 1? What is the major factor controlling the accuracy of the measurements?

3. Consider the percentage differences in Calculations Table 2. Are the high-impedance voltmeter readings a reasonable estimate of the emf?

4. Are there significant differences between the high-impedance voltmeter measurements and the low-impedance voltmeter measurements? Can you conclude anything about the internal resistance of any of the unknown dry cells from this data? State what you can about the internal resistance of each unknown dry cell.

5. The idea that a dry cell can be represented as a pure emf in series with a pure resistance R_i is a model. If that model is accurate for the dry cell for which you determined R_i in Calculations Table 3, the value of R_i should be approximately the same for the three determinations. Comment on the extent to which such a model seems appropriate to your measurements of the dry cell.

Laboratory 33

The RC Time Constant

PRELABORATORY ASSIGNMENT

Read carefully the entire description of the laboratory and answer the following questions based on the material contained in the reading assignment. Turn in the completed prelaboratory assignment at the beginning of the laboratory period prior to the performance of the laboratory.

1. In a circuit such as the one in Figure 33.1 with the capacitor initially uncharged, the switch S is thrown to position A at $t = 0$. The charge on the capacitor (a) is initially zero and finally $C\mathcal{E}$, (b) is constant at a value of $C\mathcal{E}$, (c) is initially $C\mathcal{E}$ and finally zero, or (d) is always less than $\mathcal{E}/R$.

2. In a circuit such as the one in Figure 33.1 with the capacitor initially uncharged, the switch S is thrown to position A at $t = 0$. The current in the circuit is (a) initially zero and finally $\mathcal{E}/R$, (b) constant at a value of $\mathcal{E}/R$, (c) is equal to $C\mathcal{E}$, or (d) initially $\mathcal{E}/R$ and finally zero.

3. In a circuit such as the one in Figure 33.2, the switch S is first closed to charge the capacitor, and then it is opened at $t = 0$. The expression $V = \mathcal{E}\, e^{-t/RC}$ gives the value of (a) the voltage on the capacitor but not the voltmeter, (b) the voltage on the voltmeter but not the capacitor, (c) both the voltage on the capacitor and the voltage on the voltmeter, which are the same, or (d) the charge on the capacitor.

4. For a circuit such as the one in Figure 33.1, what are the equations for the charge Q and the current I as functions of time when the capacitor is charging?

    $Q =$ _____ $I =$ _____

5. For a circuit such as the one in Figure 33.1, what are the equations for the charge Q and the current I as functions of time when the capacitor is discharging?

    $Q =$ _____ $I =$ _____

6. If a 5.00-μF capacitor and a 3.50-MΩ resistor form a series RC circuit, what is the RC time constant? (Give proper units for RC.)

    $RC =$ _____

7. Assume that a 10.0 μF capacitor, a battery of emf $\varepsilon = 12.0$ V, and a voltmeter of 10.0 MΩ input impedance are used in a circuit such as that in Figure 33.2. The switch S is first closed, and then the switch is opened. What is the reading on the voltmeter 35.0 s after the switch is opened? Show your work.

$V = $ _____ V

8. Assume that a circuit is constructed such as the one shown in Figure 33.3 with a capacitor of 5.00 μF, a battery of 24.0-V, a voltmeter of input impedance 12.0 MΩ, and a resistor $R_U = 10.0$ MΩ. If the switch is first closed and then opened, what is the voltmeter reading 25.0 s after the switch is opened? Show your work.

$V = $ _____ V

9. In the measurement of the voltage as a function of time performed in this laboratory, the voltage is measured at fixed time intervals. (a) true (b) false

OBJECTIVES

When a direct-current source of emf is suddenly placed in series with a capacitor and a resistor, there is current in the circuit for whatever time it takes to fully charge the capacitor. In a similar manner, there is a definite time needed to discharge a capacitor that has previously been charged. There is a characteristic time associated with either of these processes, called the "RC time constant," whose value depends on the value of the resistance R and the capacitance C. In this laboratory, series combinations of a power supply, a capacitor, and resistors will be used to accomplish the following objectives:

1. Demonstration of the finite time needed to discharge a capacitor
2. Measurement of the voltage across a resistor as a function of time
3. Determination of the RC time constant of two series RC circuits
4. Determination of the value of an unknown capacitor from measurements made on a series RC circuit using a voltmeter as the resistance
5. Determination of the value of an unknown resistor from measurements made on a second RC circuit with an unknown resistance in parallel with the voltmeter

EQUIPMENT LIST

1. Voltmeter (at least 10-MΩ input impedance, preferably digital readout)
2. Direct-current power supply (20 V)
3. Laboratory timer
4. High-quality capacitor (5–10 μF to serve as an unknown)
5. Resistor (approximately 10 MΩ to serve as an unknown)
6. Single-pole, double-throw switch
7. Assorted connecting leads

THEORY

Consider the circuit shown in Figure 33.1 consisting of a capacitor C, a resistor R, a source of emf $\mathcal{E}$, and a switch S. If the switch S is thrown to point A at time $t = 0$ when the capacitor is initially uncharged, charge begins to flow in the series circuit consisting of $\mathcal{E}$, R, and C and flows until the capacitor is fully charged. It can be shown that the current I starts at an initial value of $\mathcal{E}/R$ and decreases exponentially with time. The charge Q on the capacitor, on the other hand, begins at zero and increases exponentially with time until it becomes equal to $C\mathcal{E}$. The equations that describe those events are

$$Q = C\mathcal{E}\,(1 - e^{-t/RC}) \quad \text{and} \quad I = \mathcal{E}/R\,e^{-t/RC} \tag{1}$$

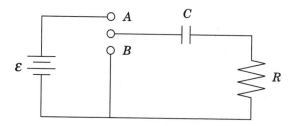

Figure 33.1 Simple Series RC Circuit.

The quantity RC is called the "time constant of the circuit," and it has units of seconds if R is expressed in ohms and C is expressed in farads. After a period of time that is long compared to the time constant RC, the term $e^{-t/RC}$ becomes negligibly small. When this is true the equations above predict that the charge Q is equal to $C\mathcal{E}$, and the current in the circuit is zero.

If switch S is now thrown to position B, which effectively takes $\mathcal{E}$ out of the circuit, the capacitor discharges through the resistor. Therefore, the charge on the capacitor and the current in the circuit both decay exponentially while the capacitor is discharging. The equations that describe this discharging process are

$$Q = C\mathcal{E}\,e^{-t/RC} \quad \text{and} \quad I = \mathcal{E}/R\,e^{-t/RC} \tag{2}$$

The equation for the current could be written with a negative sign because the current in the discharging case will be in the opposite direction from the current in the charging case. The magnitude of the current is the same in both cases. Although the above discussion has included both the case of charging and discharging a capacitor, this laboratory will only investigate the process of discharging a capacitor.

Consider the circuit shown in Figure 33.2 consisting of a power supply of emf $\mathcal{E}$, a capacitor C, a switch S, and a voltmeter whose input impedance is R. If initially the switch S is closed, the capacitor is charged almost immediately to $\mathcal{E}$, the voltage of the power supply. When the switch is opened the capacitor discharges through the resistance of the meter R with a time constant given by RC. With the switch open the only elements in the circuit are the capacitor C and the voltmeter resistance R; thus the voltage across the capacitor is equal to the voltage across the voltmeter. The voltage across the capacitor is given by Q/C, and the voltage across the voltmeter is given by IR. Solving 2 for those quantities leads in both cases to

$$V = \mathcal{E}e^{-t/RC} \tag{3}$$

Equation 3 stands either for the voltage across the voltmeter or the voltage across the capacitor as a function of time. Dividing both sides of equation 3 by $\mathcal{E}$ and taking the reciprocal of both sides of the equation leads to

$$\mathcal{E}/V = e^{-t/RC} \tag{4}$$

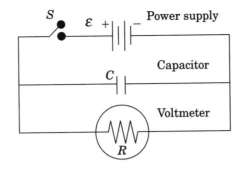

Figure 33.2 An RC circuit using a voltmeter as the resistance.

Taking the natural logarithm of both sides of equation 4 leads to the following:

$$\ln (\mathcal{E}/V) = (1/RC)\, t \qquad (5)$$

Equation 5 states that there is a linear relationship between the quantity $\ln(\mathcal{E}/V)$ and the time t with the quantity $(1/RC)$ as the constant of proportionality. Therefore, if the voltage across the capacitor is determined as a function of time, a graph of $\ln(\mathcal{E}/V)$ versus t will give a straight line whose slope is $(1/RC)$. Thus, RC can be determined, and if the voltmeter resistance R is known, then C can be determined.

  If an unknown resistor is placed in parallel with the voltmeter, it produces a circuit like that shown in Figure 33.3. The capacitor can again be charged and then discharged, but now the time constant will be equal to R_tC, where R_t is the total resistance, which is the parallel combination of R and R_U. If the relationship between R, R_U, and R_t is solved for R_U the result is

$$R_U = \frac{R\,R_t}{R - R_t} \qquad (6)$$

Therefore, a measurement of the capacitor voltage as a function of time will produce a dependence like that given by equation 5 except that the slope of the straight line will be $(1/R_tC)$. Thus, if C is known and R_tC is found from the slope, then R_t can be determined. Using equation 6, R_U can be found from R and the value just determined for R_t.

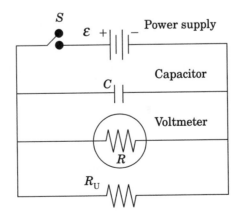

Figure 33.3 RC circuit using voltmeter and R_U in parallel as resistance.

EXPERIMENTAL PROCEDURE—UNKNOWN CAPACITANCE

1. Construct a circuit such as the one in Figure 33.2 using the capacitor supplied, the voltmeter, and the power supply. Have the circuit approved by your instructor before turning on any power. Obtain from your instructor the value of the input impedance of the voltmeter and record it in Data Table 1 as R.

2. Close the switch, and while reading its voltage on the voltmeter, adjust the power supply emf $\mathcal{E}$ to the value chosen by your instructor. Record the value of $\mathcal{E}$ in Data Table 1.

3. Open the switch and simultaneously start the timer.

4. The voltmeter reading will fall as the capacitor discharges. Let the timer run continuously, and for eight predetermined values of the voltage, record the time t on the timer when the voltmeter reads these voltages. A convenient choice for voltages at which to measure the time would be increments of 10%. For example, if $\mathcal{E} = 20.0$ V, record the time when the voltage is 18.0, 16.0, 14.0, etc. Record the values of the voltage at which the time is to be read in Data Table 1 as V.

5. Record the values of t at which each value of V occurs in Data Table 1 under Trial 1.

6. Repeat steps 2 through 4 two more times, recording the values of t under Trials 2 and 3 in Data Table 1.

EXPERIMENTAL PROCEDURE—UNKNOWN RESISTANCE

1. Construct a circuit such as the one in Figure 33.3 using the same capacitor used in the last circuit and the unknown resistor supplied. Close the switch and adjust the power supply voltage to the same value used in the last procedure.

2. Repeat steps 2 through 6 of the procedure above, but record all values in the appropriate places in Data Table 2.

CALCULATIONS—UNKNOWN CAPACITANCE

1. Calculate the values of $\ln(\mathcal{E}/V)$ and record them in Calculations Table 1.

2. Calculate the mean $\bar{t}$ and the standard error α_t for the three trials of the time t at each voltage and record them in Calculations Table 1.

3. Perform a linear least squares fit of the data with $\ln(\mathcal{E}/V)$ as the ordinate and $\bar{t}$ as the abscissa.

4. The value of the slope is equal to $1/RC$. Record the value of the slope in Calculations Table 1. (The units of $1/RC$ are s^{-1}.)

5. Calculate RC as the reciprocal of the slope. Record the value of RC in Calculations Table 1. (The units of RC are s.)

6. Using the value of RC and the value of R, calculate the value of the unknown capacitor C and record it in Calculations Table 1.

CALCULATIONS—UNKNOWN RESISTANCE

1. Calculate the values of $\ln(\mathcal{E}/V)$ and record the values in Calculations Table 2.

2. Calculate the mean $\bar{t}$ and the standard error α_t for the three trials of the time t at each voltage and record them in Calculations Table 2.

3. Perform a linear least squares fit to the data with $\ln(\mathcal{E}/V)$ as the ordinate and $\bar{t}$ as the abscissa.

4. The value of the slope of this fit is equal to $1/R_tC$. Record the value of the slope in Calculations Table 2. (The units of $1/R_tC$ are s^{-1}.)

5. Calculate the value of R_tC as the reciprocal of the slope. Record the value of R_tC in Calculations Table 2. (The units of R_tC are s.)

6. Using the value of the capacitance C determined in the first procedure and the value of R_tC, calculate the value of R_t and record it in Calculations Table 2.

7. Using equation 6, calculate the value of the unknown resistance R_U from the values of R_t and R. Record the value of R_U in Calculations Table 2.

GRAPHS

1. For the data of Part 1, graph the quantity $\ln(\mathcal{E}/V)$ as the ordinate versus $\bar{t}$ as the abscissa. Use the values of the standard errors as error bars. Also, show on the graph the straight line obtained from the linear least squares fit to the data.

2. For the data of Part 2, graph the quantity $\ln(\mathcal{E}/V)$ as the ordinate versus $\bar{t}$ as the abscissa. Use the values of the standard errors as error bars. Also, show on the graph the straight line obtained from the linear least squares fit to the data.

LABORATORY REPORT

Data Table 1

V (V)	t_1 (s)	t_2 (s)	t_3 (s)
$\mathcal{E} =$		V	
$R =$		Ω	

Calculations Table 1

$\ln(\mathcal{E}/V)$	$\bar{t}$ (s)	α_t (s)
Regression coeff. =		
Intercept =		
Slope =		s^{-1}
$RC =$		s
$C =$		F

SAMPLE CALCULATIONS

Data Table 2

V (V)	t_1 (s)	t_2 (s)	t_3 (s)
$\mathcal{E} =$		V	
$R =$		Ω	

Calculations Table 2

$\ln (\mathcal{E}/V)$	$\bar{t}$ (s)	α_t (s)
Regression coeff. =		
Intercept =		
Slope =		s^{-1}
$R_t C =$		s
$R_t =$		Ω
$R_U =$		Ω

SAMPLE CALCULATIONS

QUESTIONS

1. Evaluate the linearity of each of the graphs. Do they confirm the linear dependence between the two variables that is predicted by the theory?

2. Ask your instructor for the values of the unknown capacitor and resistor. Calculate the percentage error of your measurement compared to the values provided.
   On this basis, evaluate the accuracy of your measurement of the capacitance and
   resistance.

3. Show that RC has units of seconds if R is in Ω and C is in F.

4. A 5.60-μF capacitor and a 4.57-MΩ resistor form a series RC circuit. If the capacitor is initially charged to 25.0 V, how long does it take for the voltage on the capacitor to reach 10.0 V? Show your work.

Laboratory **34**

Kirchhoff's Rules

PRELABORATORY ASSIGNMENT

Read carefully the entire description of the laboratory and answer the following questions based on the material contained in the reading assignment. Turn in the completed prelaboratory assignment at the beginning of the laboratory period prior to the performance of the laboratory.

1. Consider the circuit in Figure 34.1. Choose any of the following statements about the circuit that are true. More than one may be correct. (a) It is a single-loop circuit. (b) It is a multiloop circuit. (c) Assuming R_1 and R_2 were known, the currents could be determined, but only if Kirchhoff's rules were used. (d) Assuming R_1 and R_2 were known the currents could be determined without the use of Kirchhoff's rules.

2. In the circuit of Figure 34.1, if I_1 = 2.00 A and I_2 = 0.75 A, what is the value of I_3?

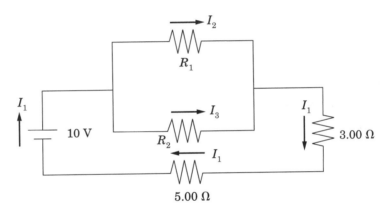

Figure 34.1 Circuit for questions 1 to 4.

For questions 3 and 4 assume that the value of R_1 and R_2 in Figure 34.1 are both 4.00 Ω.

3. What is the equivalent resistance of the circuit?

4. What is the current in the 3.00-Ω resistor?

5. Consider the circuit of Figure 34.2. Apply Kirchhoff's rules to the circuit and write three equations in terms of known circuit elements and the unknown currents shown in the figure.

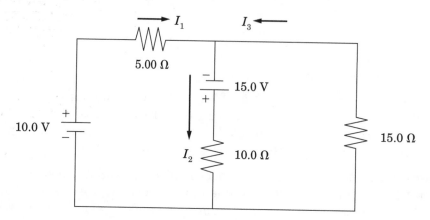

Figure 34-2 Multiloop circuit.

6. Solve the three equations that you wrote in question 5 for the values of the currents.

7. If any of the current values obtained in the solution to question 6 are negative, explain the significance of a negative value for a current.

OBJECTIVES

Multiloop circuits containing several sources of emf and several resistors will be constructed, and measurements of the voltage and current for each of the elements of the circuit will be used to achieve the following objectives:

1. Demonstration of the type of circuit to which Kirchhoff's rules must be applied
2. Theoretical application of Kirchhoff's rules to several circuits and solution for the expected currents in the circuit
3. Comparison between the theoretical values predicted by Kirchhoff's rules and the measured experimental values of the currents in multi-loop circuits

EQUIPMENT LIST

1. DC voltmeter (preferably digital, 0–20 V)
2. DC ammeter (preferably digital, 0–1000 mA)
3. Two sources of emf (DC power supplies, up to 12 V)
4. Three or four resistors or resistor boxes (range, 500–1000 Ω)
5. Digital ohmmeter (one for the class)
6. Connecting wires

THEORY

Consider the circuit in Figure 34.3. The circuit is labeled with all of the currents. The 2-Ω resistor, 8-Ω resistor, and the 12-V power supply have current I_1, while the 6-Ω resistor has current I_2 and the 3-Ω resistor has current I_3. This circuit is called a single-loop circuit because it can be solved by noting that the 6-Ω resistor and the 3-Ω resistor are in parallel. This means that their equivalent resistance is 2 Ω. That equivalent 2-Ω resistance is then in series with the 12-V power supply and the other two resistors, reducing the circuit to a single-loop circuit. The total resistance of the circuit is 12 Ω, and this resistance is across the 12-V power supply. The current in the 12-Ω equivalent resistance is therefore $I_1 = 1$ A. Application of Ohm's law to the remaining part of the circuit gives $I_2 = 1/3$ A and $I_3 = 2/3$ A.

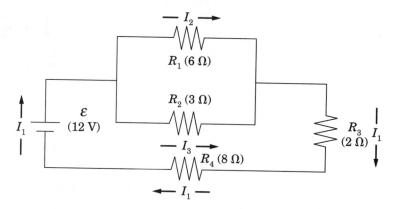

Figure 34.3 Single-loop circuit.

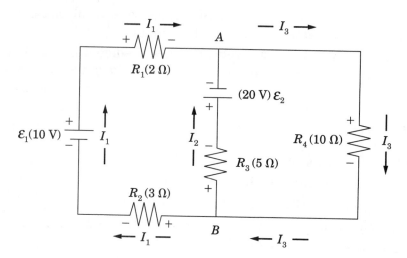

Figure 34.4 Multiloop circuit.

Consider now the circuit of Figure 34.4. This laboratory is concerned with the fundamental difference between circuits of the type depicted in Figure 34.3 and circuits of the type depicted in Figure 34.4. Note that the circuit in Figure 34.4 cannot be reduced to a single-loop circuit, but instead must be analyzed by other means. In the analysis of this circuit it is necessary to first define some terms. A point at which at least three possible current paths intersect is defined as a junction. For example, points A and B in Figure 34.4 are junctions. A closed loop is defined to be any path that starts at some point in a circuit and passes through elements of the circuit (in this case resistors and power supplies), and then arrives back at the same point without passing through any circuit element more than once. By this definition there are three loops in the circuit of Figure 34.4: (1) starting at B, going through the 10-V power supply to A, and then down through the 20-V power supply back to B; (2) starting at B, up through the 20-V power supply, and then around the outside through the 10-Ω resistor and back to B; and (3) completely around the outside part of the circuit. One can always traverse any loop in one of two directions, but regardless of which direction is chosen, the resulting equations are equivalent.

The solution for the currents in a multiloop circuit depends on analysis using two rules developed by a German professor named Gustav Robert Kirchhoff. The first of these rules is Kirchhoff's current rule (KCR) and can be stated in the following way:

KCR—The sum of currents into a junction = the sum of currents out of the junction

This rule actually amounts to a statement of conservation of charge. In effect it states that charge does not accumulate at any point in the circuit. The second rule is Kirchhoff's voltage rule (KVR) and can be stated as:

KVR—The algebraic sum of the voltage changes around any closed loop is zero.

This rule is essentially a statement of the conservation of energy, which recognizes that the energy provided by the power supplies is absorbed by the resistors.

In a multiloop circuit, the values of the resistors and the power supplies are known. As a first step it is necessary to determine how many independent currents are in the circuit, to label them, and then to assign a direction to each current. Application of Kirchhoff's rules to the circuits, treating the assigned currents as unknowns, will produce as many independent equations as there are unknown currents. Solution of those equations will determine the values of the currents.

In the application of KVR to a circuit, care must be taken to assign the proper sign to a voltage change across a particular element. The value of the voltage change across an emf $\mathcal{E}$ can be either $+\mathcal{E}$ or $-\mathcal{E}$ depending on which direction it is traversed in the loop. If the emf is traversed from the $(-)$ terminal to the $(+)$ terminal, the change in voltage is $+\mathcal{E}$. However, when going from the $(+)$ terminal to the $(-)$ terminal, the change in voltage is $-\mathcal{E}$. In the laboratory the terminal voltage of the sources of emf will be measured. It will be assumed that those values approximate the emf. In effect, this amounts to assuming that the source of emf has no internal resistance. This is a good approximation, but it is not strictly true, and thus will cause a small error.

Similarly, when a resistor R in which there is an assumed current I, is traversed in the loop in the same direction as the current, the voltage change is $-IR$. Conversely, if the resistor is traversed in the direction opposite that of the current, the voltage change is $+IR$. It is very important to understand that the sign of the voltage change across an emf is not affected by the direction of the current in the emf. The sign of the voltage change across a resistor, however, is completely determined by the current direction.

Consider the application of Kirchhoff's rules to the multiloop circuit of Figure 34.4. At the junction point A, currents I_1 and I_2 are going into the junction, and current I_3 is going out of the junction and KCR states that

$$I_1 + I_2 = I_3 \tag{1}$$

It might appear that applying KCR to the junction B would produce an additional useful equation, but in fact it would result in an equation that is equivalent to equation 1.

Applying KVR to the loop that starts at point B, goes through the 10-V power supply to A, and then down through the 20-V power supply back to B, gives the following equation.

$$-R_2 I_1 + \mathcal{E}_1 - R_1 I_1 + \mathcal{E}_2 + R_3 I_2 = 0 \tag{2}$$

Although normally the voltage change across a resistor is written as IR, it has been written here as RI in order to emphasize that I is the unknown, and R is essentially the coefficient of I. Replacing the symbols in equation 2 with the values of the resistances and emfs of the circuit gives

$$-3\,I_1 + 10 - 2\,I_1 + 20 + 5\,I_2 = 0 \tag{3}$$

The signs used in both equations 2 and 3 are consistent with the description given above for determining the signs of voltage changes. Furthermore, the $(+)$ and $(-)$ labeled on each side of the circuit elements of the circuit diagram are also consistent with that convention.

Finally, applying KVR to the loop that starts at B and goes clockwise around the right side of the circuit gives

$$-R_3\,I_2 - \mathcal{E}_2 - R_4\,I_3 = 0 \tag{4}$$

Substituting the values of the resistances and emf for the symbols in equation 4 results in

$$-5\,I_2 - 20 - 10\,I_3 = 0 \tag{5}$$

Equations 1, 3, and 5 are the three desired equations in the three unknowns I_1, I_2, and I_3. The solution of these equations gives values for the currents of $I_1 = 2.800$ A, $I_2 = -3.200$ A, and $I_3 = -0.400$ A. The fact that the currents I_2 and I_3 are negative is an indication that the original assumption of direction for these two currents was incorrect. The interpretation of the results of the solution is that there is a current of 2.800 A in the direction indicated in the figure for I_1, a current of 3.200 A in a direction opposite to that indicated in the figure for I_2, and a current of 0.400 A in a direction opposite to that indicated for I_3. This is a general feature of solutions for the currents in a multiloop circuit using Kirchhoff's rules. Even if the original assumption of the direction of a current is wrong, the solution of the equations leads to the correct understanding of the proper direction by virtue of the sign of the current.

EXPERIMENTAL PROCEDURE

1. Choose three resistors with the following values: $R_1 = 500\ \Omega$, $R_2 = 750\ \Omega$, and $R_3 = 1000\ \Omega$. If using resistance boxes these values can be chosen exactly, but if using standard resistors choose values as close as possible to the values listed. Using the ohmmeter measure the precise value of the resistors and record those values in Data Table 1.

2. Using two power supplies and the resistors R_1, R_2, and R_3, construct a circuit like that shown in Figure 34.5 with $\mathcal{E}_1 = 10$ V and $\mathcal{E}_2 = 5$ V. (*Caution*: If the laboratory is performed with batteries instead of power supplies make certain that no battery is subject to large reverse currents. Some batteries might be damaged or even explode if subjected to large reverse currents. This should be of minimal danger for the low currents that will result using voltages and resistors of the magnitude suggested.)

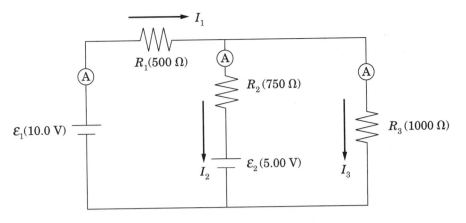

Figure 34.5 Experimental multiloop circuit with three unknown currents.

3. Measure the currents I_1, I_2, and I_3. Assuming that only one ammeter is available, the currents will have to be measured one at a time by placing the ammeter in the positions shown in the circuit diagram as a circle. Note that placing the ammeter in the circuit with the polarity shown in the circuit diagram will give positive readings when the current is in the direction assumed. If a modern digital ammeter is used for the measurements, the ammeter will give a positive reading if the current is in the direction assumed, and will give a negative reading if the current is in the opposite direction. If, however, an ammeter is used that properly deflects in only one direction, the meter could be damaged if the current is in the opposite direction from that assumed. In this case the polarity of the meter must be reversed. If there is any concern about damage to the ammeter, it may be prudent to perform the calculations (described in the next section) for the unknown currents and thus determine their magnitude and direction in advance.

4. Measure the values of the emfs $\mathcal{E}_1$ and $\mathcal{E}_2$ with a voltmeter and record those values in Data Table 1. As previously noted, these values are really the terminal voltages of the power supplies which are assumed to approximate the emfs if the internal resistance of the sources are negligible.

5. Construct the circuit of Figure 34.6, which also has two power supplies but has four resistors. Choose values of $\mathcal{E}_1 = 5$ V, $\mathcal{E}_2 = 10$ V, $R_1 = 1000$ Ω, $R_2 = 800$ Ω, $R_3 = 600$ Ω, and $R_4 = 500$ Ω or values as close to those as possible. Following similar procedures to those described above, measure the values of the resistors and record those values in Data Table 2. Following similar procedures to those described above, measure each of the four currents and the emf of the power supplies and record those values in Data Table 2.

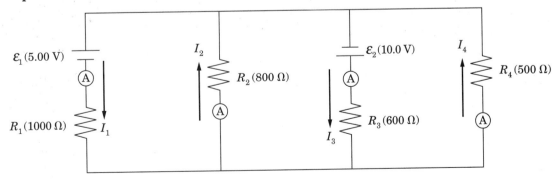

Figure 34-6 Experimental multiloop circuit with four unknown currents.

CALCULATIONS

1. Apply Kirchhoff's rules to the circuit of Figure 34.5 with the actual values that were used in your circuit. Three equations in the three currents I_1, I_2, and I_3 will result. One equation will be a KCR equation, and two will be KVR equations. Record those three equations in the appropriate place in Calculations Table 1.

2. Solve the three equations that you have written for the values of I_1, I_2, and I_3. Record those values in Calculations Table 1.

3. Calculate the percentage error of the experimental values of the current compared to the theoretical values for each of the currents. Record the results in Calculations Table 1.

4. Apply Kirchhoff's rules to the circuit of Figure 34.6 with the actual values that were used in your circuit. Four equations in the four currents I_1, I_2, I_3, and I_4 will result. One equation will be a KCR equation, and three will be KVR equations. Record those four equations in the appropriate place in Calculations Table 2.

5. Solve the four equations that you have written for the values of I_1, I_2, I_3 and I_4. Record those values in Calculations Table 2.

6. Calculate the percentage error of the experimental values of the current compared to the theoretical values for each of the currents. Record the results in Calculations Table 2.

LABORATORY REPORT

Data Table 1

Power Supply Voltages
$\varepsilon_1 = $ _____ V
$\varepsilon_2 = $ _____ V

Resistor Values (Ω)	Experimental Current (mA)
$R_1 = $	$I_1 = $
$R_2 = $	$I_2 = $
$R_3 = $	$I_3 = $

Calculations Table 1

Kirchoff's rules for the circuit	
(1) KCR —	
(2) KVR1 —	
(3) KVR2 —	

Theoretical Current (mA)	% Error of Experimental Current compared to the Theoretical Current
$I_1 = $	
$I_2 = $	
$I_3 = $	

Data Table 2

Power Supply Voltages
$\varepsilon_1 = $ _____ V
$\varepsilon_2 = $ _____ V

Resistor Values (Ω)	Experimental Current (mA)
$R_1 = $	$I_1 = $
$R_2 = $	$I_2 = $
$R_3 = $	$I_3 = $
$R_4 = $	$I_4 = $

Calculations Table 2

Kirchoff's rules for the circuit	
(1) KCR —	
(2) KVR1 —	
(3) KVR2 —	
(4) KVR3 —	

Theoretical Current (mA)	% Error of Experimental Current compared to the Theoretical Current
$I_1 = $	
$I_2 = $	
$I_3 = $	
$I_4 = $	

QUESTIONS

1. In Figure 34.5 what is the equation that relates the currents I_1, I_2, and I_3?

2. State quantitatively how well your experimental results for the circuit of Figure 34.5 agree with the expected relationship given in question 1. (*Hint*: Calculate the percentage difference between the experimental values of the two sides of the equation given in question 1.)

3. In Figure 34.6, what is the equation that relates the currents I_1, I_2, I_3, and I_4?

4. State quantitatively how well your experimental results for the circuit of Figure 34.6 agree with the expected relationship given in question 3.

5. Are the experimental values of the currents for the entire laboratory generally larger or smaller than the theoretical values expected for the currents?

6. It was pointed out in the laboratory that some error might be caused by neglect of the internal resistance of the emf. Would the internal resistance cause an error in the direction shown in your answer to question 5? State your reasoning for the direction of any error caused by the internal resistance.

7. An ideal ammeter has zero resistance. Real ammeters have small but finite resistance. Would ammeter resistance cause an error in the proper direction to account for the direction of your error indicated in question 5? State your reasoning.

8. The connecting wires in the experiment are assumed to have no resistance, but in fact have a finite resistance. Would this error be in the proper direction to account for the direction of the error stated in your answer to question 5? State your reasoning.

Laboratory 35

Magnetic Induction of a Current Carrying Long Straight Wire

PRELABORATORY ASSIGNMENT

Read carefully the entire description of the laboratory and answer the following questions based on the material contained in the reading assignment. Turn in the completed prelaboratory assignment at the beginning of the laboratory period prior to the performance of the laboratory.

1. State the right-hand rule that relates the direction of the **B** field near a long straight wire to the direction of the current in the wire.

2. The direction of current is defined to be the direction in which _____ charges would flow.

3. State the equation that relates the magnitude of the **B** field near a long straight wire to the current I in the wire and the distance r from the wire.

   $B = $ _____

4. There is a current of 10.0 A in a long straight wire. What is the magnitude of the **B** field 5.00 cm from the wire? Show your work.

5. When a current that is constant in time passes through a wire, the **B** field that is produced around the wire is (a) time-varying, (b) constant in time, (c) negative, or (d) zero.

6. Why are measurements using the inductor coil not taken close to the wire? In other words, why do the measurements start 3.00 cm away from the wire?

7. Consider an imaginary line drawn perpendicular to a long straight wire carrying an alternating current. An inductor coil is placed near the wire to measure the voltage induced in the coil. The axis of the coil should be (a) parallel to the imaginary line, (b) parallel to the wire, (c) perpendicular to the imaginary line, or (d) parallel to the imaginary line and to the wire.

8. If an inductor coil is placed near a long wire carrying a current that is constant in time, the voltage induced in the coil is (a) positive, (b) negative, (c) zero, or (d) nonzero.

Magnetic Induction of a Current-Carrying Long Straight Wire

OBJECTIVES

When current exists in an infinitely long straight wire, a **B** field will exist in the region surrounding the wire. If the current is constant in time, the **B** field that exists will be constant in time at a given point. This constant **B** field can be detected by its effect on a small compass. If the current in the wire is time-varying, the **B** field that exists will also be time-varying. This time-varying **B** field can be detected by the electric field that it induces in a small inductor coil placed near the wire. In this laboratory, measurements on an apparatus with a long straight current-carrying wire will be used to accomplish the following objectives:

1. Determination of the direction of the **B** field surrounding a long straight wire using a compass

2. Confirmation that the direction of the **B** field near the wire is consistent with a right-hand rule for relating the current direction to the direction of the **B** field

3. Determination of the induced voltage in a small inductor coil placed near the long straight wire as a relative measurement of the **B** field

4. Demonstration that the magnitude of the **B** field surrounding a long straight wire decreases as $1/r$, where r is the perpendicular distance from the wire

EQUIPMENT LIST

1. Direct-current power supply (low voltage, 2 A), direct-current ammeter (2A)

2. Sine-wave generator (variable frequency up to 100 kHz, 5 V peak to peak)

3. Alternating-current digital voltmeter (measuring frequencies up to 100 kHz)

4. A 100-mH inductor coil (length $\cong$ 1 cm and inside diameter $\cong$ 1 cm)

5. Small compass; long straight wire apparatus. (Consists of a frame on which a continuous strand of wire is wrapped for 10 loops. The 10 strands are taped together over a length of approximately 40 cm to approximate a wire whose current is 10 times the current in a single strand of the wire. The apparatus can be placed with the long straight section parallel to the laboratory table or perpendicular to the table.)

THEORY

When a current I exists in an infinitely long straight wire, the lines of magnetic induction **B** are concentric circles surrounding the wire. At a perpendicular distance r from the wire, the **B** field is tangent to the circle as shown in Figure 35.1. The direction of the current I is perpendicular to the plane of the page and directed out of the page. The direction of the current is by definition the direction that positive charge would flow. The magnitude of the **B** field as a function of I and r is given by

$$B = \frac{\mu_0 I}{2\pi r} \tag{1}$$

where $\mu_0 = 4\pi \times 10^{-7}$ weber/amp-m, I is in amperes, and r is in meters. The units of B are weber/m², which has been given the name Tesla.

The direction of the **B** field relative to the current direction is given by the following right-hand rule. If the thumb of the right hand points in the direction of the current, the four fingers of the right hand curl in the direction of the **B** field. It is important to realize that the knowledge that the **B** field forms circles is assumed by this rule, and the rule only determines in which direction to take the tangent to the circles. Note that the tangents to the circles shown in Figure 35.1 representing the **B** field agree with this rule. Also note in Figure 35.1 that the length of the **B** vectors are drawn shorter for the larger circles to show that the **B** field decreases with distance from the wire as predicted by equation 1.

In a strict sense, the above statements apply only to an infinitely long straight wire. In this laboratory the straight portion of the wire is of some finite length L. For measurements made in the center of the wire within a perpendicular distance of L/4 from the wire, the finite wire will approximate an infinite wire.

If the current in the long straight wire is constant in time, the **B** field created by that current will be constant in time. For that case, the direction of the **B** field can be determined by observing the effect of the **B** field on a small compass placed in the vicinity of the long straight wire. A portion of the laboratory will use a compass to investigate the direction of the **B** field in this manner.

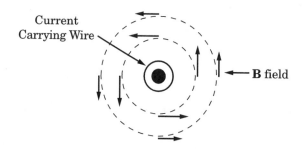

Figure 35.1 B field near wire carrying current perpendicular to page out of the page.

If the current in the long straight wire is an alternating current produced by a sine-wave generator, the **B** field surrounding the wire will also be time-varying, and it will alternate in direction and magnitude. If a small coil of 100-mH self-inductance is placed next to the wire, an alternating voltage will be induced in the coil. According to Faraday's law of induction, this induced voltage in the coil is proportional to the rate of change of the magnetic flux through the coil, and hence to the magnitude of the time-varying **B** field.

Therefore, a measurement of the voltage induced in the coil, as the coil is placed at different distances from the wire, provides a relative measure of the magnitude of the **B** field at different distances from the wire. Note carefully that the quantity actually measured is an alternating electric voltage, but its magnitude is proportional to the **B** field and will be taken to be a *relative* measurement of the **B** field at a given point.

EXPERIMENTAL PROCEDURE—DIRECTION OF THE *B* FIELD

1. Connect the circuit shown in Figure 35.2 using the direct-current power supply and the direct-current ammeter. Arrange the long-wire apparatus so that the outside long wire is in a horizontal plane along a north–south axis. Ask your instructor the direction of north in the laboratory room. Arrange the wire so that the direction of the current is from north to south. Determine the direction of the current by tracing the wires from the (+) terminal of the power supply. Have the circuit approved by your instructor to ensure that the current is in the proper direction.

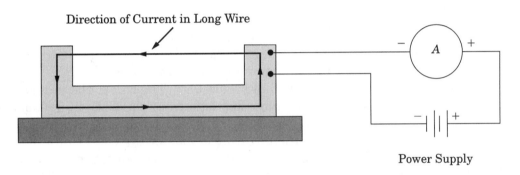

Figure 35.2 Long wire apparatus connected to Direct Current Supply.

2. Turn on the power supply and turn up the voltage until a current of 2.00 A is read on the ammeter. Do not exceed a current of 2.00 A.

3. Place the compass in the middle of the long-wire section directly above the wire as close to the wire as possible. State the direction (north, south, east, northeast, etc.) that the compass needle points. Record your answer in Data Table 1.

4. Place the compass in the middle of the long-wire section directly below the wire as close to the wire as possible. State the direction (north, south, east, northeast, etc.) that the compass needle points. Record your answer in Data Table 1.

5. Stand the long-wire apparatus on its end so that the current in the outside long wire is vertically downward. Place the compass next to the wire at the four positions indicated by the open circles in Figure 35.6 in the Laboratory Report section. The ⊗ represents the downward current viewed from above. In the open circles representing the four compass positions, draw an arrow showing the direction that the compass needle points.

EXPERIMENTAL PROCEDURE—*B* FIELD AS A FUNCTION OF DISTANCE

1. Connect the circuit shown in Figure 35.3 using the long-wire apparatus and the sine-wave generator. Turn the generator to maximum amplitude. Stand the long-wire apparatus on its end so that the outside long wire is vertical. Place in the apparatus the platform that serves to hold the the inductor coil.

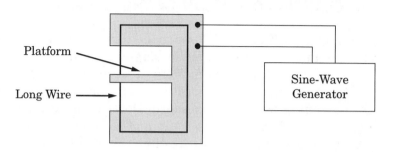

Figure 35.3 Long wire apparatus connected to the sine wave generator.

2. Connect the inductor coil to the digital voltmeter. Twist the leads about 10 to 15 times before connecting them between the inductor coil and the voltmeter. This is extremely important because it will minimize the voltage induced in the leads themselves and ensure that the voltage induced is in the inductor coil. Place the inductor coil on the platform as shown in Figure 35.4. The axis of the inductor coil should be perpendicular to an imaginary line (shown as the dotted line labeled 1 in the figure) that is perpendicular to the current-carrying wire. The inductor coil was shown in three different positions with the axis of the coil at different distances r_1, r_2, and r_3 from the wire. At each position of the inductor coil shown, the **B** field will alternate in opposite directions along the axis of the coil. The coil is chosen to be short ($\cong$ 1 cm) and of small cross section (diameter $\cong$ 1 cm) because for that choice, the **B** field lies approximately along the coil axis and is approximately uniform over the cross section of the coil.

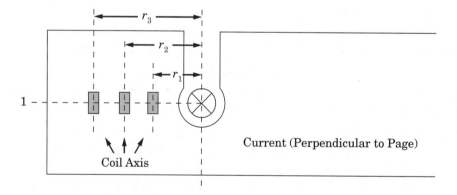

Figure 35.4 View of the platform looking down from above. The current is perpendicular to the page alternating into and out of the page.

3. The amplitude of the induced voltage on the digital voltmeter will depend on the frequency of the sine-wave generator. With the inductor about 3 cm from the wire, with its axis positioned as shown in Figure 35.4, vary the frequency of the

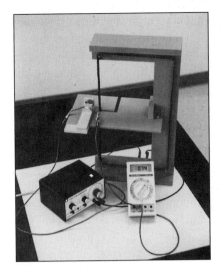

Figure 35.4 Homemade long-wire apparatus with inductor coil in position.

generator until the maximum voltage is read on the digital voltmeter. Once this frequency is found, do not change the frequency. Make all measurements at this frequency.

4. Measure the voltage induced in the inductor coil as a function of r (Figure 35.5). The quantity r is the distance from the center of the coil to the center of the wire. Take data from $r = 3.0$ cm to $r = 9.0$ cm in increments of 1 cm. The reason that data are not taken for r less than 3 cm is the fact that at distances close to the wire, the **B** field is extremely nonuniform over the coil cross section. Record the values of the voltage in the Data Table under the column labeled B (Trial 1). If this were a true measure of the **B** field, the units would be Tesla. Since the measured quantity is really a voltage proportional to B, no units are stated.

5. Repeat step 4 two more times, measuring the induced voltage as a function of distance and recording the values in the Data Table under Trial 2 and Trial 3.

CALCULATIONS

1. Calculate the mean and standard error for the three trials of B and record them as $\overline{B}$ and αB in the Calculations Table.

2. Calculate the percent standard error at each point by calculating $\alpha_B/\overline{B}$ and expressing it as a percentage. Record the values in the Calculations Table.

3. Calculate the value of $1/r$ for each of the values of r and record them in the Calculations Table.

4. Perform a linear least squares fit to the data of $\overline{B}$ versus $1/r$ with $\overline{B}$ as the ordinate and $1/r$ as the abscissa. Record the value of the slope, intercept, and the correlation coefficient r.

GRAPHS

Make a graph of the data for $\overline{B}$ versus $1/r$ with $\overline{B}$ as the ordinate and $1/r$ as the abscissa. Also show on the graph the straight line obtained from the least squares fit.

Laboratory 35

Magnetic Induction of a Current-Carrying Long Straight Wire

LABORATORY REPORT

Data Table 1

With compass above wire compass direction =
With compass below wire compass direction =

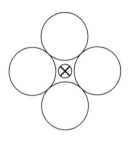

Figure 35.6 Indicate the compass direction at the positions shown.

Data Table 2

r (cm)	B Trial 1	B Trial 2	B Trial 3
3.00			
4.00			
5.00			
6.00			
7.00			
8.00			
9.00			

Calculations Table

$1/r$ (cm^{-1})	$\overline{B}$	α_B	$\% \ \alpha_B$

Slope = _____

Intercept = _____

r = _____

SAMPLE CALCULATIONS

QUESTIONS

1. Are your answers to the questions in Data Table 1 about the direction in which the compass needle points consistent with the right-hand rule for direction of the **B** field?

2. Evaluate the precision of the measurements of the induced voltage as a function of distance from the wire. Consider the percent standard error of the measurements in your evaluation.

3. State the extent to which your measurements confirm the expectation that B is proportional to $1/r$. Give the evidence for your evaluation of this question.

4. When the direct current is 2.00 A in a single wire of the bundle of 10 wires, the total current in the bundle of wire that approximates the long straight wire is 20.0 A. What is the magnitude of the **B** field 3.00 cm from this long straight wire carrying a current of 20.0 A? What is the magnitude of the **B** field 9.00 cm from the wire carrying 20.0 A?

5. A constant current is in a long straight wire in the plane of the paper in the direction shown below by the arrow. Point X is in the plane of the paper above the wire, and point Y is in the plane of the paper but below the wire. What is the direction of the **B** field at point X? What is the direction of the **B** field at point Y?

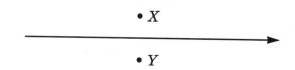

Direction at $X =$ _____

Direction at $Y =$ _____

Laboratory 36

Alternating-Current LR Circuits

PRELABORATORY ASSIGNMENT

Read carefully the entire description of the laboratory and answer the following questions based on the material contained in the reading assignment. Turn in the completed prelaboratory assignment at the beginning of the laboratory period prior to the performance of the laboratory.

1. For a resistor in a series alternating-current circuit, the phase relationship between the current in the resistor and the voltage across the resistor is (a) the current leads the voltage by 90°, (b) the voltage leads the current by 90°, (c) the current is in phase with the voltage, or (d) the current is at some phase angle ϕ relative to the voltage (ϕ is dependent on the circuit parameters).

2. For an inductor in a series alternating-current circuit, the phase relationship between the current in the inductor and the voltage across the inductor is (a) the current leads the voltage by 90°, (b) the voltage leads the current by 90°, (c) the current is in phase with the voltage, or (d) the current is at some phase angle ϕ relative to the voltage (ϕ is dependent on the circuit parameters).

3. For a generator in a series alternating-current circuit, the phase relationship between the generator voltage and the current in the generator is (a) the current leads the voltage by 90°, (b) the voltage leads the current by 90°, (c) the current is in phase with the voltage, or (d) the current is at some phase angle ϕ relative to the voltage (ϕ is dependent on the circuit parameters).

4. If a generator has a maximum voltage of 5.00 V, what is the root-mean-square voltage of the generator? Show your work.

   $V_{rms} = $ _____

5. A 2.50-mH inductor has an rms voltage of 15.0 V across it at a frequency $f = 200$ Hz. What is the rms current in the inductor? Show your work.

6. A pure inductor L and a pure resistor R are in series with a generator of voltage V. The voltage across the inductor is $V_L = 10.0$ V. The voltage across the resistor is 15.0 V. What is the voltage V of the generator? Show your work.

$V = $ _____

7. A 500-Ω resistor and a real inductor whose pure inductance is L and whose internal resistance is r are in series with a generator whose voltage is $V = 10.0$ V and whose angular frequency is $\omega = 1000$ rad/s. The voltage across the real inductor is measured to be 4.73 V, and the voltage across the 500-Ω resistor is measured to be 6.57 V. What is the value of L and r? (*Hint*: This is the measurement to be performed in this laboratory exercise. Use equation 7 to find ϕ, use equations 8 and 9 to find V_L and V_r, and then use equations 11 and 12 to find ωL and r. Finally find L from the known value of ω.) Show your work.

OBJECTIVES

If a sinusoidally varying source of emf with a frequency f is placed in series with a resistor and a pure inductor, the current I varies with time but is the same in each element of the circuit at any given instant. This current I will vary with the same frequency f as the generator, but it will be shifted in phase by an angle ϕ relative to the generator. The voltage across each of the circuit elements has its own characteristic phase relationship with the current. The resistor voltage V_R is in phase with the current I, the inductor voltage V_L leads the current by a phase angle of 90°, and the generator voltage leads the current by an angle ϕ, whose value is dependent on the circuit parameters. Measurements of the voltage across each element in a series circuit of an inductor, a resistor, and a sine-wave generator will be used to accomplish the following objectives:

1. Demonstration of the relative phase angle of the voltage across each component in the circuit

2. Determination of the phase angle of the generator current relative to the generator voltage

3. Demonstration that real inductors consist of both inductance and resistance, and that they can be represented by a pure inductor L in series with a pure resistance r

4. Determination of the value of L and r for an unknown inductor

EQUIPMENT LIST

1. Sine-wave generator (variable frequency, 5 V peak-to-peak amplitude)

2. Resistance box

3. A 100-mH inductor (resistance $\cong 350\ \Omega$ to serve as an unknown)

4. Alternating-current voltmeter (digital readout, capable of measuring high frequency)

5. Compass and protractor

THEORY

Consider first the two circuits shown in Figure 36.1 in which a sine-wave generator of frequency f is connected separately to resistor R and then to a pure inductance L. The generator is assumed to have a maximum voltage of V and will thus produce a maximum voltage of V across the resistor in circuit (a). It will also produce a maximum voltage of V across the inductor in circuit (b). The voltage across the resistor is related to the current by a relationship like that for direct current circuits which is

$$V_R = IR \tag{1}$$

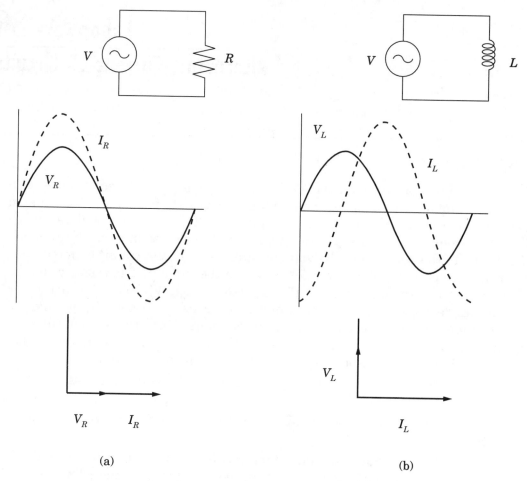

Figure 36.1 Generator and Resistor and Generator and Inductor. Phase relationships between the voltage and current and phasor diagrams of the phase relationships.

If L is the inductance (in units of H) and ω is the angular frequency of the generator ($\omega = 2\pi f$) in rad/s, then the following relationship exists between the voltage V and the current I

$$V_L = I\,\omega L \tag{2}$$

The quantity ωL is called the "inductive reactance," and it has units of Ω. For example, if a 1.50-mH inductor has a voltage of 10.0 V across it at a frequency of 100 Hz, the current is $I = V_L/\omega L = 10.0/2\pi(100)(1.50 \times 10^{-3}) = 10.6$ A.

When an alternating current or voltage is measured in the laboratory on a meter, the number read for the current or voltage must be some time-averaged value. Meters are normally calibrated in such a way that they respond to the root-mean-square value of the current or voltage. A root-mean-squared value of voltage can be designated as V_{rms}. The relationship between V_{rms} and V the maximum voltage is

$$V_{\text{rms}} = \frac{\sqrt{2}}{2} V = 0.707\,V \tag{3}$$

Note that equations 1 and 2 refer to either maximum values or rms values. If the voltage is expressed as a maximum value of voltage in those equations, then the current will be a maximum value. If the voltage is an rms value, then the current will be an rms value. In this laboratory the only quantities measured will be voltage, and all the measurements will be rms values. No subscripts will be put on the symbols, but it is understood that all measured values of voltage are rms values.

Also shown in Figure 36.1 below each circuit is a graph of the current and voltage across the element for one full period. The graph for the case of the resistor indicates that the resistor current I_R and the resistor voltage V_R are in phase. For the inductor the graph shows that the inductor current I_L and the inductor voltage V_L are 90° out of phase, with the voltage leading the current by 90°. The graph below also shows that the voltage reaches its maximum value one quarter of a period before the current reaches its maximum value.

Shown at the bottom of Figure 36.1 is a diagram called a "phasor diagram," whose purpose is also to show the phase relationship. The phasors are vectors drawn with length proportional to the value of the represented quantity, and they are assumed to be rotating counterclockwise with the frequency of the generator. At any time, a projection of one of the rotating vectors on the y-axis is the instantaneous value of that quantity. Since the resistor current and voltage are in phase, they are shown in the same direction, and therefore their projection of the y-axis is always in phase. For the inductor the vector representing the inductor voltage is 90° ahead of the vector representing the current when they are both assumed to be rotating counterclockwise.

Consider now the circuit obtained by placing a pure inductance L having no resistance and a resistor R in series with a sine-wave generator of voltage V shown in Figure 36.2. For this circuit, the current I is the same at every instant of time in all three circuit elements. Also given in Figure 36.2 is a phasor diagram in which only the voltages are shown. The phasor representing the current (which is not shown) would be in the direction of the phasor labeled V_R because the current and the resistor voltage are in phase. Note that the inductor voltage V_L is 90° ahead of the resistor voltage V_R, and the generator voltage is angle ϕ ahead of V_R. This phasor diagram shows that the generator voltage V is the vector sum of V_R and V_L. In equation form the phasor diagram states

$$V = \sqrt{V^2{}_L + V^2{}_R} \tag{4}$$

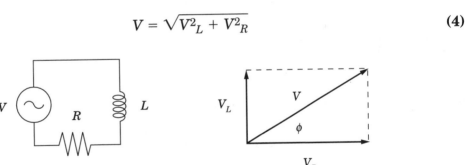

Figure 36.2 Series circuit of Resistor and Inductor and associated phasor diagram.

From an examination of the phasor diagram it is also clear that the phase angle ϕ is related to the voltages V_L and V_R, and thus to the resistance R and ωL through equations 1 and 2. The relationship is given by

$$\tan \phi = \frac{V_L}{V_R} = \frac{\omega L}{R} \tag{5}$$

Note that equation 5 is strictly valid only for a pure inductor that has no resistance, but it can be used as an approximation if the resistance of the inductor is negligible compared to the resistor R.

In fact, real inductors do have both an inductance L and an internal resistance r. Essentially, a real inductor can be represented by a pure inductance L in series with a pure resistance r. In Figure 36.3 a real inductor is shown in series with a resistor R and a generator of voltage V. The points A, B, and C in the figure represent the three points between which a voltmeter can be placed in the circuit in order to measure a voltage. The voltage between points A and B is the generator voltage V, and the voltage between A and C is the resistor voltage V_R. Between the points B and C is the combined voltage across the inductance L and the internal resistance r. This voltage will be referred to as V_{ind}. There is, of course, some voltage V_L across L alone, and some voltage V_r across r alone. However, there can be no direct measurement of V_L or V_r. The only quantity that can be measured is V_{ind}, which is the vector sum of V_L and V_r.

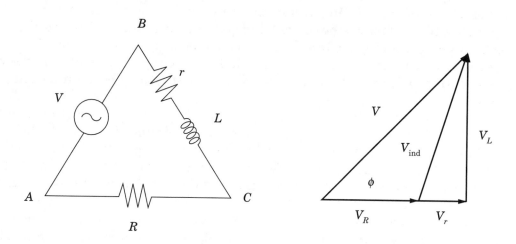

Figure 36.3 Series circuit of real inductor with internal resistance r, a resistor R, and a generator of voltage V. Also shown is the associated phasor diagram of the voltages.

A phasor diagram for the circuit is also shown in Figure 36.3. Applying the law of cosines to the triangle formed by V, V_R, and V_{ind} leads to

$$V^2_{ind} = V^2 + V^2_R - 2VV_R \cos \phi \qquad (6)$$

Solving equation 6 for $\cos\phi$ leads to the following:

$$\cos \phi = \frac{V^2 + V^2_R - V^2_{ind}}{2VV_R} \qquad (7)$$

Equation 7 states that the angle ϕ can be determined by a measurement of V, V_R, and V_{ind}.

An examination of the phasor diagram in Figure 36.3 shows that the unknown voltages V_L and V_r can be determined from V, V_R, and ϕ by the following:

$$V_L = V \sin\phi \qquad (8)$$

$$V_r = V \cos\phi - V_R \qquad (9)$$

The current I is the same in all the elements of the circuit, and it can be related to the voltage across each element by the following equations:

$$V_L = I\omega L \qquad V_R = IR \qquad V_r = Ir \qquad (10)$$

Once V_L and V_r have been determined from equations 8 and 9, the equations in 10 can be used to solve for ωL and r by eliminating I to get

$$\omega L = R\frac{V_L}{V_R} \tag{11}$$

$$r = R\frac{V_r}{V_R} \tag{12}$$

EXPERIMENTAL PROCEDURE

1. Connect the inductor in series with the sine-wave generator and a resistance box to form a circuit like that of Figure 36.3. Set the generator to maximum voltage and a frequency of 800 Hz. Set the resistance box to a value of 400 Ω and record that value as R in the Data Table.

2. Using the AC voltage scale on the voltmeter, very carefully measure the generator voltage V, the inductor voltage V_{ind}, and the voltage across the resistor V_R. Record these values in the Data Table.

3. Repeat steps 1 and 2 for values of R of 600, 800, and 1000 Ω. Do not assume that the generator voltage stays the same. Even though the voltage setting is left at the maximum setting, the generator output might change slightly in response to the changes in R. Therefore, be sure to measure all three voltages for each value of R.

4. Make careful note of the particular inductor you used. You may need to identify it and use it again in other laboratory exercises involving an inductor in an alternating current circuit. The values that are determined for L and r should be noted and saved along with the means of identification of the inductor.

CALCULATIONS

1. From the known value of the frequency f, calculate and record in the Calculations Table the value of the angular frequency ω ($\omega = 2\pi f$).

2. Using equation 7, calculate the value of $\cos\phi$ and then of ϕ for each case and record the values in the Calculations Table.

3. Using equations 8 and 9, calculate V_L and V_r for each case and record the values in the Calculations Table.

4. Using equations 11 and 12 calculate ωL and r for each of the four cases and record the values in the Calculations Table.

5. Calculate the four values for L from the four values of ωL and the known value of ω and record them in the Calculations Table.

6. Calculate the mean and standard errors for the four values of r and the four values of L and record them in the Calculations Table as $\bar{r}$, $\bar{L}$, α_r, and α_L.

GRAPHS

Construct to scale a phasor diagram like the one shown in Figure 36.4 for each of the four cases. Use one sheet of graph paper and make four separate diagrams on the one sheet of paper. Choose some scale (for example, 1.00 V = 1.00 cm) so that the diagrams are as large as possible, and that each one fits on one fourth of the sheet of paper. First, construct a vector along the x-axis whose length is scaled to the magnitude of V_R as shown in Figure 36.4. Using a compass construct an arc from the end of V_R whose radius is the length of the scaled value of V_{ind}. Finally, construct an arc from the beginning of V_R whose radius is the length of the scaled value of V. The intersection of the two arcs is the intersection of V_{ind} and V, and those two vectors can then be drawn in their proper direction as shown in part (b) of the figure. Finally V_L and V_r can be constructed as shown in part (c) of the figure by dropping a perpendicular from the intersection of the arcs to the x-axis and extending a vector from the end of V_R.

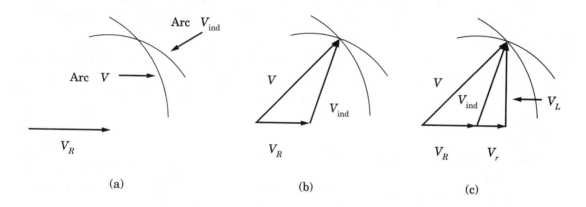

Figure 36.4 Phasor diagram construction.

LABORATORY REPORT

Data Table

	$\omega =$		rad/s	
R (Ω)				
V (V)				
V_{ind} (V)				
V_R (V)				

Calculations Table

$\cos\phi$				
ϕ (degrees)				
V_L (V)				
V_r (V)				
ωL (Ω)				
r (Ω)				
L (H)				

$\bar{r} =$ _____ Ω $\alpha_r =$ _____ Ω $\bar{L} =$ _____ H $\alpha_L =$ _____ H

SAMPLE CALCULATIONS

QUESTIONS

1. Comment on the precision of your measurement of L and r. State the evidence for your comments.

2. Examine the phasor diagrams that you have constructed. Using a protractor measure the angle ϕ of the constructed triangle of V, V_R, and V_{ind}. Compare it with the calculated value of ϕ for each of the phasor diagrams. Calculate the percentage error in the value of ϕ from the diagram compared to the calculated value.

3. If your inductor was used in a series circuit with a resistance of $R = 10,000\ \Omega$ and a generator of $\omega = 100,000$ rad/s, what would be the phase angle ϕ? (*Hint:* The resistance of the inductor would be negligible.)

4. Consider the circuit that you measured with $R = 600\ \Omega$. Calculate the value of the current from each of the three equations 10 and compare their agreement.

Laboratory 37

Alternating Current RC and LCR Circuits

PRELABORATORY ASSIGNMENT

Read carefully the entire description of the laboratory and answer the following questions based on the material contained in the reading assignment. Turn in the completed prelaboratory assignment at the beginning of the laboratory period prior to the performance of the laboratory.

1. In a series RC circuit such as the one in Figure 37.1, the following phase relationship exists between the generator voltage V, the capacitor voltage V_C, and the resistor voltage V_R. (a) V and V_R are in phase, V_C lags V_R by 90°; (b) V_C and V_R are at angle ϕ, V leads V_C by 90°; (c) V_C lags V_R by 90°, V lags V_R by angle ϕ; or (d) V_R lags V_C by ϕ, V_C leads V by 90°.

2. If a series RC circuit has $V_R = 12.6$ V and $V_C = 10.7$ V, the generator voltage must be (a) 12.6 V, (b) 23.3 V, (c) 1.9 V, or (d) 16.5 V. Show your work.

3. A series RC circuit has $\omega = 2000$ rad/s and $R = 300$ Ω. The voltage V_C is measured to be 4.76 V, and V_R is measured to be 6.78 V. What is the value of C? Show your work. (a) 2.37 μF, (b) 5.00 μF, (c) 1.17 μF, or (d) 4.67 μF.

4. A series RC circuit of $R = 500$ Ω and $C = 3.00$ μF is measured to have $V_R = 8.07$ V and $V_C = 6.68$ V. What is the current I, and what is the value of the angular frequency ω? Show your work.

   $I = $ _____ A $\omega = $ _____ rad/s

5. A series *LCR* circuit consists of an inductor of inductance L and resistance r, a capacitor C, a resistor R, and a generator of voltage V. Mark as true or false the following statements concerning the relative phase of V, V_L, V_r, V_C, and V_R.

    —————— 1. V_r and V_R are in phase.

    —————— 2. V_L leads V_C by 90°.

    —————— 3. V is at angle ϕ relative to V_R.

    —————— 4. $V_L - V_C$ is in phase with V_r.

    —————— 5. V_C lags V_R by 90°.

6. Measurements on the circuit described in question 5 give $V_L = 10.76$ V, $V_C = 5.68$ V, $V_R = 6.32$ V, and $V_r = 3.75$ V. What is the generator voltage V? Show your work.

$V = $ _____ V

7. An inductor with $L = 150$ mH and $r = 200\ \Omega$ is in series with a capacitor, a resistor, and a generator of $\omega = 1000$ rad/s. The voltage across the inductor V_{ind} is measured to be 10.87 V, V_R is measured to be 4.65 V, and V_C is measured to be 5.96 V. What is the generator voltage V? (*Hint*: This is the measurement to be performed in this laboratory for *LCR* circuits. Use equations 7 and 8 to find V_L and V_r, and then use equation 4 to find V.) Show your work.

Alternating-Current RC and LCR Circuits

This laboratory assumes that Laboratory 36 has previously been performed, and that the measurements made of the inductance and resistance of an inductor coil have been recorded and retained for use in this laboratory. That same inductor is to be used in the *LCR* circuit, which is to be constructed in this laboratory.

OBJECTIVES

If a sinusoidally varying source of emf with a frequency f is placed in series with a resistor and a capacitor, the current I varies with time but is the same in each element of the circuit at any given instant. The current I will vary with the same frequency f as the generator, but it will be shifted in phase by an angle ϕ relative to the generator. The voltage across each of the circuit elements has its own characteristic phase relationship with the current. The resistor voltage V_R is in phase with the current I, the capacitor voltage V_C lags the current by a phase angle of 90°, and the generator voltage V lags the current by the angle ϕ whose value is dependent on the circuit parameters.

If an inductor of inductance L and resistance r is now added to the above circuit to form a series *LCR* circuit, the phase relationship for the resistor and capacitor are the same, and the inductor voltage has two components. One component is the voltage V_r across the resistance r, and it is in phase with the current. The other component V_L is the voltage across the inductance L, and it leads the current by 90°.

Measurements of the voltage across each element in a series *RC* circuit and of the voltage across each element in a series *LCR* circuit will be used to accomplish the following objectives:

1. Demonstration that the voltage across the resistor V_R and the voltage across the capacitor V_C are 90° out of phase in an *RC* circuit

2. Determination of the value of the capacitance of a capacitor in an *RC* circuit

3. Verification of the phase relationships among the voltages across the resistor, the capacitor, and the inductor in an *LCR* circuit

EQUIPMENT LIST

1. Sine-wave generator (variable frequency, 5 V peak-to-peak amplitude)
2. Resistance box
3. A 100-mH inductor (same one used in Laboratory 36 whose value of L and r have already been determined)
4. A 1.00-μF capacitor
5. AC voltmeter (digital readout, capable of measuring high frequency)
6. Compass and protractor

THEORY—RC CIRCUIT

Consider a series circuit consisting of a capacitor C, a resistor R, and a sine-wave generator of frequency f as shown in Figure 37.1. Also shown in the figure is a phasor diagram for the generator voltage V, the voltage across the resistor V_R, and the capacitor voltage V_C. It is assumed that all voltages discussed in this laboratory are root-mean-squared values. Since the voltages V_R and V_C are 90° out of phase, the voltages V, V_R, and V_C form a right triangle as shown in Figure 37.1. Therefore, the equation relating the magnitudes of the measured voltages in an RC circuit is

$$V = \sqrt{V^2_C + V^2_R} \tag{1}$$

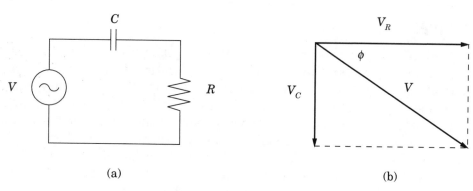

(a)   (b)

Figure 37.1 *RC circuit and the phasor diagram for the voltage across each element.*

Note that equation 1 is valid only for a pure capacitor with no resistive component. If measurements on a real capacitor show agreement with equation 1, it would indicate that the capacitor has no significant resistive component.

In an RC circuit, the current I is the same in each element of the circuit, and the relationships between the voltage and the current for the resistor and capacitor are

$$V_R = IR \quad \text{and} \quad V_C = I\left(\frac{1}{\omega C}\right) \tag{2}$$

The quantity $1/\omega C$ is called the "capacitive reactance," and it has units of ohms. If the current is eliminated between the two equations in 2, an equation for C is given by

$$C = \left(\frac{1}{\omega R}\right)\left(\frac{V_R}{V_C}\right) \tag{3}$$

Thus, a value for the capacitance of an unknown capacitor can be determined from equation 3 if ω and R are known and V_R and V_C are measured.

EXPERIMENTAL PROCEDURE—RC CIRCUIT

1. Connect the capacitor provided in series with the sine-wave generator and a resistance box to form a circuit like that shown in Figure 37.1. Set the generator output to maximum voltage and set the frequency to 250 Hz. Record the value of f in Data Table 1. Set the resistance box to a value of 300 Ω and record that value as R in Data Table 1.

2. Using the AC voltage scale on the voltmeter, very carefully measure the generator voltage V, the capacitor voltage V_C, and the voltage across the resistor V_R. Record those values in Data Table 1.

3. Repeat the above procedure for three other cases using $R = 500$, 700, and $900\ \Omega$. Do not assume that the generator voltage stays the same. Even though the voltage setting is left at the maximum setting, the generator output might change slightly in response to changes in R. Therefore, be sure to measure all three voltages for each value of R.

4. Obtain a value for the capacitance of your capacitor from your instructor. Record this known value as C_k in Data Table 1.

CALCULATIONS—RC CIRCUIT

1. From the known value of the frequency f, calculate the value of the angular frequency ω ($\omega = 2\pi f$). Record the value of ω in Calculations Table 1.

2. Calculate the quantity $\sqrt{V^2_C + V^2_R}$ for each case and record the values in Calculations Table 1.

3. Calculate the percentage error in the quantity $\sqrt{V^2_C + V^2_R}$ compared to each of the measured values of the generator voltage V. Record these percentage errors in Calculations Table 1.

4. Using equation 3, calculate the values of C from the measured values of V_R and V_C for each value of R. Record each value of C in Calculations Table 1. Calculate the mean $\overline{C}$ and the standard error α_C for the four values of C and record them in Calculations Table 1.

5. Calculate the percentage error in the value of $\overline{C}$ compared to the known value of the capacitance C_k.

THEORY—LCR CIRCUIT

Consider a series LCR circuit shown in Figure 37.2, with a generator of voltage V, a resistor R, a capacitor C, and an inductor having inductance L and resistance r. Note that the capacitor is assumed to have no resistance. Also shown in Figure 37.2 is the phasor diagram for the voltages V, V_R, V_C, V_L, and V_r. The figure shows that V_L and V_C are 180° out of phase, and V_R and V_r are in phase. Therefore, the quantities $V_L - V_C$, V, and $V_R + V_r$ form a right triangle, and thus they obey the following relationship:

$$V = \sqrt{(V_L - V_C)^2 + (V_R + V_r)^2} \tag{4}$$

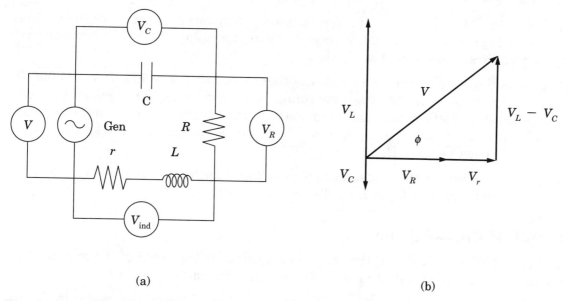

(a)   (b)

Figure 37.2 *LCR* circuit with voltmeter in the four positions to measure voltage across each element of the circuit. Also shown is a phasor diagram of all the relevant voltages.

It is not possible to measure either V_L or V_r directly. The only voltage associated with the inductor that can be experimentally measured is shown in Figure 37.2 as V_{ind}, and it is the vector sum of the voltages V_L and V_r. The relationship between V_{ind}, V_L, and V_r is shown in Figure 37.3. It is clear from that figure that the voltages V_{ind}, V_L, and V_r obey the following relationship:

$$V^2_{ind} = V^2_L + V^2_r \tag{5}$$

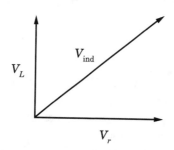

Figure 37.3 Phasor diagram for the voltages acrosss the components of the inductor.

Since the current I is the same in each element of the circuit, the voltages V_L and V_r can be expressed in terms of the current I as

$$V_L = I\,(\omega L) \quad \text{and} \quad V_r = Ir \tag{6}$$

The quantity ωL is called the "inductive reactance," and it has units of ohms. If equations 5 and 6 are combined, and the current I is eliminated it can be shown that

$$V_L = \frac{V_{ind}\,\omega L}{\sqrt{(\omega L)^2 + r^2}} \tag{7}$$

$$V_L = \frac{V_{ind}\,r}{\sqrt{(\omega L)^2 + r^2}} \tag{8}$$

Assuming that ω, L, and r are known, equations 7 and 8 can be used to determine V_L and V_r if V_{ind} is measured. These values of V_L and V_r combined with measured values of V_C and V_R can be used in equation 4 to verify the relationship between these quantities and the measured generator voltage V.

EXPERIMENTAL PROCEDURE—LCR CIRCUIT

1. Construct a series *LCR* circuit like the one shown in Figure 37.2 using the same capacitor used previously, the inductor used in Laboratory 36 (whose L and r were determined in that laboratory), a resistance box, and the sine-wave generator. Record the values of L and r in Data Table 2.

2. Set the generator output to maximum voltage and set the frequency to 800 Hz. Set the resistance box to a value of 200 Ω. Record the values of f and R in Data Table 2.

3. Using the AC voltage scale on the voltmeter, very carefully measure the generator voltage V, the capacitor voltage V_C, the resistor voltage V_R, and the inductor voltage V_{ind}. Record these values in Data Table 2.

4. Repeat the procedure of steps 1 through 3 with the generator frequency set to $f = 600$ Hz and the resistance box set to $R = 200$ Ω.

5. Repeat the procedure of steps 1 through 3 two more times, once with $R = 300$ Ω and $f = 600$ Hz, and again with $R = 300$ Ω and $f = 800$ Hz.

CALCULATIONS—LCR CIRCUIT

1. From the known value of f for each case, calculate the angular frequency ω ($\omega = 2\pi f$) and record the values in Calculations Table 2.

2. Using equations 7 and 8, calculate the four values of V_L and V_r and record them in Calculations Table 2.

3. Calculate and record in Calculations Table 2 the four values of $V_L - V_C$ and $V_R + V_r$.

4. For each of the four cases calculate and record in Calculations Table 2 the value of the quantity $\sqrt{(V_L - V_C)^2 + (V_R + V_r)^2}$.

5. Calculate the percentage error in the quantity $\sqrt{(V_L - V_C)^2 + (V_R + V_r)^2}$ compared to the measured value of the generator voltage V. Record the values of those percentage errors in Calculations Table 2.

LABORATORY REPORT

Data Table 1

$f =$ _____ Hz		$C_k =$ _____ F	
$R\ (\Omega)$			
$V_C\ (V)$			
$V_R\ (V)$			
$V\ (V)$			

Calculations Table 1

$\omega =$ _____ rad/s			
$\sqrt{V^2_R + V^2_C}$			
% error			
$C\ (F)$			

$\overline{C} =$	$\alpha_C =$	% Error $\overline{C} =$

Data Table 2

$r =$ _____ Ω		$L =$ _____ H	
$f\ (Hz)$			
$R\ (\Omega)$			
$V\ (V)$			
$V_{ind}\ (V)$			
$V_C\ (V)$			
$V_R\ (V)$			

Calculations Table 2

ω (rad/s)				
$V_L\ (V)$				
$V_r\ (V)$				
$V_L - V_C\ (V)$				
$V_R + V_r\ (V)$				
$\sqrt{(V_L - V_C)^2 + (V_R + V_r)^2}$				
% error				

QUESTIONS

1. Comment on the agreement between the measured generator voltage V and the quantity $\sqrt{V^2{}_C + V^2{}_R}$ for the RC circuit data.

2. Do your results for question 1 confirm that the capacitor has no resistance? State specifically how the data either do or do not confirm this expectation.

3. Evaluate the precision of the measurements of the value of the capacitor in the RC circuit.

4. Considering the given value of C_k as the true value, comment on the accuracy of your measurements of the capacitance.

5. Comment on the agreement between the measured generator voltage V and the quantity $\sqrt{(V_L - V_C)^2 + (V_R + V_r)^2}$ in the LCR circuit.

6. Do your results confirm the phasor diagram of Figure 37.2 as a correct model for the addition of the voltages in an LCR circuit?

PRELABORATORY ASSIGNMENT

Read carefully the entire description of the laboratory and answer the following questions based on the material contained in the reading assignment. Turn in the completed prelaboratory assignment at the beginning of the laboratory period prior to the performance of the laboratory.

1. Describe the components that make up the electron gun in a cathode-ray tube.

2. Describe the voltage waveform that produces a linear time scale when applied to the horizontal plates of a cathode-ray tube.

3. When the electron beam strikes the fluorescent screen the phospor glow that results has a persistence. Approximately how long does the glow persist?

4. A function generator outputs a sine wave of $f = 200$ Hz. It is input to an oscilloscope set at 1 ms/DIV. How many complete cycles of the sine wave are displayed on the oscilloscope? (*Hint:* The period of the sine wave T is related to the frequency f of the wave by $T = 1/f$, and there are 10 divisions on the time display of the oscilloscope.)

5. A typical student oscilloscope on its least sensitive calibrated scale can display a voltage up to a maximum of approximately (a) 1 V, (b) 5 V, (c) 20 V, or (d) 200V.

6. A typical student oscilloscope on its most sensitive calibrated scale can display a voltage up to a minimum of approximately (a) 1 mV, (b) 5 μV, (c) 20 V, or (d) 200 mV.

7. Which statement best describes the function of the trigger level and trigger slope settings of an oscilloscope? (a) The trigger level sets a slope that must be exceeded to start the sweep. (b) The trigger level sets a voltage above which the sweep is started if the slope is positive, and below which the sweep is started if the slope is negative. (c) The trigger slope sets a voltage above which the sweep is triggered. (d) The trigger slope sets a voltage above which the sweep is started for positive level, and below which the sweep is started for negative level.

8. A sawtooth wave whose period is 100 ms is applied to an oscilloscope whose screen is 10 cm wide. What time is represented by 1 cm on the screen?

OBJECTIVES

In this laboratory signals from a function generator will be observed on an oscilloscope to accomplish the following objectives:

1. Introduction of students to the fundamental principles and practical operation of the oscilloscope

2. Measurement of sine and other waveform signals of varying voltage and frequency

3. Comparison of voltage measurements with the oscilloscope to voltage measurements using an AC voltmeter

EQUIPMENT LIST

1. Oscilloscope (typically, bandwidth DC to 20 Mhz)

2. Function generator (sine wave plus additional waveform such as a square wave or triangular wave)

3. AC voltmeter (capable of measuring high frequency)

4. Appropriate connecting wires (probably BNC to banana plug)

THEORY

The fundamental working part of an oscilloscope is a cathode-ray tube (CRT). Its components include: (1) a heated filament to emit a beam of electrons, (2) a series of electrodes to accelerate, focus, and control the intensity of the emitted electrons, (3) two pairs of deflection plates that deflect the electron beam when there is a voltage between the plates (one pair for deflection in the horizontal direction and one pair for deflection in the vertical direction), (4) a fluorescent screen that emits a visible spot of light at the point where the beam of electrons strikes the screen. Together the heated filament and series of electrodes in (1) and (2) above are called an "electron gun." The electron gun and deflecting plates are arranged linearly inside an evacuated glass tube, and the fluorescent screen coats the glass tube at the opposite end of the tube from the electron gun as shown in Figure 38.1.

When there is no voltage between either pair of deflection plates, the electron beam will travel straight down the evacuated tube and strike the center of the fluorescent screen. If a constant voltage is applied between either the horizontal or vertical deflection plates the beam will be displaced a constant amount on the fluorescent screen in either the horizontal (x) or vertical (y) direction. The direction of the displacement depends on the sign of the voltage, and the magnitude of the displacement is proportional to the voltage. If a time-varying voltage is applied to either set of deflecting plates, the displacement of the beam will vary with time as the applied voltage varies with time, and the electron beam spot will move on the screen as a function of time. When the beam strikes the screen the phosphor glow persists for approximately 0.1 s.

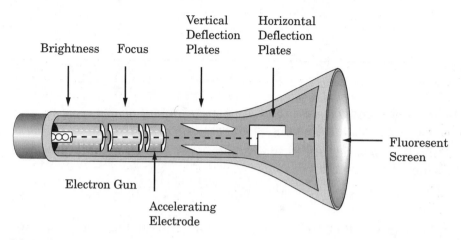

Figure 38.1 Cathode-ray tube.

Deflections of the electron beam in the horizontal (x) direction can be made to represent a time scale by applying a time-varying sawtooth voltage waveform as shown in Figure 38.2. When a voltage of that waveform and the appropriate maximum voltage is applied to the horizontal plates, the beam spot will sweep across the fluorescent screen once each time the voltage linearly increases from its minimum up to its maximum. At the end of sweep of the beam across the screen, the sawtooth voltage goes back to zero, which returns the beam back to the left of the screen. The time this takes will equal the period T of the sawtooth waveform. Since this sawtooth waveform sweeps the beam across the screen it is commonly called the "sweep generator." The finite time it takes to return the beam back to the left side of the screen could cause a visible trace. This is prevented by briefly applying a negative grid voltage in order to blank out the retrace.

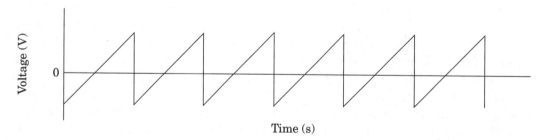

Fig 38.2 Sawtooth voltage waveform.

If the period T of the sweep generator is 1 s, the beam will clearly be recognizable as a spot that moves at constant speed across the tube face. If the period is as short as 0.1 s the beam is no longer recognizable as a spot, but instead appears to be a somewhat pulsating line. This is caused by the persistence of the phosphor, which causes the trace to still be glowing from one pass of the beam when another pass of the beam begins. For periods T of 0.02 s or less the beam is moving across the screen so often that the persistence of the phosphor makes the trace appear as a steady line.

The oscilloscope is designed so that a series of specific sweep generator periods can be applied to the horizontal plates by the selection of the position of a multi-position switch. The width of oscilloscope screen is fixed, usually 10 cm. Each different choice of period T therefore represents a specific time per length of scale division in the horizontal direction. Typically these are chosen to decrease in a series of scales that are in the ratio 2:1:0.5. For a typical student-type oscilloscope the times scales would be 19 settings ranging from 0.2 s/cm to 0.2 μs/cm. Note that since the screen is 10 cm wide there is a factor of 10 difference between the period T and the time scale. If the period of the sweep generator is 10 ms, that represents a time scale of 1 ms/cm. Also note that $t = 0$ is assumed to occur at the left of screen, and time is assumed to increase to the right.

In the vertical direction the screen is typically smaller, usually about 8 cm total. The vertical input is calibrated directly in volts. The input voltage scale is also variable by the choice of a multiposition switch that selects the appropriate amplification of the input voltage over some chosen voltage range. Typical ranges of possible voltage scales are from 5 V/cm to 5 mV/cm, again chosen in the ratio 2:1:0.5. This choice of voltage scales allows a range of input voltages to be displayed with deflections on the oscilloscope screen that are large enough to be easily visible. Since the voltage can be of either positive or negative polarity, the vertical scale has its zero in the center of the screen in order to be able to display both positive and negative voltages. Note that for the voltage scales given, input voltages must be equal to or less than 20 V either positive or negative from zero since a 20-V signal would deflect the beam 4 cm from the center of the screen even if the scale chosen were the maximum 5 V/cm.

The most common use of the oscilloscope is to use the time scale provided by the sweep generator to display the time variation of a voltage signal that is applied to the vertical plates. Usually this is some specific waveform that is repeated with a fixed frequency. For example, if a simple sine-wave voltage is applied to the vertical plates, a display of the voltage versus time will be directly displayed on the oscilloscope screen as a sine-wave trace of the beam whose maximum amplitude is proportional to the maximum voltage of the signal and whose period on the time scale of the oscilloscope is equal to the period of the signal. If the voltage waveform applied to the vertical plates is a more complex waveform, the resulting trace on the screen will represent the shape of that complex waveform.

The discussion so far has ignored one important point, which involves the means to coordinate the starting time of the sweep generator with the starting point of the voltage signal that is to be displayed. This is accomplished by using some waveform as a "trigger" to start the sweep generator. The triggering waveform can be the same signal that is input to the vertical plates for analysis, a secondary external signal, or the 60-Hz line voltage. When the signal itself is used as the trigger for the sweep generator, the signal is observed on the oscilloscope as a steady display that is constant in time because the sweep generator is initiated at the same point on the repetitive vertical signal for each pass of the sweep generator. On most oscilloscopes this is referred to as internal triggering and is the mode to be used in this laboratory. The trigger function of the oscilloscope has both a level and slope control. The level control sets a voltage level at which the trigger pulse is initiated, and the slope control refers to whether the trigger pulse will be generated as the signal goes above the set level or when it goes below that level. For example, if the level control is set at 1 V and the slope set at (+), the sweep will be triggered when the signal goes above 1 V. With the same level setting and the slope set on (−), the sweep will be trigger when the signal goes below 1 V.

TYPICAL OSCILLOSCOPE CONTROLS

A description is given below of the controls for a particular oscilloscope, the Hitachi model V-212 shown in Figure 38.3. It is a typical student-type oscilloscope. Although the description given is for that specific instrument, it is intended to show what will be available in the instruction manual of essentially any oscilloscope that is being used. Using the instruction manual for the oscilloscope, familiarize yourself with the operation of the controls that are the same as or similar to those described below.

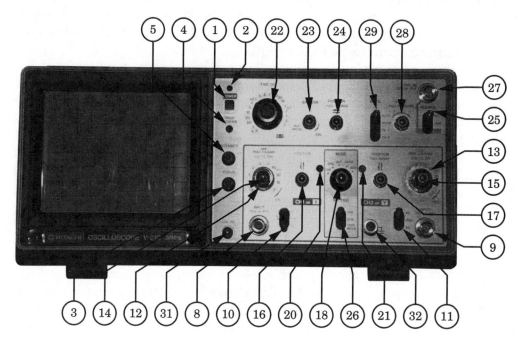

Figure 38.3 The Hitachi model V-212 Oscilloscope.

Power Supply and CRT

1. POWER switch—The power is on at the pushed-in position and off at the released position.

2. POWER LAMP—The lamp glows red when the power switch is on.

3. FOCUS—After obtaining an appropriate brightness by operating the intensity control, adjust the focus until the display is sharpest.

4. TRACE ROTATION—Used to align the trace of the CRT with horizontal scale.

5. INTENSITY—This knob adjusts the brightness. Brightness increases as the intensity is rotated clockwise. Perform all measurements at the lowest intensity level that is comfortable to the eyes. Avoid leaving the intensity set to a high level for extended time periods to prevent permanent damage to the phosphor.

6. VOLTAGE SELECTOR—This control is on the back of the instrument. It is used to select the power source, either 110 V or 220 V. (not pictured)

7. AC inlet—This is an inlet for the detachable AC power cord. (not pictured)

Controls of the Vertical Deflection System

8. CH1 INPUT—BNC connector for the vertical axis input. The signal input to this terminal becomes the x-axis signal when the oscilloscope is used as an x-y oscilloscope.

9. CH2 INPUT—BNC connector for a second vertical axis input. This input becomes the y-axis signal when the oscilloscope is used as an x-y oscilloscope.

10 & 11. (AC-GND-DC)—These input coupling switches are used to select the coupling system between the input signal and the vertical axis amplifier. (a) AC—At this setting the signal is connected through a capacitor that blocks any DC component of the input signal and displays only the AC component. (b) GND—At this setting the input to the vertical axis amplifier is grounded. (c) DC—At this setting the input signal is directly connected to the vertical axis amplifier and displayed unchanged, including any DC component.

12 & 13. VOLTS/DIV—A step attenuator that selects the vertical deflection factor. Set it to an easily observable range corresponding to the amplitude of the input signal. Ranges available from 5 V/DIV to 5 mV/DIV.

14 & 15. VAR (PULL x 5 GAIN)—When this control is fully rotated clockwise the vertical scale is calibrated to the ranges described above in 12 & 13. When rotated counterclockwise a continuously variable vertical attenuation up to a factor of 2.5 is introduced. When the knob is pulled out a fixed gain of 5 is introduced.

16 & 17. VERTICAL POSITION—This knob is used to adjust the position of the vertical axis. The display rises with clockwise rotation of this knob and falls with counterclockwise rotation.

18. MODE—This switch is used to select the operation mode of the vertical deflection. (a) CH1—Only the signal that has been applied to CH1 appears on the screen (b) CH2—Only the signal that has been applied to CH2 appears on the screen (c) Alt—Signals applied to CH1 and CH2 appear on the screen alternately at each sweep. This is used when the sweep time is short in two-channel observation. (d) Chop—At this setting the input signals applied to CH1 and CH2 are switched at about 250 kHz independent of the sweep and at the same time appear on the screen. This setting is used when the sweep time is long in two-channel observation. (e) Add—The algebraic sum of the input signals applied respectively to CH1 and CH2 appears on the screen.

19. CH1 OUTPUT—An output connector providing a sample of the signal applied to the CH1 connector. It is on the back of the oscilloscope. (not pictured)

20 & 21. DC BAL—These adjustment controls are used for the attenuator balance adjustment.

Controls of the Horizontal Deflection System

22. TIME/DIV (x-y)—(a) Sweep time ranges are available in 19 steps from 0.2 μs/cm to 0.2 s/DIV (b) x-y—This position is used when using the instrument as an x-y oscilloscope. In the position the x (horizontal) signal is connected to the input of CH1 and the y (vertical) signal is applied to the input of CH2 and has a deflection range from less than 1 mV to 5 V/DIV at a reduced bandwidth of 500 khZ.

23. SWP VAR—When this control is rotated fully in the clockwise direction of the arrow, the CAL state is produce and the sweep time is calibrated to value indicated by the time/DIV. Counterclockwise rotation of the control produces a continuously variable delay of the sweep time up to a factor of 2.5.

24. HORIZONTAL POSITION (PULL × 10 MAG)—This knob is used to move the display in the horizontal directions. The display is moved right when the knob is rotated clockwise and moved left with counterclockwise rotation. The sweep is magnified 10 times by pulling out the POSITION knob. In this case the sweep time is 1/10 of the value indicated by the time/DIV.

Synchronization System

25. (INT-LINE-EXT)—This switch is used to select the triggering signal source. (a) INT—The input signal applied to CH1 or CH2 becomes the triggering signal (b) LINE—This setting is used when observing a signal with the power supply line frequency. (c) EXT—An external triggering signal applied to the TRIG INPUT becomes the triggering signal.

26. INT TRIG—This switch is used to select the internal triggering signal source. (a) CH1—The input signal applied to CH1 becomes the triggering signal (b) CH2—The input signal applied to CH2 becomes the triggering signal. (c) VERT MODE—For observing two waveforms, the sync signal changes alternately corresponding to the signals on CH1 and CH2 to trigger the signal.

27. EXT TRIG—Input terminal for use for external triggering signal.

28. LEVEL/PULL (−) SLOPE—This knob sets the voltage level at which the sweep generator is triggered. The zero level is with the knob at the 12 o'clock position. Turning clockwise from there sets a positive level and counterclockwise from there sets a negative level. The slope of the trigger is positive (normally set there) when the knob is pushed in and negative when the knob is pulled out.

29. TRIGGER MODE—(a) AUTO—In this mode a sweep is always conducted. In the presence of a triggered signal, normal triggered sweep is obtained and the waveform stands still. In the case of no signal or out of triggering the sweep line will appear automatically. This setting is convenient in most cases and used almost all the time. (b) NORM—Triggered sweep is obtained and sweep is conducted only when triggering is effected. No sweep line will appear in the case of no signal or out of synchronization. Use this mode when effecting synchronization to a very-low-frequency signal (25 Hz or less).

Experimental Procedure

In several of the instructions below you are asked to draw what is on the oscilloscope display on the grids provided. In each of those cases assume the VOLTS/DIV and TIME/DIV are properly calibrated, and fill in the blank given for the values of VOLTS/DIV and TIME/DIV for the exercise associated with each set of grids. On the vertical scale, 0 V is labeled. Label the full-scale voltage both positive and negative. The time scale is labeled with 0 s. Label the value of the full-scale time on the horizontal axis. Do this for each grid.

1. Turn on the power to the oscilloscope and let it come to thermal equilibrium for at least 10 minutes. Set the oscilloscope mode setting to CH1, the trigger source

to INT, the trigger level to zero (center of range), trigger SLOPE to + (level knob pushed in), trigger MODE to AUTO, the INT TRIG to CH1, and CH1 to AC.

2. Set the TIME/DIV control to 1ms/DIV, the SWP VAR control rotated fully clockwise to the CAL position, the VOLTS/DIV control to 1 V/DIV, and the VAR (PULL × 5 GAIN) control rotated fully clockwise to the CAL position.

3. Turn on the power to the function generator and let it come to thermal equilibrium for at least 10 minutes. Select a sine-wave voltage, set the frequency $f = 100$ Hz, and connect the output of the function generator to the CH1 INPUT of the oscilloscope. Adjust the amplitude control of the function generator to zero. Adjust the VERTICAL POSITION control of the oscilloscope until the flat trace is exactly on the center line of the vertical display.

4. (a) Adjust the amplitude control of the function generator until the display on the oscilloscope is full-scale positive on the positive part of the cycle and full-scale negative on the negative part of the cycle. In the laboratory report section carefully draw on the grid labeled 1A what is displayed on the screen. (b) Leaving all other parameters fixed, set the VOLT/DIV control to 2 V/DIV, and draw on the grid labeled 1B what is now displayed on the screen. (c) Leaving all other parameters fixed, set the VOLT/DIV control to 5 V/DIV, and draw on the grid labeled 1C what is now displayed on the screen.

5. (a) Leaving all other parameters fixed, set the VOLT/DIV control to 1 V/DIV, and select $f = 200$ Hz from the function generator. Draw on the grid labeled 2A what is now displayed on the screen. (b) Leaving all other parameters fixed select $f = 400$ Hz from the function generator, and draw on the grid labeled 2B what is now displayed on the screen. (c) Leaving all other parameters fixed, select $f = 600$ Hz from the function generator and draw on the grid labeled 2C what is now displayed on the screen.

6. (a) Leaving all other parameters fixed, set the VOLT/DIV control to 1 V/DIV, the TIME/DIV control to 2ms/DIV, and select $f = 100$ Hz from the function generator. Note that the trigger slope control is still set at (+). Draw on the grid labeled 3A what is now displayed on the screen. (b) Leaving all other parameters fixed, pull out the trigger level control, which sets the trigger slope to (−). Draw on the grid labeled 3B what is now displayed on the screen.

7. (a) Leaving all other parameters fixed, push in the trigger level control, which sets the trigger slope to (+), leaving the trigger level still set at zero. Draw on the grid labeled 4A what is now displayed on the screen. (b) Leaving all other parameters fixed, slowly turn the trigger level control clockwise, increasing the trigger level. Increase it only so long as the display remains triggered. At the maximum level that the display is triggered, draw on the grid labeled 4B what is displayed on the screen. (c) Leaving all other parameters fixed, slowly turn the trigger level control counterclockwise, decreasing the trigger level. Decrease it only so long as the display remains triggered. At the minimum level that the display is triggered, draw on the grid labeled 4C what is displayed on the screen.

8. Push the trigger level control in for (+) slope and turn the level back to zero. Set the TIME/DIV to 2 ms/DIV and set the function generator to a sine wave of $f = 100$ Hz. Use the AC voltmeter to set the output of the function generator to 1.00 V as read on the voltmeter. Input this sine wave to the oscilloscope and measure the peak voltage of the sine wave. To measure the peak voltage of the sine wave you are free to adjust the VOLT/DIV control to give the most accurate measurement

possible. Generally this means adjusting the scale for as large a deflection as possible. Record the peak voltage of the sine wave as read from the oscilloscope in Data Table 1. Complete all the measurements in Data Table 1 from 1.00 V to 5.00 V. For each voltage, set the output from the generator using the voltmeter, and then read the voltage from the oscilloscope, each time choosing the VOLT/DIV that will allow the most accurate reading from the oscilloscope.

9. Set the function generator to output a triangular wave with $f = 1000$ Hz, and the TIME/DIV on the oscilloscope to 1 ms/DIV. Use the AC voltmeter to set the output of the function generator to 1.00 V as read on the voltmeter. Input this triangular wave to the oscilloscope and measure the peak voltage of the wave. Proceed as done for the sine wave above, this time measuring the voltages between 1.00 V and 5.00 V as read on the voltmeter. Record the results in Data Table 2.

10. The goal of this laboratory is to introduce students to the oscilloscope. At this time simply experiment for yourself with the features of the oscilloscope. Input as many different frequencies and waveforms as you have time for and attempt to learn everything you can about the operation of the oscilloscope by simply trying different settings of all of the oscilloscope controls.

CALCULATIONS

1. Perform a linear least squares fit to the data in Data Table 1 with the peak voltage read on the oscilloscope as the ordinate and voltage as read on the voltmeter as the abscissa. Determine the slope, intercept, and regression coefficient. Record those values in Calculations Table 1.

2. Perform a linear least squares fit to the data in Data Table 2 with the peak voltage read on the oscilloscope as the ordinate and voltage as read on the voltmeter as the abscissa. Determine the slope, intercept, and the regression coefficient. Record those values in Calculations Table 2.

LABORATORY REPORT

1A. TIME/DIV = _____   1B. TIME/DIV = _____   1C. TIME/DIV = _____

1A. VOLTS/DIV = _____   1B. VOLTS/DIV = _____   1C. VOLTS/DIV = _____

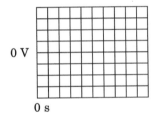
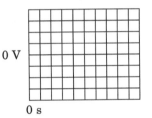
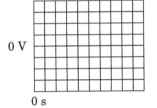

2A. TIME/DIV = _____   2B. TIME/DIV = _____   2C. TIME/DIV = _____

2A. VOLTS/DIV = _____   2B. VOLTS/DIV = _____   2C. VOLTS/DIV = _____

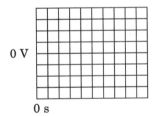
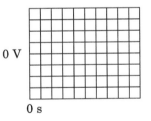
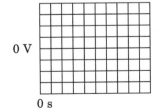

3A. TIME/DIV = _____   3B. TIME/DIV = _____

3A. VOLTS/DIV = _____   3B. VOLTS/DIV = _____

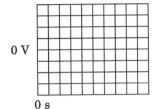
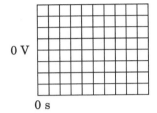

4A. TIME/DIV = _____   4B. TIME/DIV = _____   4C. TIME/DIV = _____

4A. VOLTS/DIV = _____   4B. VOLTS/DIV = _____   4C. VOLTS/DIV = _____

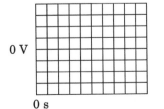
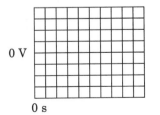
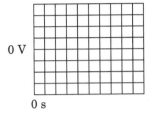

Data Table 1

Voltmeter (V)	Oscilloscope (V)
1.00	
2.00	
3.00	
4.00	
5.00	

Data Table 2

Voltmeter (V)	Oscilloscope (V)
1.00	
2.00	
3.00	
4.00	
5.00	

Calculations Table 1

Intercept = _____

Slope = _____

r = _____

Calculations Table 2

Intercept = _____

Slope = _____

r = _____

SAMPLE CALCULATIONS

QUESTIONS

1. In the grid labeled 2A, how many complete cycles are sketched in your figure? From your sketch, what is period of the wave? Using this period, calculate the frequency of the wave for this sketch. Is it in agreement with the frequency used for this part of the experiment?

2. In your own words, explain why the two sketches in 3A and 3B look like they do. They both have the trigger level zero, but one has a positive trigger slope and the other has a negative trigger slope.

3. Explain the appearance of sketches 4A, 4B, and 4C. They all have a positive trigger slope, but the trigger level of 4A is zero, the trigger level of 4B is positive, and the trigger level of 4C is negative.

4. For a sine wave, an AC voltmeter measures a root-mean-square value that is 0.707 of the peak value of the sine wave. Therefore, the peak value measured on the oscilloscope should be 1/.707 or 1.414 times the voltmeter readings. The slope of the data in Data Table 1 that you calculated and recorded in Calculations Table 1 should be approximately 1.414. Calculate the percentage error between you slope for this data and 1.414.

5. For a triangular wave, an AC voltmeter measures a root-mean-square value that is 0.576 of the peak value of the triangular wave. Therefore, the peak value measured on the oscilloscope should be 1/.576 or 1.736 times the voltmeter readings. The slope of the data in Data Table 2 that you calculated and recorded in Calculations Table 2 should be approximately 1.736. Calculate the percentage error between your slope for this data and 1.736.

6. An oscilloscope is set on a TIME/DIV setting of 50 μs. There are 10 divisions on the time scale. A sine wave on the oscilloscope display has exactly three full cycles of the sine wave that fit on the 10 divisions. What is the frequency of the wave?

PRELABORATORY ASSIGNMENT

Read carefully the entire description of the laboratory and answer the following questions based on the material contained in the reading assignment. Turn in the completed prelaboratory assignment at the beginning of the laboratory period prior to the performance of the laboratory.

1. What is the equation for the power P dissipated by a resistor of resistance R, current I, and voltage V?

2. What is the constant ratio between electrical energy (in joules) when it is converted completely to heat (in calories)?

3. A resistor has a current of 3.75 A when its voltage is 6.75 V. What is the resistance of the resistor? What power does it dissipate?

4. A resistor has a resistance of 1.50 Ω and a voltage of 6.00 V across it. What is the current in the resistor? What power does it dissipate?

5. If the resistor in question 4 is immersed in water, how much energy does it deliver to the water in 350 s? Express your answer in joules and in calories.

6. A resistor has a voltage of 6.65 V and a current of 4.45 A. It is placed in a calorimeter containing 200 g of water at 24.0°C. The calorimeter is aluminum (specific heat = 0.220 cal/g-°C), and its mass is 60.0 g. The heat capacity of the resistor itself is negligible. What is the temperature of the system 500 s later if all the electrical energy goes into heating the water and the calorimeter?

7. Two 2.00-Ω resistors are placed in (a) series (b) parallel across a 10.0-V battery as shown. What is the total power dissipated in each case?

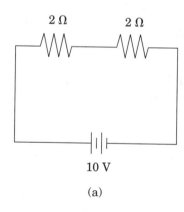

10 V

(a)

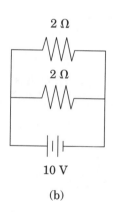

10 V

(b)

Joule Heating of a Resistor

OBJECTIVES

The power P absorbed in an electrical resistor of resistance R, current I, and voltage V is given by $P = I^2R = V^2/R = VI$. Despite the fact that it has units of power, it is commonly referred to as joule heat. A given amount of electrical energy absorbed in the resistor (in units of joules) produces a fixed amount of heat (in units of calories). The constant ratio between the two has the value $J = 4.186$ J/cal. In this laboratory, an electric coil will be immersed in water in a calorimeter, and a known amount of electrical energy will be input to the coil. Measurements of the heat produced will be used to accomplish the following objectives:

1. Demonstration that the rise in temperature of the system is proportional to the electrical energy input

2. Determination of an experimental value for J and comparison of that value with the known value

EQUIPMENT LIST

1. Immersion heater coil to fit standard calorimeter
2. Calorimeter and thermometer
3. Direct-current power supply (5 A at 6 V)
4. Ammeter (0–5 A), Voltmeter (0–10 V)
5. Laboratory timer
6. Laboratory balance and calibrated masses

THEORY

When a resistor of resistance R has a current I at voltage V, the power absorbed in the resistor is

$$P = I^2R = V^2/R = VI \tag{1}$$

Power is energy per unit time, and if the power is constant, the energy U delivered in time t is given by

$$U = Pt \tag{2}$$

Substituting equation 1 into equation 2 gives the following expression for the electrical energy:

$$U = VIt \tag{3}$$

When a resistor absorbs electrical energy, it dissipates this energy in the form of heat Q. If the resistor is placed in a calorimeter, the amount of heat produced can be measured when it is absorbed in the calorimeter. Consider the experimental arrangement shown in Figure 39.1, in which a resistor coil (also called an "immersion heater") is immersed in the water in a calorimeter. The heat Q produced in the resistor is absorbed by the water, calorimeter cup, and the resistor coil itself. This heat Q produces a rise in temperature ΔT. The heat Q is related to ΔT by

$$Q = (m_\mathrm{w} c_\mathrm{w} + m_\mathrm{c} c_\mathrm{c} + m_\mathrm{r} c_\mathrm{r})\, \Delta T \qquad (4)$$

where the ms and cs are the masses and specific heats of the water, calorimeter, and the resistor. Let mc stand for the sum of the product of mass and specific heat for the three objects that absorb the heat. In those terms the heat Q is given by the following:

$$Q = mc\Delta T \qquad (5)$$

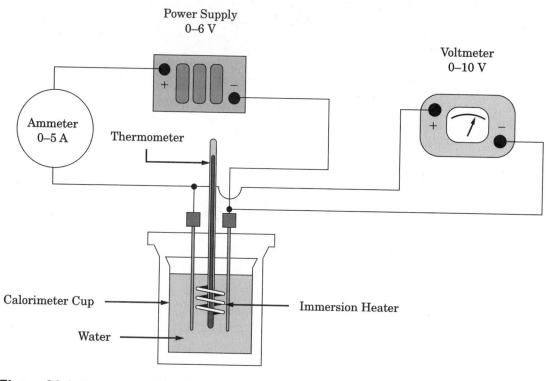

Figure 39.1 Experimental arrangement and circuit diagram for the calorimetric measurement of the heat produced in an immersion heater by an electrical current.

The electrical energy absorbed in the resistor is completely converted to heat. The equality of those two energies is expressed as

$$U(\mathrm{J}) = J(\mathrm{J/cal})\, Q(\mathrm{cal}) \qquad (6)$$

where J represents the conversion from joules to calories. Using the expression for U and Q from equations 3 and 5 in equation 6 leads to

$$VIt(\mathrm{J}) = J(\mathrm{J/cal})\, mc\Delta T(\mathrm{cal}) \qquad (7)$$

Suppose that a fixed current and voltage are applied to the resistor in a calorimeter, and the temperature rise ΔT is measured as a function of the time t. Equation 7 predicts that a graph with $VIt(\mathrm{J})$ as the ordinate and $mc\Delta T(\mathrm{cal})$ as the abscissa should produce a straight line with J as the slope.

EXPERIMENTAL PROCEDURE

1. Determine the mass m_c of the calorimeter cup and record it in the Data Table. Obtain from your instructor the specific heat of the calorimeter cup c_c, the mass of the resistor coil m_r, and its specific heat c_r and record them in the Data Table.

2. Place enough water in the calorimeter cup to immerse the resistor coil completely. The water temperature should be a few degrees below room temperature. Be sure that the coil is completely covered by the water, but do not use any more water than is necessary. Determine the mass of the water plus the calorimeter cup and record it in the Data Table. Determine the mass of the water by subtraction and record it as m_w in the Data Table.

3. Place the immersion heater in the calorimeter cup and construct the circuit shown in Figure 39.1. Check again that the immersion heater is below the water level. If it is not, add some more water and determine the mass of the water again.

4. Turn on the power supply and adjust the current between 4.0 and 5.0 A. Do this quickly and then turn off the supply with the output level still adjusted to the setting that produced the desired current. Do not allow the supply to stay on long enough to appreciably heat the water. Stir the system several minutes to allow it to come to equilibrium.

5. Determine the initial temperature T_i and record it in the Data Table. Estimate the thermometer readings to the nearest 0.1 C° for all temperature measurements.

6. With the power supply still set to the output level required to produce 4.0 A, turn on the power supply and simultaneously start the laboratory timer. Record the initial values of the current I and the voltage V in the Data Table. Let the timer run continuously and stir the system often. Measure and record the temperature T, the current I, and the voltage V every 60 s for 8 minutes. Record all data in the Data Table.

CALCULATIONS

1. Calculate the quantity mc, where $mc = m_w c_w + m_c c_c + m_r c_r$ and record it in the Calculations Table.

2. Calculate the temperature rise ΔT above the initial temperature T_i from $\Delta T = T - T_i$ for each of the measured values of T and record the results in the Calculations Table.

3. Calculate the quantity $mc\Delta T$ for each case and record the results in the Calculations Table.

4. For each measurement of the voltage V and current I calculate the product VI and record the results in the Calculations Table.

5. Calculate the mean $\overline{VI}$ and standard error α_{VI} for the values of VI and record the results in the Calculations Table.

6. Calculate the quantity $\overline{VI}t$ for each time t and record the results in the Calculations Table.

7. Perform a linear least squares fit to the data with $\overline{VI}t$ as the ordinate and $mc\Delta T$ as the abscissa. Determine the slope J_{exp}, the intercept A, and the correlation coefficient r and record them in the Calculations Table.

8. Calculate the percentage error in the value of J_{exp} compared to the known value of $J = 4.186$ J/cal.

GRAPHS

Graph the data with $\overline{VI}t$ as the ordinate and $mc\Delta T$ as the abscissa. Also show on the graph the straight line obtained from the linear least squares fit.

LABORATORY REPORT

Data Table

Mass calorimeter + water =		g	$T_i =$	°C
Mass calorimeter =		g	c for calorimeter =	cal/g-°C
Mass water =		g	c for water =	cal/g-°C
Mass coil =		g	c for coil =	cal/g-°C

t (s)	V (V)	I (A)	T (°C)
0			
60			
120			
180			
240			
300			
360			
420			
480			

SAMPLE CALCULATIONS

Calculations Table

$mc =$ _____ cal/°C

ΔT (°C)	$mc\Delta T$ (cal)	VI (W)	$\overline{VI}t$ (J)
0	0		0

$\overline{VI} =$ W	$\alpha_{VI} =$ W	
$J_{exp} =$ J/cal	$A =$ J	$r =$
Percentage error in experimental $J =$		

SAMPLE CALCULATIONS

QUESTIONS

1. When the electrical power input is approximately constant, the temperature rise of the system should be proportional to the elapsed time. Do your data confirm this expectation? State the evidence for your answer.

2. What is the accuracy of your experimental value for J?

3. Does the correlation coefficient of the least squares fit to the data indicate strong evidence for linearity in this data?

4. How long from the original starting time would it have taken to achieve a temperature of 50.0°C with the experimental arrangement you used? Show your work.

5. Assuming that one used the same heating coil and that its resistance did not change, how much would the power be increased if the voltage were increased by 50%? Show your work.

6. Suppose that some liquid other than water were used in the calorimeter. If the liquid had a specific heat of 0.25 cal/g-°C, would that tend to improve the results, make them worse, or have no effect on them? Explain clearly the reasoning behind your answer.

Laboratory 40

Reflection and Refraction with the Ray Box

PRELABORATORY ASSIGNMENT

Read carefully the entire description of the laboratory and answer the following questions based on the material contained in the reading assignment. Turn in the completed prelaboratory assignment at the beginning of the laboratory period prior to the performance of the laboratory.

1. State the law of reflection. Use a diagram to define the angles involved.

2. Define the index of refraction.

3. State Snell's law. Define terms and angles using a diagram.

4. A light ray is incident on a plane interface between two media. The ray makes an incident angle with the normal of 25.0° in a medium of $n = 1.25$. What is the angle that the refracted ray makes with the normal if the second medium has $n = 1.55$? Show your work.

5. A 60.0° prism has an index of refraction of 1.45 as shown below. A ray is incident as shown at an angle of 60.0° to the normal of one of the prism faces. Trace the ray on through the prism and find the angles θ_2, θ_3, and θ_4 as defined in the laboratory instructions. Show your work.

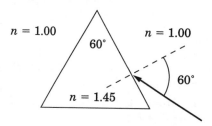

$n = 1.00$ 60° $n = 1.00$ 60° $n = 1.45$

Reflection and Refraction with the Ray Box

OBJECTIVES

In geometrical optics an assumption is made called the "ray approximation." It assumes that light consists of plane waves that propagate in straight lines. Rays are lines perpendicular to the wavefronts, which define the direction in which the light travels. A light ray that is traveling in some medium is either reflected or transmitted when it comes to a boundary into a second medium. The rays that are transmitted into the second medium are changed in direction. They are said to be refracted. A quantity called the "index of refraction" for a given medium is defined as the ratio of the speed of light in a vacuum to the speed of light in the medium. In this laboratory light rays from a ray box will be used to accomplish the following objectives:

1. Demonstration that for reflection from a plane surface the angle of incidence is equal to the angle of reflection

2. Demonstration that rays going from air into a transparent plastic medium are refracted at a plane boundary

3. Determination of the index of refraction of a plastic prism from direct measurement of incident and refracted angles of a light ray

4. Demonstration of the focal properties of spherical reflecting and refracting surfaces

EQUIPMENT LIST

1. Ray box, 60.0° prism, planoconvex lens, circular metal reflecting surfaces, converging lens, diverging lens

2. Protractor, straight edge, compass, sharp hard-lead pencil

3. Black tape, several sheets of white paper

THEORY

Reflection

The reflection of light from a plane surface is described by the law of reflection, which states that the angle of incidence θ_i is equal to the angle of reflection θ_r. By convention these angles are measured with respect to a line perpendicular or normal to the plane surface. Reflection from a plane mirror or a flat transparent plane surface of a piece of glass or plastic are the most easily demonstrated examples of the law of reflection.

In Figure 40.1(a) several rays are shown incident on a plane surface, and in each case the reflected ray is also shown. For each ray the angle of incidence θ_i is seen to be equal to the angle of reflection θ_r.

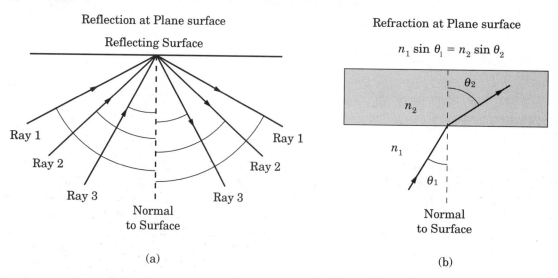

Figure 40.1 Illustration of reflection and refraction of light rays at a plane surface.

Refraction

A light ray incident on a plane surface that forms the boundary between two transparent media can be reflected at the surface or transmitted into the second medium. In general, light rays incident on a plane interface will be partially reflected and partially transmitted. The ray that is transmitted does not continue in the same direction. Instead it undergoes a change in direction. The ray is said to be refracted. This is illustrated in Figure 40.1(b). The angle of incidence is θ_1, and the angle of refraction is θ_2. The change in direction of the light ray is caused by the fact that the speed of light is different for different media.

The speed of light in a vacuum is c, and it represents the maximum possible speed of light. For any material, the speed of light is v, where $v \leq c$. A quantity called the "index of refraction," or n, for any medium is defined by $n = c/v$. Note that since $v \leq c$ then the only allowed values of n are $n \geq 1$.

The relationship between the angle of incidence θ_1 and the refracted angle θ_2 is an expression involving the indices of refraction of the two media n_1 and n_2. The relationship is known as Snell's law, and it is

$$n_1 \sin\theta_1 = n_2 \sin\theta_2 \tag{1}$$

When $n_1 > n_2$ equation 1 implies that $\theta_1 < \theta_2$. This states that a ray going from a medium of a given index of refraction to one of a smaller index of refraction is bent away from the normal. If $n_1 < n_2$ then it must be true that $\theta_1 > \theta_2$, and thus a ray going into a medium of larger index of refraction is bent toward the normal.

Focal Properties of Reflection and Refraction

Descriptions of the focal properties of reflection from spherical mirrors are shown in Figure 40.2. When reflection takes place from a concave spherical surface, incident parallel rays are converging and come to an approximate point focus. If R is the radius of curvature of the spherical surface, the focal point is a distance f from the vertex of the spherical mirror, where $f = R/2$. The distance f is called the "focal length" of the mirror and is positive by convention for a concave converging mirror. Incident parallel rays on a convex spherical mirror are diverging, but they appear to have come from a point. The distance from the vertex of the mirror to that point is called the "focal length," and its magnitude is given by $f = R/2$. By convention, the focal length is negative for a convex diverging mirror.

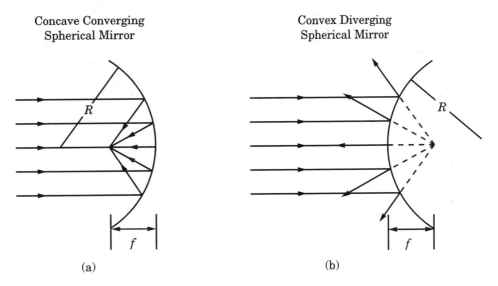

Figure 40.2 The focal properties of spherical mirrors for incident parallel rays.

EXPERIMENTAL PROCEDURE

Reflection

1. Using black tape, cover all the slits from the ray box except the central one. This will produce a single ray that will be used to examine reflection and refraction from a plane surface.

2. Place the 60.0° prism on a piece of white paper with one of its prism faces down. In this position the surface on the paper will be in the form of a rectangle. This will produce two parallel plane surfaces as shown in Figure 40.3(a). Draw a straight line along the face of the prism for one of these surfaces where it touches the paper as shown in the figure. Place a small dot in the center of the line as shown.

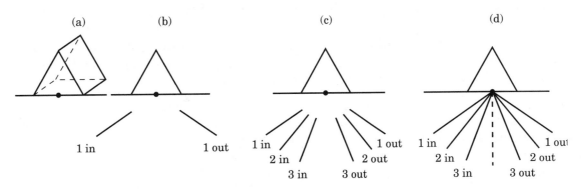

Figure 40.3 Tracing incident and reflected rays from a plane surface.

3. Place the ray box about 0.15 m away from the prism and adjust the single ray to strike the plane surface at the position of the dot at an angle of incidence estimated to be about 60°. With a straight edge, draw a line in the direction of the incident ray and one in the direction of the reflected ray. This will produce the lines shown in Figure 40.3(b). (In parts b, c, and d of the figure, no perspective is shown, and the view is straight on to the plane surface.) Repeat this process two more times, once for an incident ray of about 45° and once for an incident ray of about 30°. This procedure will produce the lines shown in Figure 40.3(c).

4. At the point of the dot, construct a perpendicular to the face of the prism. Extend all six of the lines showing the ray directions until they intersect at the point of the dot. This will produce the lines shown in Figure 40.3(d). Using a protractor, measure the incident angles θ_{1i}, θ_{2i}, and θ_{3i} and the reflected angles θ_{1r}, θ_{2r}, and θ_{3r} for each of the rays. Record all these angles (to the nearest 0.1°) in the Data Table.

Refraction

1. Place the prism on the paper as shown in Figure 40.4(a). In this position, the surface of the prism on the paper is in the form of an equilateral triangle. Draw straight lines on the paper along two adjacent faces of the prism as shown in Figure 40.4(b).

2. Place the ray box about 0.15 m away from the prism. Adjust the direction of the ray box so that the incident ray strikes one face of the prism at an angle of about 50° relative to normal to the prism face. Using a straight edge, draw a line in the direction of the incident ray and one in the direction of the refracted ray as shown in Figure 40.4(b).

(a) (b)

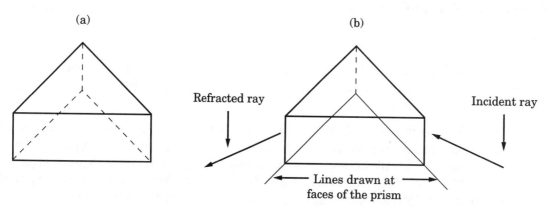

Refracted ray   Incident ray

Lines drawn at faces of the prism

Figure 40.4 Refraction of a ray incident on one face of a 60° prism.

3. Using a separate sheet of paper for each ray, repeat steps 1 and 2 for two other rays, one incident at an angle of about 60° with respect to the normal and the other incident at about 70° with respect to the normal.

4. Construct the lines tracing the path of each of the incident rays through the prism in the order of the steps shown in Figure 40.5. This will produce a figure from which the angles θ_1, θ_2, θ_3, and θ_4 can be determined with a protractor. Measure these angles for each of the three cases and record the values of all four angles (to the nearest 0.1°) in the Data Table.

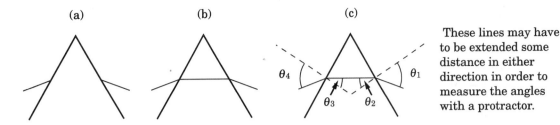

Figure 40.5 Step by step process to trace the rays and determine the four angles.

Focal Properties of Reflection and Refraction

1. Remove the black tape from the ray box slits. Place the plastic planoconvex lens next to the slits to produce five parallel rays from the ray box. The lens may have to be rotated slightly to produce the best set of parallel rays.

2. Place the circularly shaped metal reflector on a piece of white paper and trace its outline on the paper. Place the ray box about 0.15 m away on the concave side of the reflector. Align the five parallel rays with the center of the reflector to produce a pattern like the one in Figure 40.2(a). Make a tracing of this pattern, and from it, measure the focal length of the concave reflector. Record it in the Data Table as f_{con}.

3. Turn the reflector around and repeat step 2 on another piece of paper with the reflector now acting as a convex mirror. Trace the pattern, which should look like that of Figure 40.2(b). Extend the reflected rays back to the point from which they appear to come. Measure the focal length and record it in the Data Table as f_{div}.

4. Using a compass, construct a circular arc that is the same radius of curvature as the reflector. Record in the Data Table the radius of that constructed circle as R, the radius of curvature of the reflector.

5. Place the plastic converging and diverging lenses on separate pieces of paper and trace the ray pattern produced by the parallel beam of rays from the ray box. Patterns like those of Figure 40.6 should be observed. Measure the focal length of the converging lens and record it in the Data Table as f_{con}. Measure the diverging lens focal length and record it in the Data Table as f_{div}.

CALCULATIONS

Reflection

Calculate the difference $(|\theta_i - \theta_r|)$ between the measured values of the incident angle and the reflected angle for each of the three rays and record them in the Calculations Table.

Refraction

1. According to Snell's law (equation 1), for refraction at the first surface the relationship (1) $\sin\theta_1 = n \sin\theta_2$ should be true. The value of $n = 1$ has been used for air, and n stands for the index of refraction of the prism. At the second surface the equation is $n \sin\theta_3 = (1) \sin\theta_4$. Solving these two equations for n gives

$$n = \frac{\sin\theta_1}{\sin\theta_2} \quad \text{and} \quad n = \frac{\sin\theta_4}{\sin\theta_3} \tag{2}$$

2. Using these two equations, calculate two values of n for each of the incident rays. It turns out that these are not independent measurements because the errors made in drawing the rays to determine the angle tend to produce two values of n with errors in the opposite direction. Therefore, take the average of the two values calculated by equations 2 as a single measurement with that incident ray. Record in the Calculations Table these averaged values of n as the measurements for n for each ray.

3. Calculate the mean $\bar{n}$ and standard error α_n for these three measurements of n. Record these values in the Calculations Table.

FOCAL PROPERTIES OF REFLECTION AND REFRACTION

1. Calculate the percentage difference between the value of f_{con} and the value of f_{div} for the reflector.

2. According to theory, the value of the focal length for the reflector should be equal to $R/2$. Calculate the percentage difference between the measured value of $R/2$ and the focal lengths f_{con} and f_{div} for the reflector.

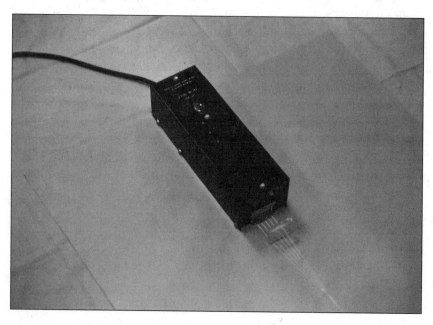

Figure 40.6 Ray box showing focus of incident parallel rays.

LABORATORY REPORT

Data Table

Calculations Table

Reflection

Ray	θ_i	θ_r
1		
2		
3		

Difference in Angle

Refraction

Ray	θ_1	θ_2	θ_3	θ_4
1				
2				
3				

n	$\bar{n}$	α_n

Mirrors

	f (cm)	R (cm)
Concave Mirror f_{con}		
Convex Mirror f_{div}		

% diff fs	% diff $R/2$

Lenses

	f (cm)
Converging Lens	
Diverging Lens	

QUESTIONS

1. Are your data consistent with the law of reflection? State your answer as quantitatively as possible.

2. State as quantitatively as possible the precision of your value for n, the index of refraction of the prism.

3. State how your data for the prism are evidence for the validity of Snell's law.

4. How well do your data for the focal lengths of the concave and convex mirror agree with the expectation that $f = R/2$? State your answer as quantitatively as possible.

5. Using the value of n determined for the prism, find the speed of light in the prism.

6. Inside the prism the wavelength of the light must change as well as the speed. Is a given wavelength longer or shorter inside the prism? Consider specifically light whose wavelength is 500 nm in air. What is the wavelength of this light inside the prism?

PRELABORATORY ASSIGNMENT

Read carefully the entire description of the laboratory and answer the following questions based on the material contained in the reading assignment. Turn in the completed prelaboratory assignment at the beginning of the laboratory period prior to the performance of the laboratory.

1. Mark the following statements about lenses as true or false.

_____a. Incident parallel light rays converge if the lens focal length is negative.

_____b. If the path of converging light rays are traced backwards, they appear to have come from a point called the "focal point."

_____c. A double convex lens has a negative focal length.

_____d. The focal length of a lens is always positive.

2. A double convex lens is made from glass whose index of refraction is $n = 1.50$. The *magnitudes* of its radii of curvature R_1 and R_2 are 10.0 cm and 15.0 cm, respectively. What is the focal length of the lens? Show your work.

$f =$ _____ cm

3. What is a real image? What is a virtual image?

4. For a diverging lens, state what kinds of images can be formed and the conditions under which those images can be formed.

5. For a converging lens, state what kinds of images can be formed and the conditions under which those images can be formed.

6. A lens has a focal length of $f = +10.0$ cm. If an object is placed 30.0 cm from the lens, where is the image formed? Is the image real or virtual? Show your work.

7. An object is 16.0 cm from a lens. A real image is formed 24.0 cm from the lens. What is the focal length of the lens? Show your work.

8. One lens has a focal length of $+15.0$ cm. A second lens of focal length $+20.0$ cm is placed in contact with the first lens. What is the equivalent focal length of the combination of lenses? Show your work.

9. Two lenses are in contact. One of the lenses has a focal length of $+10.0$ cm when used alone. When the two are in combination, an object 20.0 cm away from the lenses forms a real image 40.0 cm away from the lenses. What is the focal length of the second lens? Show your work.

Focal Length of Lenses

OBJECTIVES

When a ray of light is incident upon the interface between two media in which the speed of light is different, the ray changes direction as it passes from one medium into the other. The process is called "refraction," and the different media are characterized by a constant called the "index of refraction." Lenses are devices that can cause parallel rays of light to converge or diverge by appropriate shaping of the interface between media of differing index of refraction. This laboratory will use an optical bench and several lenses either alone or in combination to accomplish the following objectives:

1. Demonstration that converging lenses form real images and diverging lenses form virtual images

2. Measurement of the focal length of converging lenses by forming a real image of a very distant object

3. Determination of the equivalent focal length of two lenses in contact in terms of their individual focal lengths

4. Measurement of the focal length of a diverging lens by using it in combination with a converging lens to form a real image

EQUIPMENT LIST

1. Optical bench with holders for lenses, a screen, and an object

2. Lamp with object painted on its face to serve as an illuminated object

3. Three lenses (focal lengths approximately $+20$, $+10$, and -30 cm)

4. Meter stick, screen on which to form image, masking tape

THEORY

When a beam of light rays parallel to the central axis of a lens is incident upon a converging lens, the rays are brought together at a point called the "focal point" of the lens. The distance from the center of the lens to the focal point is called the "focal length" of the lens, and it is a positive quantity for a converging lens.

When a parallel beam of light rays is incident upon a diverging lens the rays diverge as they leave the lens; however, if the path of the outgoing rays are traced backward, they appear to have emerged from a point called the "focal point" of the lens. The distance from the center of the lens to the focal point is called the "focal length" of the lens, and it is a negative quantity for a diverging lens.

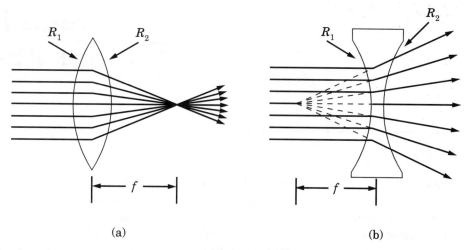

Figure 41.1 Ray diagram for converging and diverging showing the definition of the focal length for both the converging and diverging case.

In Figure 41.1, two common types of lenses are pictured. The first, shown in part (a), is a double convex lens whose focal length is positive, and thus it brings the incident parallel rays to a focus. The second type of lens, shown in part (b), is a double concave lens whose focal length is negative, and thus it causes the incident parallel rays to diverge. They appear to come from a fixed point as shown. In general, a lens is converging or diverging depending on the curvature of its surfaces. In Figure 41.1 the radii of curvature of the surfaces of the two lenses are denoted as R_1 and R_2. The general relationship that determines the focal length f in terms of the radii of curvature and the index of refraction n of the glass from which the lens is made is called the "lens makers' equation." It is given by

$$\frac{1}{f} = (n - 1)\left(\frac{1}{R_1} - \frac{1}{R_2}\right) \tag{1}$$

For the converging lens shown in Figure 41.1(a), the radius R_1 is positive and the radius R_2 is negative, but for the diverging lens of part (b), the radius R_1 is negative and the radius R_2 is positive. The signs of these radii are determined according to a sign convention that is described in most elementary textbooks.

As an example, consider a double convex lens like the one shown in Figure 41.1(a) made from glass of index of refraction 1.60 with radii of curvature R_1 and R_2 of *magnitude* 20.0 and 30.0 cm, respectively. According to the sign convention given above, that would mean $R_1 = +20.0$ cm and $R_2 = -30.0$ cm. Putting those values into equation 1 gives a value for the focal length f of $+20.0$ cm.

Essentially, equation 1 states that a lens that is thicker in the middle than at the edges is converging, and a lens that is thinner in the middle than at the edges is diverging. Therefore, a lens can be classified as converging or diverging merely by taking it between one's fingers to see if the lens is thicker at its center than at its edge.

Lenses are used to form images of objects. There are two possible kinds of images. The first type, called a "real image," is one that can be focused on a screen. For a real image, light actually passes through the points at which the image is formed. The second type of image is called a "virtual image." For a virtual image, light does not actually pass through the points at which the image is formed, and the image cannot be focused on a screen. Diverging lenses can form only virtual images, but converging lenses can form either real images or virtual images. Whether the image formed by a converging lens is real or virtual depends on how far the object is from the lens compared to the focal length of the lens. If an object is further from a converging lens than its focal length, a real image is formed. However, if the object is closer to a converging lens than the focal length, the image formed is virtual. Note that whenever a virtual image is formed it will serve as the object for some other lens system to ultimately form a real image. Often the other lens system is the human eye, and the real image is formed on the retina of the eye.

In the process of image formation, the distance from an object to the lens is called the "object distance," or p, and the distance of the image from the lens is called the "image distance," or q.

The relationship between the object distance p, the image distance q, and the focal length of the lens f is given by

$$\frac{1}{p} + \frac{1}{q} = \frac{1}{f} \tag{2}$$

Note that equation 2 is valid both for converging (positive f) and for diverging (negative f) lenses. Normally, the object distance is considered positive. In that case, a positive value for the image distance means that the image is on the opposite side of the lens from the object, and the image is real. A negative value for the image distance means that the image is on the same side of the lens as the object, and the image is virtual.

When a lens is used to form an image of a very distant object, the image distance p is very large. For that case, the term $1/p$ in equation 2 is very small and is thus negligible compared to the other terms $1/q$ and $1/f$ in that equation. Therefore, for the case of a very distant object equation 2 becomes

$$\frac{1}{q} = \frac{1}{f} \tag{3}$$

For this case, the image distance is equal to the focal length. This provides a quick and accurate way to determine the focal length of a converging lens. The method is only applicable to a converging lens because the image must be focused on a screen. Since a diverging lens cannot form a real image, this technique will not work directly for a diverging lens.

If two lenses with focal lengths of f_1 and f_2 are placed in contact, the combination of the two act as a single lens of effective focal length f_e. The effective focal length of the two lenses in contact f_e is related to the individual focal lengths of the lenses f_1 and f_2 by

$$\frac{1}{f_e} = \frac{1}{f_1} + \frac{1}{f_2} \tag{4}$$

Equation 4 holds for any combination of converging and diverging lenses. If the individual lenses f_1 and f_2 are converging, then the effective focal length f_e will, of course, also be converging. If one of the lenses is converging and the other is diverging, then the effective focal length can be either converging or diverging depending on the values of f_1 and f_2. If the converging lens has a smaller magnitude than the diverging lens, then the effective focal length will be converging. This fact can be used to determine the focal length of an unknown diverging lens if it is used in combination with a converging lens whose focal length is short enough to produce a converging combination.

EXPERIMENTAL PROCEDURE—FOCAL LENGTH OF A SINGLE LENS

1. Place one of the three lenses in a lens holder on the optical bench and place the screen in its holder on the optical bench (Figure 41.2). Place the optical bench in front of a window in the laboratory and point the bench toward some distant object. Adjust the distance from the lens to the screen until a sharp real image of the distant object is formed on the screen. You will be able to form such an image for only two of the three lenses. This experimental arrangement satisfies the conditions of equation 3. Therefore, the measured image distance is equal to the focal length of the lens. Record these measured image distances in Data Table 1 as the focal length of the two lenses for which the method works. Call the lens with the longest focal length A, the one with the shortest focal length B, and the one for which no image can be formed C.

2. Place lens B in the lens holder on the optical bench and use the lamp with the object painted on its face as an object. For various distances p of the object from the lens, move the screen until a sharp real image is formed on the screen. For each value of p measure the image distance q from the screen to the lens. Make sure that the lens, the object, and the screen are at the center of their respective holders. Try values for p of 20, 30, 40, and 50 cm determining the value of q for each case. If these values of p do not work for your lens, try other values until you find four values that differ by at least 5 cm. Record the values for p and q in Data Table 2.

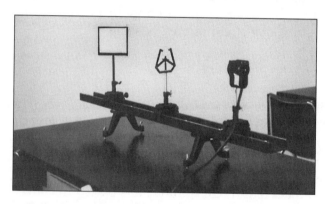

Figure 41.2 Optical bench with object, lens, and screen on which a real image is formed.

CALCULATIONS—FOCAL LENGTH OF A SINGLE LENS

1. Using equation 2, calculate the values of the focal length f for each of the four pairs of object and image distances p and q. Record them in Calculations Table 2.

2. Compute the mean $\bar{f}$ and the standard error α_f for the four values for the focal length f and record them in Calculations Table 2.

3. The mean $\bar{f}$ represents the measurement of the focal length of lens B using finite object distances. Compute the percentage difference between $\bar{f}$ and the value determined using essentially infinite object distance in Data Table 1. Record the percentage difference between the two measurements in Calculations Table 2.

EXPERIMENTAL PROCEDURE—FOCAL LENGTH OF LENSES IN COMBINATION

1. Place lens A and lens B in contact, using masking tape to hold the edges of the two lenses parallel. Measure the focal length of the combination f_{AB} both by the very distant object method and by the finite object method. For the finite object method, use just one value of the object distance p and determine the image distance q. Record the results for both methods in Data Table 3.

2. Place lens B and lens C in contact, using masking tape to hold the edges of the two lenses parallel. Repeat the measurements described in step 1 above for these lenses in combination. Record the results in Data Table 3.

CALCULATIONS—FOCAL LENGTH OF LENSES IN COMBINATION

1. From the data for lenses A and B in Data Table 3 calculate the value of f_{AB} from the value of p and q using equation 2. Record that value of f_{AB} in Calculations Table 3.

2. Record the value of f_{AB} determined by the very distant object method in Calculations Table 3.

3. Calculate the average $\overline{f_{AB}}$ of the two values for f_{AB} determined in steps 1 and 2. This average value $\overline{f_{AB}}$ will be considered to be the experimental value for the combination of these two lenses.

4. Using equation 4, calculate a theoretical value expected for the combination of lenses A and B. Use the values determined in Data Table 1 by the distant object method for the values of f_A and f_B in the calculation. Record this value as $(f_{AB})_{theo}$ in Calculations Table 3.

5. Calculate the percentage difference between the experimental value and the theoretical value for f_{AB}. Record it in Calculations Table 3.

6. From the data for lenses B and C in Data Table 3, calculate the value of f_{BC} from the values of p and q using equation 2. Record that value of f_{BC} in Calculations Table 3.

7. Record the value of f_{BC} determined by the very distant object method in Calculations Table 3.

8. Calculate the average $\overline{f_{BC}}$ of the two values for f_{BC} determined in steps 6 and 7. This average value $\overline{f_{BC}}$ will be considered to be the experimental value for the combination of these two lenses.

9. Using the value of $\overline{f_{BC}}$ determined in step 8 and the value of f_B from Data Table 1 for the focal length of B, calculate the value of f_C, the focal length of lens C using equation 4. Record the value of f_C in Calculations Table 3.

LABORATORY REPORT

Data Table 1

Lens	Image Distance	Focal Length
A		$f_A =$
B		$f_B =$

Data Table 2

p (cm)	q (cm)

Calculations Table 2

f_B (cm)	$\overline{f_B}$	α_f	% difference

SAMPLE CALCULATIONS

Data Table 3

Lenses	$q = f$ (infinite object)	p	q
A and B	$f_{AB} =$		
B and C	$f_{BC} =$		

Calculations Table 3

Lenses A and B	
f_{AB} (infinite object)	
f_{AB} (from p and q)	
$\overline{f_{AB}}$	
$(f_{AB})_{theo}$	
% difference	

Lenses B and C	
f_{BC} (infinite object)	
f_{BC} (from p and q)	
$\overline{f_{BC}}$	
f_C	

SAMPLE CALCULATIONS

QUESTIONS

1. Why is it not possible to form a real image with lens C alone?

2. Take lens C between your thumb and index finger. Is it thinner at the center of the lens or thicker? Take lens B between your thumb and index finger. Is it thinner at the center of the lens or thicker? From this information alone, what can you conclude about lenses C and B?

3. Consider the percentage difference between the two measurements of the focal length of lens B. Express α_f as a percentage of $\overline{f}_B$. Is the percentage difference between the two measurements less than the percentage standard error?

4. Compare the agreement between the experimental and theoretical values of f_{AB} the focal length of lenses A and B combined. Does this data suggest that equation 4 is a valid model for the equivalent focal length of two lenses in contact?

5. If lens A and lens C were used in contact, could they produce a real image? State clearly the basis for your answer.

Laboratory 42

Diffraction Grating Measurement of the Wavelength of Light

PRELABORATORY ASSIGNMENT

Read carefully the entire description of the laboratory and answer the following questions based on the material contained in the reading assignment. Turn in the completed prelaboratory assignment at the beginning of the laboratory period prior to the performance of the laboratory.

1. What is a continuous spectrum? What is a discrete spectrum?

2. What kind of light sources produce each type of spectrum?

3. The wavelengths produced by a hot gas of helium (a) form a discrete spectrum, (b) form a line spectrum, (c) are characteristic of the electronic structure of helium, or (d) all of the above are true.

4. A diffraction grating has a grating spacing of $d = 1500$ nm. It is used with light of wavelength 500 nm. At what angle will the first-order diffraction image be seen? Show your work.

5. For a given wavelength λ and a diffraction grating of spacing d (a) an image is formed at only one angle, (b) at least two orders are always seen, (c) the number of orders seen can be any number and depends on d and λ, or (d) there can never be more than four orders seen.

6. The grating used in this laboratory (a) can only produce images in the horizontal, (b) must be rotated in its holder until it produces the desired horizontal pattern, (c) produces images only in the vertical direction, or (d) produces images only to the left of the slit.

7. A diffraction grating with $d = 2000$ nm is used with a mercury discharge tube. At what angle will the first-order blue-green wavelength of mercury appear? What other orders can be seen, and at what angle will they appear? Show your work.

8. The diffraction grating of question 7 is used at a distance $L = 50.0$ cm from the slit. What is the distance D from the slit to the first-order image for the blue wavelength of mercury? Show your work.

9. What is the voltage and current of the spectrum-tube power supply?

Diffraction Grating Measurement of the Wavelength of Light

OBJECTIVES

Sources of visible light often produce many different wavelengths or colors. Light from sources utilizing hot solid metal filaments contain essentially a continuous distribution of wavelengths forming a white light. Light produced by a discharge in a gas of a single chemical element contains only a limited number of discrete wavelength components that are characteristic of the element. There are several methods that can be used to separate a light source into its component wavelengths. The technique that will be used in this laboratory employs a diffraction grating to accomplish the following objectives:

1. Demonstration of the difference between a continuous spectrum and a discrete spectrum
2. Demonstration of the fact that individual elements have spectra that are characteristic of the element
3. Determination of the average spacing between the lines of a diffraction grating by assuming the characteristic wavelengths of mercury are known
4. Determination of the characteristic wavelengths of helium

EQUIPMENT LIST

1. Optical bench
2. Diffraction grating (600 lines/mm replica grating)
3. Spectrum-tube power supply
4. Mercury and helium discharge tubes
5. Meter stick and slit arrangement
6. Incandescent light bulb (15 W)

THEORY

When light is separated into its component wavelengths, the resulting array of colors is called a "spectrum." If a light source produces all the colors of visible light it is called a "continuous spectrum." It is given this name because there is no distinct beginning or end to any one color, but rather one color fades continuously into another. Generally, such sources of light are produced by heated solid metal filaments. For example, an ordinary incandescent light bulb with a tungsten filament produces such a continuous spectrum.

Other light sources produce only certain discrete wavelengths of light, and the spectrum appears as mostly dark with a few discrete lines of color at the wavelengths emitted by the source. Such light sources are produced by hot discharges of gas of a single chemical element, and the wavelengths of light emitted are characteristic of the electronic structure of that element. This spectrum is called a "discrete spectrum" or a "line spectrum." The term *line spectrum* is used because the images produced are usually images of a narrow slit that is illuminated by the light source.

There are several methods that can be used to separate a light source into its component wavelengths and thus produce a spectrum. This laboratory will use a diffraction grating to produce spectra from an incandescent light bulb and from gas-discharge tubes of mercury and helium.

A transmission diffraction grating is a piece of transparent material on which has been ruled a large number of equally spaced parallel lines. The distance between the lines is called the "grating spacing," or d, and it is usually only a few times as large as a typical wavelength of visible light. Wavelengths of visible light are in the range from about 4×10^{-7} to 7×10^{-7} m. It is customary to express the wavelength of light in units of nm where 1 nm $= 1 \times 10^{-9}$ m. In those units the range of visible light is from 400 to 700 nm. Grating spacings d are thus in the range 1000 to 2000 nm.

The wavelengths of light are associated with the color of the light as seen by the human eye. Starting from short wavelength and going to long wavelength the order of colors is violet, blue, green, yellow, orange, and red. The actual range of the visible spectrum is somewhat different for individuals, and there may be a distinct difference in the ability of two laboratory partners to see the wavelengths at either end of the spectrum. It is often very difficult for some people to see the very short wavelengths.

Light rays that strike the transparent portion of the grating between the ruled lines will pass through the grating at all angles with respect to their original path. If the deviated rays from adjacent rulings on the grating are in phase, an image of the source will be formed. This will be true when the adjacent rays differ in path length by an integral number of wavelengths of the light. Thus, for a given wavelength λ there will exist a series of angles at which an image is formed. The first time an image is formed will occur when the path difference between adjacent rays from adjacent ruled lines is exactly equal to one wavelength λ. According to Figure 42.1, that condition will be true at an angle θ_1 such that the equation

$$\lambda = d \sin\theta_1 \tag{1}$$

is satisfied. At some larger angle θ_2, when the path difference between adjacent rays from adjacent ruled lines is exactly equal to 2λ, then the equation

$$2\lambda = d \sin\theta_2 \tag{2}$$

is satisfied. In general, an image will be formed at any angle θn for which the adjacent rays from adjacent rulings have a path difference equal to $n\lambda$, where n is an integer called the "order number." Thus, the general case is described by the equation

$$n\lambda = d \sin\theta_n \tag{3}$$

The number of orders that can be seen are determined by d and λ. Equation 3 is valid only for integer values of n and for angles θ_n up to 90°. For a given λ and d, the maximum n is the one that gives the largest value of $\sin\theta$ less than 1. Although it will be possible to see both first-order ($n = 1$) and second-order ($n = 2$) for the experimental arrangement used in this laboratory, measurements will be made only on the first-order images.

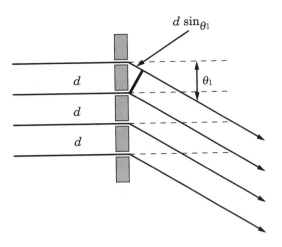

Figure 42.1 Ray diagram for the conditions of the first order diffraction image.

The experimental arrangement is shown in Figure 42.2. The discharge-tube light source is viewed through the grating as shown. The distance L from the grating to the slit is chosen at a convenient value and then kept fixed. If the source has a number of different wavelengths, the first-order image for each wavelength will occur at different angles and thus at different distances D from the slit as defined by Figure 42.2. The angle θ corresponding to each wavelength can be determined by measuring D with L fixed and known.

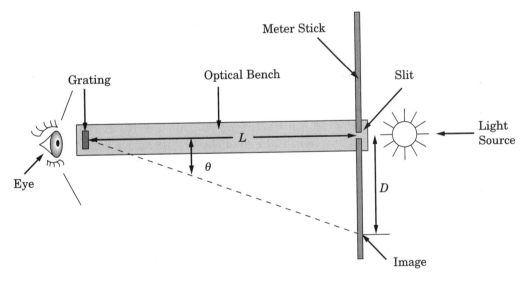

Figure 42.2 Arrangement of the diffraction grating, slit, light source, and optical bench.

If λ and n are assumed known in equation 3, then d can be determined. In the first part of the laboratory, a mercury light source will be used and its wavelengths given. A series of measurements will be made to accurately determine the value of the grating spacing d. In the second part of the laboratory, using this value of d for the grating, the wavelengths of a helium source will be determined.

EXPERIMENTAL PROCEDURE

1. Set up the experimental arrangement shown in Figure 42.2. Place the slit near one end of the optical bench just above the meter stick, which is held by the same holder that holds the slit. The meter stick should be perpendicular to the axis of the optical bench and should be level. The zero of the meter stick should be to the left, with the markings increasing to the right. The slit should be at the 50.00-cm mark on the meter stick just above the meter stick. The slit is located just above the meter stick so that its images will be just above the meter stick and can be easily located relative to a mark on the meter stick. Place the grating some distance L away from the slit with the plane of the grating perpendicular to the axis of the optical bench. Record the value (to the nearest 0.1 mm) of L in Data Table 1 and take all the data at this same value for the grating to slit distance.

2. *Use extreme caution with the discharge-tube power supply. It produces 5000 V and sufficient current to make it potentially lethal. Do not touch the supply electrodes while the supply is turned on.* With the power supply turned off and unplugged, place the mercury-discharge tube into the electrode receptacles. Place the supply behind the slit with the discharge tube as close to the slit as possible. It will probably be necessary to place blocks or books under the power supply to adjust the height of the discharge tube. The narrow portion of the discharge tube (which is the most intense) must be at the height of the slit.

3. *Turn on the power supply now. Do not accidentally touch the power supply electrodes while making the following adjustments.* While one partner looks through the grating directly at the slit, the other partner should make very fine adjustments in the position of the power supply to place the bright narrow portion of the discharge tube in alignment with the slit. Proper alignment is achieved when the slit is as bright as possible as seen by the person looking through the grating directly at the slit. It is critical that the light source be positioned such that the slit is as bright as possible. The slightest movement of the light source relative to the slit after this adjustment has been made may severely alter the brightness of the images seen.

4. Look through the grating to the right and left of the slit. Just above the meter stick there should appear a series of images of the slit in various colors. It may be necessary to rotate the grating in its holder to place the images in the horizontal. The images may originally appear at any angle to the horizontal up to the extreme case of 90°, in which case they would be in the vertical. Rotate the grating until the images are horizontal and just above the meter stick.

5. In Data Table 1 are listed seven wavelengths of mercury that should be prominent. They are listed in the order of increasing wavelength. They should appear in this order with the smallest wavelength at the smallest angle. Try to match the images that you see with the wavelengths given. It may be difficult to identify all seven of the lines. In particular, many people have difficulty seeing the violet lines clearly. Looking through the grating, locate the position of the first-order images that are to the right of the optical bench above the meter stick. One partner should locate the position of a given line by having the other partner move a small pointer (for example, a pencil point) along the meter stick until the pointer is in line with a given image. It may be helpful to use the 15-W light source to illuminate the meter stick to read the position once it has been located. Record (to the nearest 0.1 mm) the position P_R of each of the seven wavelengths in Data Table 1.

6. Repeat the process for the images on the left of the optical bench corresponding to the seven wavelengths. Record (to the nearest 0.1 mm) the position P_L of each image in Data Table 1.

7. Turn off the power supply and allow the discharge tube to cool. With the power supply still off, remove the mercury-discharge tube and replace it with the helium-discharge tube. Position the power supply with the discharge tube aligned with the slit, turn on the power supply, and adjust the position of the supply for maximum brightness as done previously for the mercury-discharge tube.

8. In Data Table 2 are listed eight wavelengths of helium that should be visible. Again try to match the images that you see with the wavelengths given. Perform the same procedure as done above for mercury, measuring the positions P_R and P_L of each image on the right and the left. Record (to the nearest 0.1 mm) the data in Data Table 2.

9. Place the 15-W light bulb behind the slit and observe the continuous spectrum. Locate the positions P_R and P_L of the following parts of the spectrum: (a) the shortest wavelength visible, (b) the division between blue and green, (c) the division between green and yellow, (d) the division between yellow and orange, (e) the division between orange and red, and (f) the longest wavelength visible. Record (to the nearest 0.1 mm) the position of these points in Data Table 3.

CALCULATIONS

1. Calculate the distance D_R from the slit to each image on the right ($D_R = P_R - 50.0$) and calculate the distance D_L from the slit to each image on the left ($D_L = 50.0 - P_L$) for the mercury data in Data Table 1. Calculate the average distance ($\overline{D} = [D_R + D_L]/2$) and calculate $\tan\theta = \overline{D}/L$, θ, and $\sin\theta$ for each image. Record those values (to four significant figures) in Calculations Table 1.

2. Each of the measurements on mercury is an independent measurement for d, the grating spacing. Using equation 1, calculate the seven values of d from the seven wavelengths and record them (to four significant figures) in Calculations Table 1.

3. Calculate the mean $\overline{d}$ and the standard error α_d for the seven values of d and record them in Calculations Table 1.

4. Calculate the values of D_R, D_L, and $\overline{D}$ for the helium data in Data Table 2 and calculate $\tan\theta = \overline{D}/L$, θ, and $\sin\theta$ for each image. Record those values (to four significant figures) in Calculations Table 2.

5. Using equation 1, calculate the wavelengths of helium from the values of $\sin\theta$ in step 4. Use the value of $\overline{d}$ determined in the mercury measurements for the grating spacing d. Record the results (to four significant figures) in Calculations Table 2.

6. From the data in Data Table 3 for the continuous spectrum, determine the wavelength corresponding to the various points in the spectrum that were located. Calculate and record all the information called for in Calculations Table 3.

Laboratory 42
Diffraction Grating Measurement of the Wavelength of Light

LABORATORY REPORT

Data Table 1

	λ (nm)	P_R (cm)	P_L (cm)
L = _____ cm			
Mercury Colors			
Violet	404.7		
Violet	407.8		
Blue	435.8		
Blue-green	491.6		
Green	546.1		
Yellow	577.0		
Yellow	579.0		

Calculations Table 1

λ (nm)	D_R (cm)	D_L (cm)	$\overline{D}$ (cm)	$\tan\theta$	θ	$\sin\theta$	d (nm)
404.7							
407.8							
435.8							
491.6							
546.1							
577.0							
579.0							
$\overline{d}$ = _____ nm				α_d = _____ nm			

Data Table 2

Helium Wavelengths	Helium Colors	P_R (cm)	P_L (cm)
	$L =$ _____ cm		
438.8 nm	Blue-violet		
447.1 nm	Deep Blue		
471.3 nm	Blue		
492.2 nm	Blue-green		
501.5 nm	Green		
587.6 nm	Yellow		
667.8 nm	Red		
706.5 nm	Red		

Calculations Table 2

D_R (cm)	D_L (cm)	$\overline{D}$ (cm)	$\tan \theta$	θ	$\sin \theta$	λ (nm)

SAMPLE CALCULATIONS

Data Table 3

Portion of Spectrum	P_R	P_L
Shortest Wavelength		
Division Blue and Green		
Division Green and Yellow		
Division Yellow and Orange		
Division Orange and Red		
Longest Wavelength		

Calculations Table 3

Portions of Spectrum	$\overline{D}$ (cm)	$\tan\theta$	θ	$\sin\theta$	λ (nm)
Shortest Wavelength					
Division Blue and Green					
Division Green and Yellow					
Division Yellow and Orange					
Division Orange and Red					
Longest Wavelength					

SAMPLE CALCULATIONS

QUESTIONS

1. Comment on the precision of your measurement of d.

2. The values for the eight wavelengths of helium are 438.8, 447.1, 471.3, 492.2, 501.5, 587.6, 667.8, and 706.5 nm. Calculate the percentage error in your measured values compared to these true values including the sign of the error. Comment on the accuracy of your measurements.

λ = 438.8 nm   % Error = _____

λ = 447.1 nm   % Error = _____

λ = 471.3 nm   % Error = _____

λ = 492.2 nm   % Error = _____

λ = 501.5 nm   % Error = _____

λ = 587.6 nm   % Error = _____

λ = 667.8 nm   % Error = _____

λ = 706.5 nm   % Error = _____

3. If all the errors in question 2 are of the same sign it might be evidence of a systematic error. Do your data show evidence of a systematic error?

4. If the grating were actually 600 lines/mm, the value of d would be 1667 nm. If you used that value of d with the values of $\sin\theta$ in Calculations Table 2 would the resulting wavelengths for helium be better or worse than the wavelength values in that table?

5. Hydrogen has known emission lines of wavelength 656.3 nm and 434.1 nm. At what distance D away from the slit would each of these lines be observed in your experimental arrangement? Show your work.

6. In the continuous spectrum what is the range of yellow wavelengths? What is the range of orange wavelengths? What is the middle of the visible spectrum according to your measured values of the range of the visible spectrum?

Laboratory 43

Bohr Theory of Hydrogen—The Rydberg Constant

PRELABORATORY ASSIGNMENT

Read carefully the entire description of the laboratory and answer the following questions based on the material contained in the reading assignment. Turn in the completed prelaboratory assignment at the beginning of the laboratory period prior to the performance of the laboratory.

1. State the Balmer formula for the visible light spectrum of hydrogen.

2. Using the Balmer formula calculate the first four wavelengths of the spectrum corresponding to $n = 3, 4, 5,$ and 6. Show your work.

3. Describe the possible orbits of an electron in a hydrogen atom that are allowed by the Bohr theory.

4. What is a stationary state of the atom in Bohr theory?

5. State Bohr's postulate about the frequency f of light emitted when an electron makes a transition from a state of energy E_i to one of E_f.

6. In the Bohr theory the Rydberg constant turns out to be equal to $me^4/8\varepsilon_0^2 ch^3$. Using accepted values for the constants m, e, ε_0, c, and h, calculate the value of the Rydberg constant. Show your work.

7. A diffraction grating has a grating constant of $d = 1.500 \times 10^{-6}$ m. At what angle θ will the first order image of light of wavelength 5.555×10^{-7} m appear?

Bohr Theory of Hydrogen—The Rydberg Constant

OBJECTIVES

The spectrum from a hot gas of an element consists of discrete wavelengths that are characteristic of the element. In 1885 in an attempt to understand these spectra, Johann Balmer published an empirical relationship that described the visible spectrum of hydrogen. Although Balmer published the relationship in a somewhat different form, the modern equivalent of the Balmer formula is

$$\frac{1}{\lambda} = R_{\text{H}}\left(\frac{1}{2^2} - \frac{1}{n^2}\right) \quad n = 3, 4, 5, 6, \ldots \tag{1}$$

where $R_{\text{H}} = 1.097 \times 10^7$ m^{-1} is a constant called the "Rydberg constant," λ stands for the wavelength, and n is an integer that takes on successive values greater than 2. In 1913, Neils Bohr was able to derive the Balmer relationship by making a series of revolutionary postulates. The Bohr theory was historically of great importance in the developments that eventually led to modern quantum theory. In this laboratory the wavelengths of the hydrogen spectrum will be determined and used to accomplish the following objectives:

1. Demonstration of the agreement between the measured values of the wavelength and those predicted by the Bohr theory
2. Determination of an experimental value for the Rydberg constant from a linear least squares fit of the measured values of wavelengths to the form of equation 1

EQUIPMENT LIST

1. Spectrometer, diffraction grating in holder (600 lines/mm or better)
2. Hydrogen gas–discharge tube, mercury-discharge tube
3. Power supply for the discharge tubes

THEORY

In his attempts to explain the spectrum of hydrogen, Neils Bohr was influenced by several theories that had recently been developed. He incorporated concepts from the quantum theory of Max Planck, from the photon description of light by Albert Einstein, and from the nuclear theory of the atom suggested by Ernest Rutherford's α-particle scattering from gold. The central ideas of Bohr theory are contained in a series of four postulates that can be stated as follows:

1. The electron is assumed to move in circular orbits around the nucleus. The Coulomb attraction between the negative electron and the positive nucleus is the force on the electron.

2. Only orbits of certain discrete radii are allowed. The allowed orbits are those for which the angular momentum of the electron relative to the nucleus is an integral multiple of $h/2\pi$, where Planck's constant $h = 6.626 \times 10^{-34}$ J-s. In equation form this is

$$mvr = \frac{nh}{2\pi} \quad n = 1, 2, 3, 4, \ldots \tag{2}$$

where m is the electron mass, v is the electron velocity, and r is the radius of the electron's orbit.

3. When the electron is in one of its allowed orbits, it does not radiate energy. In these orbits the atom is stable, and the atom is said to be in a stationary state. This postulate was a radical departure from classical physics. Classical electromagnetic theory predicts that an electron moving in a circle is accelerated and thus must radiate electromagnetic energy continuously.

4. The atom radiates energy only when an electron makes a transition from one of the allowed orbits to another of the allowed orbits. Because the radii are discrete, the energy levels of the stationary states are discrete. If E_i and E_f stand for the energies of the initial and final stationary states, then the energy radiated by the atom is in the form of a photon of energy hf, where f is the frequency of the photon. Bohr's postulate states that

$$hf = E_i - E_f \tag{3}$$

With these postulates it is possible to derive an expression for the energy of the stationary states. They are given by

$$E_\mathrm{n} = -\frac{me^4}{8\varepsilon_0{}^2 h^2} \frac{1}{n^2} \quad n = 1, 2, 3, 4, \ldots \tag{4}$$

The transitions that correspond to the visible Balmer spectrum are those transitions from all of the states above ($n = 3, 4, 5, 6, \ldots$) down to the $n = 2$ state. The frequencies of those transitions can be shown to be

$$f = \frac{me^4}{8\varepsilon_0{}^2 h^2} \left(\frac{1}{2^2} - \frac{1}{n^2} \right) \quad n = 3, 4, 5, 6, \ldots \tag{5}$$

Since $c = f\lambda$, equation 5 can be written in terms of the wavelength λ as

$$\frac{1}{\lambda} = \frac{me^4}{8\varepsilon_0{}^2 h^3} \left(\frac{1}{2^2} - \frac{1}{n^2} \right) \quad n = 3, 4, 5, 6, \ldots \tag{6}$$

Bohr was able to show that the value of the constant $me^4/8\varepsilon_0{}^2 ch^3$ was in excellent agreement with the value of the Rydberg constant in Balmer's formula. This is, of course, striking confirmation for the validity of the Bohr theory of hydrogen.

The four wavelengths of the visible hydrogen spectrum that are easily seen and measured are in excellent agreement with the first four wavelengths predicted by equation 6. This will be demonstrated by measuring the value of λ for those four wavelengths and performing a linear least squares fit with $1/\lambda$ as the ordinate and the quantity $([1/2^3] - [1/n^3])$ as the abscissa, where $n = 3, 4, 5,$ and 6 for the four measured wavelengths. The correlation coefficient of the least squares fit is a measure of the agreement of Bohr theory with the data. The slope of the fit is an experimental value of the Rydberg constant R_H.

The wavelengths will be measured with a diffraction grating spectrometer (Figure 43.1). (For details of diffraction grating theory, see the theory section of Laboratory 42.) The grating spacing d will be determined from the first-order diffraction grating equation

$$\lambda = d \sin\theta \qquad (7)$$

Initially the grating spacing d will be assumed to be unknown, and the wavelengths of mercury will be considered as known. Measurements of the angles at which the mercury spectrum occur can then be used to determine d from equation 7. Using that value of d, measurements of the angles at which the hydrogen wavelengths occur will allow the determination of those wavelengths.

Figure 43.1 Spectrometer that can be used with a prism or with a diffraction grating.

EXPERIMENTAL PROCEDURE

1. Place the diffraction grating (in its holder) on the spectrometer table as shown in Figure 43.2. Place the mercury spectrum tube between the electrodes of the spectrum tube power supply. *DO NOT TOUCH THE HIGH-VOLTAGE ELECTRODES WHILE THE POWER SUPPLY IS ON. IT PROVIDES A VOLTAGE OF 5000 V.*

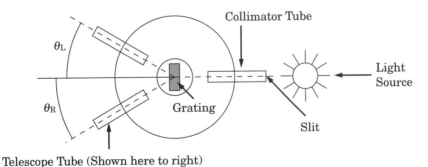

Figure 43.2 Experimental arrangement for the diffraction grating spectrometer.

2. Turn on the power supply and place the spectrum tube as close to the slit in the collimator tube as is possible. Rotate the telescope tube of the spectrometer until it is directly in line with the collimator tube. Adjust the slit in the collimator and the eyepiece of the telescope until a sharp image of the slit is obtained. The vertical cross hair of the telescope must also be in focus and in the center of the slit.

3. Rotate the telescope tube to the left or right until images of the spectral lines for mercury are located. The wavelengths of the mercury spectrum (with the relative intensities in parentheses) are violet 4.047×10^{-7} m (1800), blue 4.358×10^{-7} m (4000), blue-green 4.916×10^{-7} m (80), green 5.461×10^{-7} m (1100), yellow 5.770×10^{-7} m (240), yellow 5.790×10^{-7} m (380). Rotate the telescope tube to the other side to be sure that all the lines can be located. This is just a preliminary check to be sure that all the lines are visible. It may not be possible to resolve the two yellow lines. If not, just assume one line at 5.780×10^{-7} m. It is extremely important that the grating is never moved after it is originally positioned.

4. Carefully measure to the nearest 1 minute of arc the angle at which each of the wavelengths of mercury occurs on both sides of 180°. The spectrometer will probably have a vernier scale capable of reading to 1 minute of arc. Consult your instructor for directions in the use of the vernier scale. If the spectrometer does not have a vernier, estimate the angles with as much precision as possible. Record the two angles for each of the wavelengths in Data Table 1. Be sure that the diffraction grating does not ever move.

5. Without moving the diffraction grating, turn off the spectrum-tube power supply. Carefully remove the mercury tube and replace it with the hydrogen tube. Turn on the supply and place the hydrogen tube as close to the slit as possible. *AGAIN BE VERY CAREFUL NOT TO TOUCH THE HIGH-VOLTAGE ELECTRODES WHILE MAKING THESE ADJUSTMENTS IN THE POSITION OF THE SUPPLY.* Rotate the telescope tube back to 180° and carefully adjust the position of the hydrogen tube until a sharp image of the slit is seen directly through the grating. Everything should be in focus from the mercury measurements, and it should just be necessary to place the hydrogen tube in the correct position to give the brightest image. At all times be extremely careful not to move the grating.

6. Carefully measure to 1 minute of arc the angle at which the first four wavelengths of the visible hydrogen spectrum occur on both sides of 180°. In order of decreasing wavelength they are red, blue-green, blue, and violet. Record the two angles for each of the wavelengths in Data Table 2.

CALCULATIONS

1. For the mercury data in Data Table 1 calculate the diffraction angle θ for each wavelength from $\theta = | \theta_R - \theta_L |/2$. See Figure 43.2 for a description of θ_R and θ_L. Record these values of θ in Calculations Table 1.

2. For each of the wavelengths of mercury, calculate a value for the grating spacing d using equation 7 and record them in Calculations Table 1.

3. Calculate the mean $\overline{d}$ and standard error α_d for the measured values of d from the mercury spectrum. Record the values of $\overline{d}$ and α_d in Calculations Table 1.

4. For the hydrogen data in Data Table 2 calculate the diffraction angle θ from $\theta = | \theta_R - \theta_L |/2$ and record these values in Calculations Table 2.

5. Using the values of θ for hydrogen and the value of $\overline{d}$ in equation 7, calculate the experimental values for the wavelengths of hydrogen λ_{exp} and record them in Calculations Table 2.

6. Using these experimental values of the hydrogen wavelengths, calculate the quantity $1/\lambda$ for each wavelength and record the results in Calculations Table 2.

7. For each of the hydrogen wavelengths and its associated value of n, calculate the quantity $([1/4] - [1/n^2])$ and record the results in Calculations Table 2.

8. Perform a linear least squares fit with $1/\lambda$ as the ordinate and $([1/4] - [1/n^2])$ as the abscissa. Determine the slope R_{exp}, the intercept I, and the correlation coefficient r and record them in Calculations Table 2.

9. Calculate the percentage error for each of the experimental values of the hydrogen wavelengths and record them in Calculations Table 2.

10. Calculate the percentage error for the experimental value of the Rydberg constant R_{exp} compared to the known value of 1.097×10^7 m^{-1}. Record the results in Calculations Table 2.

GRAPHS

Make a graph with the experimental values of $1/\lambda$ as the ordinate and $([1/4] - [1/n^2])$ as the abscissa. Also show on the graph the straight line obtained from the linear least squares fit to the data.

Laboratory **43**

Bohr Theory of Hydrogen—The Rydberg Constant

LABORATORY REPORT

Data Table 1

Mercury Colors	λ (10^{-7}m)	θ_R (degrees)	θ_L (degrees)
Violet	4.047		
Blue	4.358		
Blue-green	4.916		
Green	5.461		
Yellow	5.770		
Yellow	5.790		

Calculations Table 1

Mercury Colors	θ (degrees)	d (10^{-7}m)	$\bar{d}$ (10^{-7}m)	α_d (10^{-7}m)
Violet				
Blue				
Blue-green				
Green				
Yellow				
Yellow				

Data Table 2

Hydrogen Colors	λ (10^{-7}m)	n	θ_R (degrees)	θ_L (degrees)
Red	6.563	3		
Blue-green	4.861	4		
Blue	4.341	5		
Violet	4.102	6		

Calculations Table 2

n	θ (degrees)	λ_{exp} (m)	$1/\lambda_{exp}$ (m^{-1})	$(1/4) - (1/n^2)$	% Error in λ
3					
4					
5					
6					
$R_{exp} =$		$I =$		r =	
Percentage error in the experimental value of $R_{exp} =$					

SAMPLE CALCULATIONS

QUESTIONS

1. What is the accuracy of your experimental value for the Rydberg constant R_H?

2. What is the accuracy of your experimental values for the wavelengths of hydrogen?

3. What is the precision of your determination of the value of the grating spacing d?

4. Which of the results discussed in questions 1, 2, and 3 is the best quantitative evidence for the agreement of your data with the Balmer equation?

5. The diffraction grating has some nominal value for the number of lines/mm stamped on the box in which it is stored. From it you can calculate the grating spacing d. Considering the agreement of your data for the hydrogen wavelengths, is your measured value for d more accurate than the one determined from the information about the number of lines/mm? State clearly the evidence for your answer.

6. Using the Balmer formula (equation 1), calculate the $n = 7$ wavelength for the hydrogen spectrum. Why was it not observed in the laboratory?

7. The intensity of the blue wavelength of mercury is given as about four times as great as the intensity of the green wavelength of mercury. However, the green line appears brighter to your eye. Why is this true?

Laboratory 44

Simulated Radioactive Decay Using Dice "Nuclei"

PRELABORATORY ASSIGNMENT

Read carefully the entire description of the laboratory and answer the following questions based on the material contained in the reading assignment. Turn in the completed prelaboratory assignment at the beginning of the laboratory period prior to the performance of the laboratory.

1. A typical sample of radioactive material would contain as a lower limit approximately how many nuclei? (a) 1000, (b) 10^6, (c) 10^{12}, or (d) 10^{23}

2. The theory of radioactive decay can predict when each of the radioactive nuclei in a sample will decay. (a) true (b) false

3. State the definition of the decay constant λ. What are its units?

4. A radioactive decay process has a decay constant $\lambda = 1.50 \times 10^{-4}$ s^{-1}. There are 5.00×10^{12} radioactive nuclei in the sample at $t = 0$. How many radioactive nuclei are present in the sample 1 hour later? Show your work.

5. For the radioactive sample described in question 4, what is the activity A (in decays per second) at $t = 0$? What is the activity 1 hour later? Show your work.

6. What is the half-life of the radioactive sample described in question 4? Show your work.

7. For the simulation of a radioactive decay using 20-sided dice, what are the analogous quantities to the real quantities listed below?

undecayed nucleus—

decayed nucleus—

time—

decay constant—

8. What quantity can be measured for the simulation experiment that cannot normally be directly measured in a real radioactive decay experiment?

Simulated Radioactive Decay using Dice "Nuclei"

OBJECTIVES

In a radioactive source containing a very large number of radioactive nuclei, it is not possible to predict when any one of the nuclei will decay. Although the decay time for any one particular nucleus cannot be predicted, the average rate of decay of a large sample of radioactive nuclei is highly predictable. This laboratory uses 20-sided dice with two marked faces to simulate the decay of radioactive nuclei. When a marked face of a die is up after a throw of the dice, it represents a decay. Measurements on a collection of these dice will be used to accomplish the following objectives:

1. Demonstration of the analogy between the decay of radioactive nuclei and the decay of dice "nuclei"

2. Demonstration that both the number of "nuclei" not yet decayed (N) and the rate of decay (dN/dt) both decrease exponentially

3. Determination of experimental and theoretical values of the decay probability constant λ for the dice "nuclei"

4. Determination of the experimental and theoretical values for the half-life of the dice "nuclei"

EQUIPMENT LIST

1. 20-sided dice to simulate radioactive nuclei
2. Three-cycle semilog graph paper

THEORY

One of the most noticeable differences between classical physics known prior to 1900 and modern physics since that time is the increased role that probability plays in modern physical theories. The exact behavior of many physical systems cannot be predicted in advance. On the other hand, there are some systems that involve a very large number of possible events, each of which is not predictable; and yet, the behavior of the system as a whole is quite predictable. One example of such a system is a collection of radioactive nuclei that emit α-, β-, or γ-radiation. It is not possible to predict when any one radioactive nucleus will decay and emit a particle. However, since any reasonable sample of radioactive material contains such a large number of nuclei (say at least 10^{12} nuclei), it is possible to predict the average rate of decay with high probability.

A basic concept of radioactive decay is that the probability of decay for each type of radioactive nuclide is constant. In other words, there are a predictable number of decays per second even though it is not possible to predict which nuclei among the sample will decay. A quantity called the "decay constant," or λ, characterizes this concept. It is the probability of decay per unit time for one radioactive nucleus. The fundamental concept is that because λ is constant, it is possible to predict the rate of decay for a radioactive sample. The value of the constant λ is, of course, different for each radioactive nuclide.

Consider a sample of N radioactive nuclei with a decay constant of λ. The rate of decay of these nuclei dN/dt is related to λ and N by the equation

$$\frac{dN}{dt} = -\lambda N \qquad (1)$$

The symbol dN/dt stands for the rate of change of N with time t. The minus sign in the equation means that dN/dt must be negative because the number of radioactive nuclei is decreasing. The number of radioactive nuclei at $t = 0$ is designated as N_0. The question of interest is how many radioactive nuclei N are left at some later time t. The answer to that question is found by rearranging equation 1 and integrating it subject to the condition that $N = N_0$ at $t = 0$. The result of that procedure is

$$N = N_0 e^{-\lambda t} \qquad (2)$$

Equation 2 states that the number of nuclei N at some later time t decreases exponentially from the original number N_0 that are present. A second question of interest is the value of the rate of decay dN/dt of the radioactive sample. That can be found by substituting the expression for N from equation 2 back into equation 1. The result is

$$\frac{dN}{dt} = -\lambda N_0 e^{-\lambda t} \qquad (3)$$

Furthermore, the expression for dN/dt in equation 1 can then be substituted into equation 3, leading to

$$\lambda N = \lambda N_0 e^{-\lambda t} \qquad (4)$$

The quantity λN is the activity of the radioactive sample. Since λ is the probability of decay for one nucleus, the quantity λN is the number of decays per unit time for N nuclei. Typically, λ is expressed as the probability of decay per second; so in that case, λN is the number of decays per second from a sample of N nuclei. The symbol A is used for activity ($A = \lambda N$); thus, equation 4 becomes

$$A = A_0 e^{-\lambda t} \qquad (5)$$

Equations 2 and 5 thus state that both the number of nuclei N and the activity A decay exponentially according to the same exponential factor. For measurements made on real radioactive nuclei, the activity A is the quantity that is usually measured.

An important concept associated with radioactive decay processes is the concept of half-life. The time for the sample to go from the initial number of nuclei N_0 to half that value $N_0/2$ is defined as the half-life $t_{1/2}$. If equation 2 is solved for the time t when $N = N_0/2$ the result is

$$t_{1/2} = \frac{\ln(2)}{\lambda} = \frac{0.693}{\lambda} \qquad (6)$$

This same result could also be obtained by considering the time for the activity to go from A_0 to $A_0/2$.

Figure 44.1 shows graphs of activity of a radioactive sample versus time. Figure 44.1(a) shows a semilog graph with the activity scale logarithmic and the time scale linear. The time scale is simply marked in units of the half-life. Note that the graph is linear on this semilog plot. The half-life is the time to go from any given value of activity to half that activity. Figure 44.1(b) shows the shape of the activity versus time graph if linear scales are chosen for both quantities.

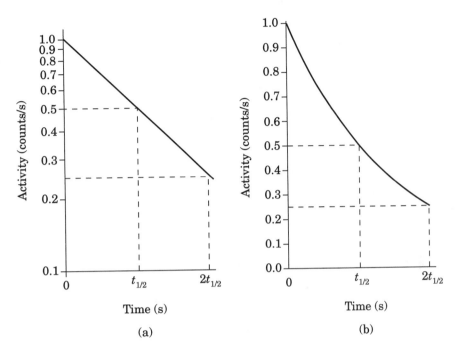

(a)   (b)

Figure 44.1 Graph of activity versus time on semilog and on linear scales.

The laboratory exercise to be performed does not involve the decay of real radioactive nuclei. Instead it is designed to illustrate the concepts described above by a simulated decay of dice "nuclei." In the exercise, radioactive nuclei are simulated by a collection of 160 dice with 20 faces. Two of the 20 faces of each die are marked with a dot. The dice are shaken and thrown, and a dice "nuclei" has decayed if a marked face is up after the throw. In this simulation, the decay constant λ is equal to the probability of 2 out of 20 of a marked face coming up. Thus, the theoretical decay constant λ is 0.100. For the analogy, each throw of the dice is one unit of time.

A unique aspect of this simulation experiment is that measurements can be taken on both the number remaining N and the number that decay. The number that decay is analogous to the activity A. For real radioactive nuclei, N cannot be measured directly but is inferred from measurements of the activity A.

EXPERIMENTAL PROCEDURE

1. Place all 160 of the dice in the square plastic tray provided with the dice. Place the clear plastic cover over the dice and shake them vigorously. Remove all dice that come to rest with a marked face up. Remove only those that have a marked face pointing directly upward. Record in Data Table 1 the number of dice that decay (are removed) on the first throw of the dice. Also record the number of dice remaining after the ones that decay are removed.

2. Place the cover on the dice and vigorously shake the remaining dice. Remove the dice that come to rest with a marked face up on the second throw. Record in Data Table 1 the number of dice removed on the second throw, and also record the number of dice remaining after the ones that decay are removed.

3. Continue this process of shaking the dice, removing the ones that have a marked face up, and recording the number of dice removed and the number of dice left for each throw. Continue this procedure for a total of 20 throws of the dice or until all of the dice have been removed.

4. Each experimental group should record its data on the blackboard so that the class results can be plotted as a set of data with better statistics. Sum the total number of dice thrown originally for the entire class and sum the number removed at each throw of the dice. Record this class data in Data Table 2.

GRAPHS

1. On the three-cycle semilog graph paper provided by the instructor, graph the results. Plot N, the number of "nuclei" remaining on the log scale, versus the number of throws on the linear scale, using x as a symbol. On the same piece of graph paper plot the activity A (number removed each throw) on the log scale versus the number of throws on the linear scale, using + for a symbol.

2. On a second piece of three-cycle semilog graph paper graph the class data for N and A. With better statistics these curves should be smoother than the individual data.

3. If the dice behaved exactly according to the theory described, all of the graphs described above would fall on a straight line on the semilog plot. The data for N will most likely show this trend better than the data for A.

4. For the individual data and the class data draw a straight line that best fits the data. Do this for both N and A.

CALCULATIONS

1. For each of the 20 shakes of the dice, calculate the ratio of the number of dice removed after a given throw to the number shaken for that throw. In the simulation, the number of dice removed is A, the activity, and the number shaken is N, the number of radioactive nuclei. Thus, these ratios amount to an experimental value for λ. Note carefully that this ratio must be calculated with data from two different rows in Data Table 1. For example, the number of dice *thrown* on the fourth throw is listed as the number of dice *remaining* after the third throw. Thus, the ratio is calculated with the number removed on each row to the number remaining in the preceding row. Record these 20 values as λ_{exp} in Calculations Table 1.

2. Calculate the mean of these 20 values for λ_{exp} and record it in Calculations Table 2 as $\overline{\lambda_{exp}}$.

3. The theoretical value of λ is 0.100. Record this value in Calculations Table 2 as λ_{theo}.

4. Calculate the percentage error in the value of $\overline{\lambda_{exp}}$ compared to λ_{theo}. Record this percentage error in Calculations Table 2.

5. Calculate the theoretical half-life from equation 6 using the value of $\lambda = 0.100$. Record that value in Calculations Table 2 as $(t_{1/2})_{theo}$. For purposes of this calculation consider a fractional throw as possible.

6. Consider the straight line drawn through the data points of your individual data for N versus number of throws. The experimental half-life is the number of throws needed to go from any point on the line to half that value. Determine the number of throws needed to go from 120 to 60 on the straight line through your data. Record that number in Calculations Table 2 as $(t_{1/2})_{exp}$. For purposes of this determination consider a fractional throw as possible.

7. Calculate the percentage error in the value $(t_{1/2})_{exp}$ compared to the value of $(t_{1/2})_{theo}$. Record that percentage error in Calculations Table 2.

Simulated Radioactive Decay using Dice "Nuclei"

LABORATORY REPORT

Data Table 1

Throw Number	Dice Removed (A)	Dice Remaining (N)
0	0	160
1		
2		
3		
4		
5		
6		
7		
8		
9		
10		
11		
12		
13		
14		
15		
16		
17		
18		
19		
20		

Calculations Table 1

λ_{exp}

Calculations Table 2

$\lambda_{theo} =$	$\overline{\lambda_{theo}} =$	% Error =
$(t_{1/2})_{theo} =$	$(t_{1/2})_{exp} =$	% Error =

Data Table 2

Throw Number	Dice Removed (A)	Dice Remaining (N)
0	0	
1		
2		
3		
4		
5		
6		
7		
8		
9		
10		
11		
12		
13		
14		
15		
16		
17		
18		
19		
20		

SAMPLE CALCULATIONS

QUESTIONS

1. Do your data for N as a function of the number of throws give a reasonably straight line? Would a line with the same slope as you drew through the N versus number of throws fit reasonably well through the A versus number of throws plot?

2. Comment on the agreement between your experimental value for λ and the theoretical value for λ.

3. Comment on the agreement between your experimental value for the half-life and the theoretical value of the half-life.

4. Are the graphs for the class data smoother and more nearly a straight line than your individual data?

5. Calculate the half-life from the class data graph. Compare it to the theoretical value for the half-life. Would you expect it to show better agreement, and if so, why?

Laboratory 45

Geiger Counter Measurement of the Half-Life of ^{137}Ba

PRELABORATORY ASSIGNMENT

Read carefully the entire description of the laboratory and answer the following questions based on the material contained in the reading assignment. Turn in the completed prelaboratory assignment at the beginning of the laboratory period prior to the performance of the laboratory.

1. What do each of the symbols stand for in the following notation for an isotope?
$$^{A}_{Z}X_{N}$$

2. What are the different types of radioactive decay? What kind of particles are produced in each type of decay?

3. Which radioactive decay process essentially amounts to either a neutron being turned into a proton or a proton being turned into a neutron? (a) α-decay, (b) β-decay, (c) γ-decay, or (d) neutron decay

4. Which radioactive decay process is analogous to the process in which atoms lose energy when they emit visible light? (a) α-decay, (b) β-decay, (c) γ-decay, or (d) neutron decay

5. What is the basis for the detection of the particles from any radioactive decay?

6. The inside of a Geiger tube is filled with a (a) gas, (b) liquid, (c) solid, or (d) plasma.

7. The pulse that is counted in a Geiger tube is caused by a (a) rise in voltage from ions arriving at the anode, (b) drop in the voltage from ions arriving at the anode, (c) rise in voltage from electrons arriving at the anode, (d) drop in voltage from electrons arriving at the anode.

8. In a Geiger tube, the total time taken to create a pulse and then let the electrons recombine with the ions to again form a neutral state is about (a) 1 μs, (b) 30 μs, (c) 300 μs, or (d) 10,000 μs.

9. A typical operating voltage for a Geiger counter is about (a) 5 V, (b) 50 V, (c) 500 V, or (d) 5000 V.

10. In this laboratory, the half-life of ^{137}Ba will be determined by measuring (a) the number of atoms left as a function of time, (b) the activity, which is constant, (c) the activity, which decreases as a function of time, or (d) the activity, which increases as a function of time.

Geiger Counter Measurement of the Half-Life of ^{137}Ba

OBJECTIVES

When radioactive nuclei decay they emit either α-, β-, or γ-radiation. The process of detecting such radiation involves the interaction of the radiation with matter in some form and the subsequent measurement of that interaction. This laboratory will use a short-lived ^{137}Ba isotope and a Geiger counter to accomplish the following objectives:

1. Measurement of the count rate versus voltage for a Geiger counter in order to determine its characteristics
2. Determination of the appropriate operating voltage for the Geiger counter based on the shape of the count rate versus voltage data
3. Measurement of the activity of the ^{137}Ba source as a function of time
4. Determination of the half-life of ^{137}Ba from the activity-versus-time data

EQUIPMENT LIST

1. Geiger counter and scaler with instruction manual
2. Timer
3. ^{90}Sr β-radiation source
4. ^{137}Ba source (minigenerator which produces ^{137}Ba from the decay of a parent nuclide of ^{137}Cs)
5. Disposable plachet (very thin small metal plate to contain a radioactive sample)
6. Three-cycle semilog graph paper

THEORY

Atoms are composed of a nucleus containing neutrons and protons surrounded by a cloud of electrons. A given chemical element contains a fixed number of protons in the nucleus but possibly differing numbers of neutrons. The forms of a given element having different numbers of neutrons are called "isotopes" of that element. The complete description of all aspects of a given isotope is given by the symbols $^{A}_{Z}X_{N}$, where X stands for the chemical symbol of the element; Z is the number of protons, which is called the atomic number; N is the number of neutrons; and A is called the mass number, where $A = Z + N$. As an example the symbol, $^{226}_{88}Ra_{138}$ stands for the isotope of radium with 88 protons and 138 neutrons, and thus 226 neutrons plus protons. Other isotopes of radium must also have 88 protons but can have a different number of neutrons. Usually the number of neutrons is not displayed explicitly and the symbol is given as $^{226}_{88}$Ra. It is inferred from this symbol that the number of neutrons is $226 - 88 = 138$.

Some nuclei are unstable and decay spontaneously into other nuclei. There are three different types of processes that occur naturally and produce natural radioactivity. In addition, there are a large number of radioactive isotopes that are produced by nuclear reactions for isotopes that no longer appear naturally. Many such isotopes are available commercially for many practical uses. Presumably, all the man-made radioactive isotopes were naturally occurring at one time, but they decayed away because their half-lives were much shorter than the age of the earth. For all types of radioactivity, the process is identified by the type of particle produced in the decay.

α-Decay is a process by which an unstable nucleus ejects a particle called an α-particle. In fact an α-particle is really just a twice ionized $^{4}_{2}$He atom. Thus, an α-particle is just the nucleus of a $^{4}_{2}$He atom consisting of the two protons and two neutrons left when the two electrons have been removed from a neutral atom. The process in which a radioactive nucleus $^{A}_{Z}X$ decays by α-decay is described by

$$^{A}_{Z}X \rightarrow {}^{4}_{2}He + {}^{A-4}_{Z-2}Y \quad \text{or} \quad {}^{A}_{Z}X \rightarrow \alpha + {}^{A-4}_{Z-2}Y \tag{1}$$

where α stands for the α-particle or $^{4}_{2}$He nucleus produced. As an example of α-decay consider the case when the radioactive nucleus is $^{226}_{88}$Ra. According to equation 1, the decay produces an α-particle and a daughter nucleus of $^{222}_{86}$Rn.

β-decay is a process taking place inside the nucleus that essentially amounts to either a neutron being turned into a proton or a proton being turned into a neutron. When this process happens, either an electron or an anti-electron (called a "positron") is produced. When the particle produced is an ordinary electron, it is called a β^- and when the particle produced is a positron, it is called a β^+. The process in which a radioactive nucleus $^{A}_{Z}X$ decays by one of the β-decay events is described by one of the equations shown below:

$$^{A}_{Z}X \rightarrow \beta^- + {}^{A}_{Z+1}Y \quad \text{and} \quad {}^{A}_{Z}X \rightarrow \beta^+ + {}^{A}_{Z-1}Y \tag{2}$$

Actually equations 2 are not complete. In each case there is in fact another particle created called a "neutrino" or an "antineutrino." These particles interact very weakly with matter and are extremely difficult to detect. They could not be detected by the detection process that will be used in this laboratory. As an example of equations 2, the radioactive nuclide $^{40}_{19}$K decays both by β^- and by β^+. In the case of β^- decay, the daughter nucleus is $^{40}_{20}$Ca, and in the case of β^+ decay the daughter nucleus is $^{40}_{18}$Ar. Many nuclides that undergo β-decay do so by only one of the processes.

γ-Decay is a process in which a nucleus that has an excess of energy lowers its energy state by the emission of a high-energy photon—a γ-ray. The process is very much analogous to the process whereby excited atoms lose energy when they emit visible light photons. The decay is described by

$$^{A}_{Z}X^* \rightarrow {}^{A}_{Z}X + \gamma \tag{3}$$

where the * indicates the excess energy, and γ is the symbol for a γ-ray or photon. Note that when gamma emission occurs there is no change in the nucleus as far as the number of protons or neutrons. Only the energy of the nucleus is altered.

A basic concept of radioactive decay is that the probability of decay for each type of radioactive nuclide is constant. In other words, there are a predictable number of decays per second even though it is not possible to predict which nuclei among the sample will decay. A quantity called the "decay constant," or λ, characterizes this concept. It is the probability of decay per unit time for one radioactive nucleus. The fundamental concept is that because λ is constant, it is possible to predict the rate of decay for a radioactive sample. The value of the constant λ is, of course, different for each radioactive nuclide.

Consider a sample of N radioactive nuclei with a decay constant of λ. The rate of decay of these nuclei dN/dt is related to λ and N by the equation

$$\frac{\mathrm{d}N}{\mathrm{d}t} = -\lambda N \qquad (4)$$

The symbol dN/dt stands for the rate of change of N with time t. The minus sign in the equation means that dN/dt must be negative because the number of radioactive nuclei is decreasing. The number of radioactive nuclei at $t = 0$ is designated as N_0. The question of interest is how many radioactive nuclei N are left at some later time t. The answer to that question is found by rearranging equation 4 and integrating it, subject to the condition that $N = N_0$ at $t = 0$. The result of that procedure is:

$$N = N_0 e^{-\lambda t} \qquad (5)$$

Equation 5 states that the number of nuclei N at some later time t decreases exponentially from the original number N_0 that are present. A second question of interest is the value of the rate of decay dN/dt of the radioactive sample. That can be found by substituting the expression for N from equation 5 back into equation 4. The result is

$$\frac{\mathrm{d}N}{\mathrm{d}t} = -\lambda N_0 e^{-\lambda t} \qquad (6)$$

Furthermore, the expression for dN/dt in equation 4 can then be substituted into equation 6 leading to

$$\lambda N = \lambda N_0 e^{-\lambda t} \qquad (7)$$

The quantity λN is called the "activity" of the radioactive sample. Since λ is the probability of decay for one nucleus, the quantity λN is the number of decays per unit time for N nuclei. Typically, λ is expressed as the probability of decay per second, and so in that case, λN is the number of decays per second from a sample of N nuclei. The symbol A is used for activity ($A = \lambda N$); thus, equation 7 becomes

$$A = A_0 e^{-\lambda t} \qquad (8)$$

Equations 5 and 8 state that both N, the number of nuclei, and A, the activity, decay exponentially according to the same exponential factor. For measurements made on real radioactive nuclei the activity A is the quantity that is usually measured. Another useful form of equation 8 can be derived by inverting that equation and taking the natural logarithm of both sides of the equation leading to,

$$\ln(A_0/A) = \lambda t \qquad (9)$$

An important concept associated with radioactive decay processes is the concept of half-life. The time for the sample to go from the initial number of nuclei N_0 to half that value $N_0/2$ is defined as the half life $t_{1/2}$. If equation 5 is solved for the time t when $N = N_0/2$ the result is

$$t_{1/2} = \frac{\ln(2)}{\lambda} = \frac{0.693}{\lambda} \qquad (10)$$

This same result could also be obtained by considering the time for the activity to go from A_0 to $A_0/2$.

In order to study these radioactive processes, it is necessary to be able to detect the presence of these particles that are the product of the decay. Although there are many forms in which devices can be constructed to accomplish the detection, they all have one feature in common. Every practical device that detects radiation does so by allowing the particles to interact with matter, and then uses that interaction as the basis for detection. The particular device that will be used in this laboratory is the Geiger counter. It consists of a tube in which the incident particle interacts and a scaling circuit to count the pulses. A diagram of a Geiger tube is shown in Figure 45.1.

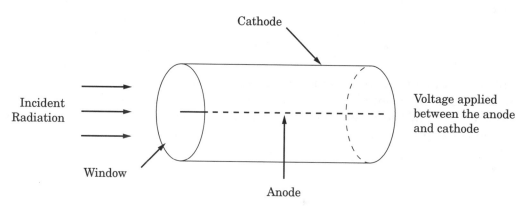

Figure 45.1 Diagram of the essential elements of a Geiger tube.

The Geiger tube is a small metal cylinder with a thin self-supporting wire along the axis of the cylinder. The wire is insulated from the cylinder. The cylindrical wall of the tube serves as the negative electrode (cathode), and the wire along the axis is the positive electrode (anode). At the entrance end of the tube there is a thin "window" formed by a very thin piece of fragile mica. Inside the counter is a special gas mixture that is ionized by any radiation that penetrates the "window."

In operation, a voltage is applied across the electrodes. The particular voltage for each tube must be determined experimentally. The applied voltage creates a large electric field in the tube, and the field is especially large in the region near the central wire. When radiation passes through the window and ionizes the gas, the large electric field cause an acceleration of the free electrons. These accelerated electrons cause additional ionizations creating what is referred to as an "avalanche" effect. The total number of ion-electron pairs created by a single incident particle is on the order of one million.

The electrons are more mobile and drift toward the positive central wire. When they arrive at the wire their negative charge causes the voltage of the wire to be lowered, and this sudden drop in voltage creates a pulse that is counted by the electronic circuitry. Each pulse counted signifies the passage of a particle through the counter. The ions then recombine with electrons, leaving the gas neutral again and ready for the passage of another particle. The whole process takes a time on the order of 300 μs, and during that time if another particle goes through the counter, it may not be counted. Thus, one disadvantage of Geiger counters is this "dead time" during which counts may be missed. This is a negligible effect unless the count rate is very high.

The count rate of a Geiger counter is a function of the voltage applied across the electrodes. Therefore, the counter should be operated in a region where the rate at which the count rate changes with voltage is a minimum. This is accomplished experimentally by measuring the count rate of some fixed source of radiation as a function of the voltage applied to the tube. A graph of the count rate versus the voltage will be made from the data and the operating point will be chosen to be some voltage at which the count rate versus voltage curve is as nearly flat as possible. This is referred to as the plateau region.

The counter has a minimum voltage needed in order to produce pulses at all. Both this minimum voltage and the operating voltage is quite variable and depends on the dimensions of the Geiger tube as well as the particular gas used in the tube. Therefore, the exact nature of the count rate versus voltage curve is dependent on the particular tube used. In general, the student-type Geiger tubes used in most undergraduate laboratories operate at a voltage of about 500 V and do not have a very flat count rate versus voltage curve anywhere. Thus, it is difficult to identify much of a plateau region. Typically, research-grade Geiger tubes operate in the neighborhood of 1000 V, and they have a much flatter count rate versus voltage curve. A typical count rate versus voltage curve for a student-type Geiger tube is shown in Figure 45.2.

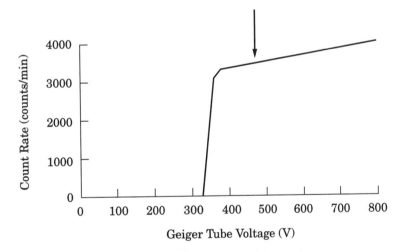

Figure 45.2 Typical count rate versus voltage for a Geiger tube.

EXPERIMENTAL PROCEDURE—COUNT RATE VERSUS VOLTAGE TO DETERMINE OPERATING VOLTAGE

1. All Geiger tubes have a maximum permissible operating voltage above which they are subject to breakdown. This may permanently damage the tube. Consult your instructor or the instruction manual for the specific maximum voltage of the particular Geiger tube used. *Do not ever exceed this maximum voltage.* Before plugging in the power cord of your instrument, make sure that the high-voltage control is turned to the minimum setting.

2. Place the $^{90}_{38}$Sr β-radiation source about 2 or 3 cm from the window of the Geiger tube. Once the source is positioned, take all the measurements for this procedure without changing the position of the source relative to the detector.

3. Turn on the power to the instrument, reset the scaler to zero, and start the counter in a continuous count mode with the high voltage still set to minimum. Slowly increase the high-voltage setting until counts begin to register on the scaler. Leave the voltage at this setting for which the Geiger tube just begins to count. This is the threshold voltage.

4. Reset the scaler to zero, start the counter, and let it count for 1 minute. Repeat this procedure two more times for a total of three 1-minute counts at this voltage setting. Record the voltage and the three values of the count in Data Table 1.

5. Raise the high-voltage setting by 50 V and repeat step 4 at this new high-voltage setting and record the high voltage and the counts for these three trials.

6. Continue this process up to the maximum permissible voltage of the Geiger tube. Be sure that you obtain the proper maximum voltage for your tube either from your instructor or from the instruction manual for the instrument.

CALCULATIONS—COUNT RATE VERSUS VOLTAGE

1. Calculate the mean $\overline{C}$ and standard error α_c for the three trials of the count at each voltage. Record these values in Calculations Table 1.

2. Calculate the square root of the mean count $\sqrt{\overline{C}}$ at each voltage and record it in Calculations Table 1.

GRAPHS—COUNT RATE VERSUS VOLTAGE

On linear graph paper, plot the mean count rate $\overline{C}$ (counts/min) versus voltage to produce a graph like the one shown in Figure 45.2.

EXPERIMENTAL PROCEDURE—HALF-LIFE OF ¹³⁷BA

1. The graph of count rate versus voltage should be similar in shape to Figure 45.2. The operating point of the tube should be chosen to be just beyond the shoulder on the relatively flat portion of the graph. That region is shown by an arrow in Figure 45.2. If the graph of your data does not look like you expect, consult your instructor for help in picking the proper operating voltage for your Geiger counter. Make all the measurements in this procedure at the same voltage once it is determined.

2. With no radioactive source near the Geiger tube, reset the scaler to zero and count the background for 2 minutes. The activity will be determined by counting for 15.0-s intervals. Divide the number of counts obtained in 2 minutes by eight to obtain an average background count rate for 15.0 s. Record that background rate in Data Table 2.

3. The ^{137}Ba isotope must be prepared at the time of its use because it has such a short half-life. The minigenerator contains ^{137}Cs, which decays with a 30-year half-life to produce ^{137}Ba as a daughter product. A saline solution of HCl is used to wash out the ^{137}Ba that has established an equilibrium with ^{137}Cs. Your instructor will prepare the sample when you are ready to count.

4. *Very carefully review all of the remaining steps of the procedure so that you will be ready to begin counting the sample immediately after it is prepared.*

5. Set the scaler to the stop position and reset the scaler to zero.

6. As quickly as possible after you receive your sample, place the plachet containing the sample as close to the Geiger tube window as possible in order to obtain the maximum counting rate.

7. As quickly as possible start the scaler in the continuous count mode and start the external timer simultaneously. *Do not ever stop the timer. Let it run continuously for the rest of this procedure.*

8. When the timer reaches 15.0 s stop the scaler. Record the counts obtained during the interval 0–15.0 s in the appropriate place in Data Table 2.

9. During the time interval when the timer reads between 15 and 30 seconds elapsed time, reset the scaler to zero and wait for the next counting period.

10. Start the scaler when the timer reads 30.0 s and stop the counter when the timer reads 45.0 s. Record the counts obtained in the interval 30.0–45.0 s in the appropriate place in Data Table 2.

11. Continue this process letting the timer run continuously and alternately counting for a 15.0-s interval and resetting and waiting a 15.0-s interval until Data Table 2 is completed. If a mistake is made on one of the counting periods, just skip it and start at the next counting period.

CALCULATIONS-HALFLIFE OF ^{137}BA

1. Each count period is 15.0 s long, during which time the activity changes continuously. For each of the counting periods, subtract the average background count for 15.0 s from the count in that time period. Divide that result for each period by 15.0 s and record it as the activity A in counts/s at the appropriate place in Calculations Table 2. This is actually the average activity during each 15.0-s time interval, but it will be assumed that it approximates the instantaneous activity at the beginning of each time interval.

2. Calculate the quantities (A_0/A) and $\ln (A_0/A)$ for each of activities in Data Table 2, where A_0 is taken to be the activity at $t = 0$.

3. According to equation 9 the quantity ln (A_0/A) should vary linearly with t the time, and the slope should be equal to the disintegration constant λ. Perform a linear least squares fit of the data in Calculations Table 2 with ln (A_0/A) as the ordinate and t the time as the abscissa. Record the value of the slope as the disintegration constant λ in Calculations Table 2.

4. Using equation 10, calculate the experimental value of the half-life of ^{137}Ba from the experimental value of the disintegration constant λ.

GRAPHS—HALF-LIFE OF ^{137}BA

On three-cycle semilog graph paper, plot the activity in counts/s versus the time in seconds using the log scale for the activity and the linear scale for time. Graph each activity at the time corresponding to the beginning of the time interval as given in Calculations Table 2.

Laboratory 45

Geiger Counter Measurement of the Half-Life of ^{137}Ba

LABORATORY REPORT

Data Table 1

Voltage	Count 1	Count 2	Count 3

Calculations Table 1

$\overline{C}$	α_C	$\sqrt{\overline{C}}$

SAMPLE CALCULATIONS

Data Table 2

Background Counts =	

Count Period	Counts
0–15 s	
30–45 s	
60–75 s	
90–105 s	
120–135 s	
150–165 s	
180–195 s	
210–225 s	
240–255 s	
270–285 s	
300–315 s	
330–345 s	
360–375 s	

Calculations Table 2

Background counts for 15.0 s =	

A (ct/s)	(A_0/A)	$\ln (A_0/A)$	time (s)

$\lambda =$	s^{-1}	$t_{1/2} =$	s

SAMPLE CALCULATIONS

QUESTIONS

1. For the count rate versus voltage data consider the standard error α_c and the square root of the mean count $\sqrt{\overline{C}}$. According to nuclear statistical theory, those quantities should be approximately equal. Do your data confirm that expectation?

2. State the threshold voltage (voltage at which the counter begins to operate) for your Geiger tube.

3. On the semilog graph of the activity of ^{137}Ba, draw the best straight line that you can through the data points. From the straight line determine the half-life of the sample. Indicate exactly which points on the straight line are used to determine the half-life.

4. The accepted half-life of ^{137}Ba is 2.6 min. Calculate the percentage error of each of your determinations of this half-life. Show your work.

5. A physics professor purchased in September 1970 a ^{137}Cs source that had an activity of 2.00×10^5 disintegrations per second at that time. What is the activity of that source today? (The half-life of ^{137}Cs is 30.2 years.) Show your work.

Laboratory 46

Nuclear Counting Statistics

PRELABORATORY ASSIGNMENT

Read carefully the entire description of the laboratory and answer the following questions based on the material contained in the reading assignment. Turn in the completed prelaboratory assignment at the beginning of the laboratory period prior to the performance of the laboratory.

1. For nuclear counting experiments no true value of a given count is assumed. What quantity is assumed to have a true value?

2. What is the exact statistical distribution function that describes the statistics of nuclear counting experiments?

3. What statistical distribution function approximates nuclear counting statistics and is used because it deals with continuous variables? For what values of the true mean m is this distribution valid?

4. In a nuclear counting experiment, a single measurement of C counts is obtained. What is the approximate value for σ_{n-1} for the count C?

5. According to the normal distribution function for the case when a given count is repeated 30 times, approximately how many of the results should fall in the range $\overline{C} \pm \sigma_{n-1}$? How many should fall in the range $\overline{C} \pm 2\sigma_{n-1}$?

6. A single count of a radioactive nucleus is made, and the result is 927 counts. What is the approximate value of σ_{n-1}?

7. A set of 10 repeated measurements of the count from a given radioactive sample were taken. The results were 633, 666, 599, 651, 654, 690, 660, 659, 664, and 612. What is the mean count $\overline{C}$? What is the value of σ_{n-1}? What is the value of α? Which of the counts fall outside $\overline{C} \pm \sigma_{n-1}$? Is this approximately the number of cases expected?

8. For the data in question 7, is $\sqrt{\overline{C}}$ approximately equal to σ_{n-1}?

OBJECTIVES

Consider a sample of long-lived radioactive nuclei. If the number of nuclei that decay in some fixed time interval is measured several times, there will be some variation in the count obtained for the different trials. This will be true even if all experimental errors have been eliminated. This variation is caused by the random nature of the nuclear decay process itself. The average count $\overline{C}$ can be determined with high precision by taking a large number of trials of the count for the given time period. This laboratory will use a Geiger counter to make repeated measurements of the number of counts from a long-lived radioactive isotope to accomplish the following objectives:

1. Determination of the average count in a fixed time interval for fixed experimental conditions by 50 independent determinations of the count

2. Determination of the standard deviation from the mean and standard error of the individual measurements

3. Comparison of the observed distribution of the individual counts relative to the mean compared to what is predicted by the normal distribution

4. Demonstration that $\sqrt{\overline{C}}$ is an approximation of the standard deviation from the mean for a single measurement resulting in C

EQUIPMENT LIST

1. Geiger counter (single unit containing Geiger tube, power supply, timer, and scaler)

2. Long-lived radioactive source (such as ^{137}Cs or ^{60}Co)

THEORY

If all other sources of error are removed from a nuclear counting experiment, there remains an uncertainty due to the random nature of the nuclear decay process. It is assumed that there exists some true mean value of the count that shall be designated as m. However, it is extremely important to realize that it is not assumed that there is a true value for any individual count C_i. Although m is assumed to exist, it can never be known exactly. Instead, one can approach knowledge of the true mean

m by a large number of observations. It can be shown that the best approximation to the true mean m is the mean $\overline{C}$, which is given by

$$\overline{C} = \frac{1}{n} \sum_{i=1}^{n} C_i \tag{1}$$

where C_i stands for the ith value of the count obtained in n trials. The standard deviation from the mean σ_{n-1} and the standard error α are defined in the usual manner as

$$\sigma_{n-1} = \sqrt{\sum_{i=1}^{n} \frac{1}{n-1} (\overline{C} - C_i)^2} \quad \text{and} \quad \alpha = \frac{\sigma_{n-1}}{\sqrt{n}} \tag{2}$$

These are no different from the assumptions that have been applied to essentially all of the measurements in this laboratory manual. The only difference is that in many of the cases in which these ideas have been applied, they are somewhat questionable because the random errors are not necessarily the determining factor. For nuclear counting experiments it is usually the case that the random errors are the limiting factor, and these concepts generally do apply strictly to such measurements.

The manner in which the measurements C_i are distributed around the mean $\overline{C}$ depends on the statistical distribution. The binomial distribution is the fundamental law for the statistics of all random events, including radioactive decay. Calculations are difficult with this distribution, and it is often approximated by another integral distribution called the "Poisson distribution." For cases of m greater than 20, both the binomial and the Poisson distribution can be approximated by the normal distribution. It has the advantage that it deals with continuous variables, and thus calculations are much easier with the normal distribution. The result is that for most nuclear counting problems of interest, the normal distribution predicts the same results for nuclear counting that has been assumed for measurements in general. These results are that approximately 68.3% of the measured values of C_i should fall within $\overline{C} \pm \sigma_{n-1}$, and approximately 95.5% of the measured values of C_i should fall within $\overline{C} \pm 2\sigma_{n-1}$.

There is one further statistical idea valid for nuclear counting experiments that is not true for measurements in general. For any given single measurement of the count C in a nuclear counting experiment, an approximation to the standard deviation from the mean σ_{n-1} is given by

$$\sigma_{n-1} \approx \sqrt{C} \tag{3}$$

For a series of repeated trials of a given count, the most accurate determination is given by $\overline{C} \pm \alpha$. If only a single measurement of the count is made, the most accurate statement that can be made is given by $C \pm \sqrt{C}$.

In this laboratory, a series of measurements of the same count will be made to determine the distribution of the measurements about the mean. In addition, the validity of equation 3 will be investigated.

EXPERIMENTAL PROCEDURE

1. Consult your instructor for the operating voltage of the Geiger counter. (Figure 46.1) It may be necessary to perform a quick set of measurements to determine the Geiger plateau. If so, consult Laboratory 45 for the procedure.

2. Set the Geiger counter to the proper operating voltage. Place a long-lived radioactive isotope on whichever counting shelf is necessary to produce between 500 and

Figure 46.1 Geiger counter with timer-scaler and encapsulated radioactive sources. (Photo courtesy of Sargent-Welch Scientific Company)

700 counts in a 30-s counting period. For best results, the Geiger counter should have preset timing capabilities. If it does not and a laboratory timer is used, it would improve the timing precision if 60-s counting intervals are used. For whatever time is counted, between 500 and 700 counts should be recorded.

3. Repeat the count for a total of 50 trials. Make no changes whatsoever in the experimental arrangement for these 50 trials. Record each count in the Data Table. Do not make any background subtraction. Simply record the total count for each counting period.

CALCULATIONS

1. Calculate the mean count $\overline{C}$, the standard deviation from the mean σ_{n-1}, and the standard error α for the 50 trials of the count and record the results in the Calculations Table.

2. For each count C_i calculate $| C_i - \overline{C} |/\sigma_{n-1}$ and record the results in the Calculations Table.

3. Determine what percentage of the counts C_i are further from $\overline{C}$ than σ_{n-1} by counting the number of times a value of $| C_i - \overline{C} |/\sigma_{n-1} > 1$ occurs. Express this number divided by 50 as a percentage. Count the number of times that $| C_i - \overline{C} |/\sigma_{n-1} > 2$ occurs. Express this number divided by 50 as a percentage. Record these results in the Calculations Table.

4. Calculate $\sqrt{\overline{C}}$ and record its value in the Calculations Table.

GRAPHS

1. Construct a histogram of your data on linear graph paper. Consider the range of the data and arbitrarily divide the range into about 15 segments. For counts in the range used this should give intervals of 8 or 10 counts. An example of some data is displayed in this manner in Figure 46.2. The mean of the data is 659 with $\sigma_{n-1} = 27$, and an interval of 10 has been chosen.

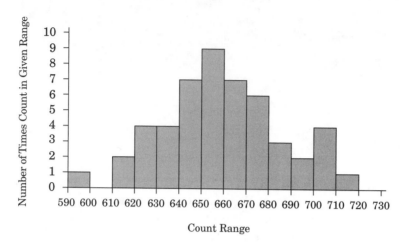

Figure 46.2 Histogram of 50 repeated counts with mean of 659.

LABORATORY REPORT

Data Table

i	C_i		i	C_i
1			26	
2			27	
3			28	
4			29	
5			30	
6			31	
7			32	
8			33	
9			34	
10			35	
11			36	
12			37	
13			38	
14			39	
15			40	
16			41	
17			42	
18			43	
19			44	
20			45	
21			46	
22			47	
23			48	
24			49	
25			50	

Calculations Table

i	$|C_i - \bar{C}|/\sigma_{n-1}$		i	$|C_i - \bar{C}|/\sigma_{n-1}$
1			26	
2			27	
3			28	
4			29	
5			30	
6			31	
7			32	
8			33	
9			34	
10			35	
11			36	
12			37	
13			38	
14			39	
15			40	
16			41	
17			42	
18			43	
19			44	
20			45	
21			46	
22			47	
23			48	
24			49	
25			50	

$\bar{C} =$		$\sigma_{n-1} =$
$\alpha =$		$\sqrt{\bar{C}} =$
% trials $> \sigma_{n-1}$ from mean =		
% trials $> 2\sigma_{n-1}$ from mean =		

QUESTIONS

1. Consider the shape of the histogram of your data. Does it show the expected distribution relative to the mean of the data?

2. Compare the percentage of trials that have $|C_i - \overline{C}|/\sigma_{n-1} > 1$ with that predicted by the normal distribution. Compare the percentage of trials that have $|C_i - \overline{C}|/\sigma_{n-1} > 2$ with that predicted by the normal distribution.

3. What is the most accurate statement that you can make about the count from the sample based on the data that you have taken?

4. Calculate the percentage difference between $\sqrt{\overline{C}}$ and σ_{n-1}. Do the results confirm the expectations of equation 3?

Suppose that you were to perform another 50 trials of the count of the same sample under the exact same conditions as the first 50 trials. Answer questions 5 to 7 about what would be expected for the total 100 trials that you now have.

5. Would the mean $\overline{C}$ for the 100 trials be expected to be significantly different from the mean $\overline{C}$ for the first 50 trials?

6. Would the standard deviation from the mean σ_{n-1} for the 100 trials be expected to be significantly different from the standard deviation from the mean σ_{n-1} for the first 50 trials?

7. Would the standard error α for the 100 trials be expected to be significantly different from the standard error α for the first 50 trials?

Laboratory 47

Absorption of β- and γ-rays

PRELABORATORY ASSIGNMENT

Read carefully the entire description of the laboratory and answer the following questions based on the material contained in the reading assignment. Turn in the completed prelaboratory assignment at the beginning of the laboratory period prior to the performance of the laboratory.

1. What are the names of the three types of natural radioactivity? Describe the nature of the particle produced in each.

2. Which kind of nuclear radiation undergoes a true exponential absorption as a function of absorber thickness? What property of its interaction with matter causes this to happen?

3. Which kind of natural radioactivity produces particles with a continuous spectrum of energy?

4. Which type of natural radioactivity produces particles that have such little penetrating power that they pose no external health hazard?

5. State the form of the equation for the absorption of γ-rays in matter. Define all the terms used in the equation.

6. What type of radiation is the most penetrating?

7. A pure γ-ray radioactive source has a count rate of 5000 counts in 1 minute with no absorber between the source and the detector. An absorber of thickness 0.375 cm is placed between the source and the detector. The number of counts in the detector in 1 minute is now 3245. What is the absorption coefficient μ of the material? Show your work.

8. In the experimental arrangement of question 7, if an additional 0.235 cm of the same material is placed between the detector and source, what is now the count in 1 minute?

Absorption of β- and γ-rays

OBJECTIVES

The products of natural radioactivity are α-particles, β-particles, and γ-particles. There is considerable difference between the three types of radiation with respect to their tendency to be absorbed when passing through matter. α-particles are so easily absorbed that they cannot penetrate the window of most Geiger counters to reach the active volume of the detector. Therefore, no measurements will be performed on α-particles in this laboratory. β-particles are the next most readily absorbed radiation, and γ-rays are the most penetrating of the three types of radiation. In this laboratory, the intensity of a γ-source and a β-source will be measured as a function of the thickness of two different types of material placed between the source and the detector. These measurements will be used to accomplish the following objectives:

1. Demonstration of the difference in relative absorption by different types of materials for different kinds of radiation
2. Determination of the absorption coefficient μ for γ-radiation in lead

EQUIPMENT LIST

1. Geiger counter (single unit with Geiger tube, power supply, timer, and scaler)
2. Absorber set (lead and polyethylene)
3. γ-radiation source (^{60}Co for example)
4. β-radiation source (^{90}Sr for example)
5. Two-cycle semilog graph paper

THEORY

The different nature of the particles produced in natural radioactivity accounts for the differences in their relative absorption in matter. The two most important characteristics are the charge and the mass of the particles. In general, the charge is the most important factor. Because γ-rays are photons and have no charge, they tend to interact least with matter. Therefore, γ-rays are the most penetrating of the three types of radiation, and it takes a relatively large thickness of matter to stop them.

γ-rays interact with matter in a fundamentally different way from the way in which charged particles interact. There are three processes responsible for the absorption of γ-rays. They are the photoelectric effect, Compton scattering, and the production of electron-positron pairs. In the photoelectric effect, the photon disappears, and an electron with essentially the total energy of the photon is ejected from an atom of the absorber. The Compton effect is a process in which the original γ-ray photon elastically scatters from an essentially free electron in the absorber atom, ejecting an energetic electron and a photon of less energy than the original photon. In pair production the photon disappears, and its energy is turned into the mass plus kinetic energy of an electron and a positron. Pair production cannot occur unless the original photon has energy greater than the sum of the rest energies of the electron and positron, which is 1.022 Mev.

Consider γ-rays from a radioactive source as a large collection of photons moving in a given direction in a beam. In each of the processes described above, a photon is completely removed from the beam in a single process if it interacts at all. Compton scattering might be thought to be an exception to this statement because it does produce another photon. However, the photon produced has very little probability of going in the original direction, so even photons that interact by the Compton effect are effectively removed from the beam. The consequence of this fact, that photons either do not interact at all or else are completely removed by an interaction, means that γ-ray intensity decreases exponentially with absorber thickness. Photons are the only particles from natural radioactivity to interact in this manner and are the only ones to show an exact exponential decrease in matter.

It is found experimentally that when a beam of γ-rays of intensity I is incident on a slab of matter of thickness Δx, the change in intensity ΔI of the beam as it passes through the slab is proportional to the thickness Δx and to the incident intensity I. In equation form this is

$$\Delta I = - \mu I \Delta x \tag{1}$$

where μ is a constant of proportionality called the "absorption coefficient." The constant μ has dimensions of inverse length and is commonly expressed in cm^{-1}. If the limit is taken so that the finite changes become differential, equation 1 can then be integrated to give

$$I = I_0 e^{-\mu x} \tag{2}$$

which is the characteristic exponential absorption described earlier. The term I_0 is the intensity at thickness $x = 0$, and I is the intensity at thickness x. Equation 2 can be rewritten in the form

$$\frac{I_0}{I} = e^{\mu x} \tag{3}$$

If the natural logarithm of both sides of equation 3 is taken, the result is

$$\ln = (I_0/I) = \mu x \tag{4}$$

Equation 4 states that the quantity $\ln(I_0/I)$ should be proportional to x with μ as the constant of proportionality. This relationship will be used to determine μ. The relationship applies only to a source of pure γ-rays. The ^{60}Co source emits both γ- and β-rays, but the β-particles will be absorbed first, and the γ-rays that are left can be assumed to be a pure source of γ-rays.

The absorption of β-particles is completely different from the absorption of γ-rays. β-particles are either electrons or positrons. In either case, they are charged particles with mass equal to the mass of the electron. As charged particles, they tend to lose their energy gradually in a series of collisions with the electrons in the atoms of the absorbing material. Generally, each collision results in a relatively small energy loss, and a larger number of collisions are necessary before all the energy of the incident β-particle is lost. As a consequence of the nature of this process, a beam of electrons all of which have the same initial energy has a definite range in a given type of absorber. Therefore, electrons do not exhibit absorption that is an exponential function of the thickness of the absorber.

The situation for the case of natural β-radiation is complicated by the fact that not all of the betas have the same energy. β-rays from a β-emitter have a spectrum of energy ranging from almost zero up to some maximum energy characteristic of the isotope. If an absorption experiment is performed for this spectrum of energies present for any β-emitter, often the intensity does in fact decrease in a nearly exponential manner. This is simply a fortuitous result of the combined effects of the initial energy distribution, of back scattering into the detector, and of the true range-energy relationship. Measurements will be performed on the absorption of β-particles from a ^{90}Sr source, which can be assumed to be a pure β-ray source.

Although no measurements will be made in this laboratory using α-particles, it is worth noting that like β-particles, α-particles also show a definite range. α-Particles are doubly ionized helium atoms. Because they are so massive and because their charge is twice as great as that of β-particles, α-particles interact very strongly with matter. All naturally occurring α-particles are completely absorbed by the thickness of a few pieces of paper. α-Particles will not even penetrate the dead layer of skin on the human body. Thus, they pose no external health hazard.

EXPERIMENTAL PROCEDURE

γ-Absorption

1. Consult your instructor for the operating voltage of the Geiger counter. It may be necessary to perform a quick set of measurements to determine the Geiger plateau. If so, consult Laboratory 45 for the procedure (Figure 47.1).

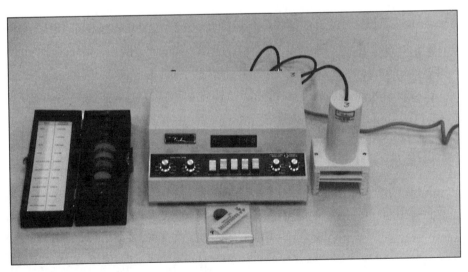

Figure 47.1 Geiger counter and set of standard absorbers described in procedure.

2. Set the Geiger counter to the proper operating voltage. Once it is set, leave the voltage unchanged for the rest of the laboratory. Place the ^{60}Co γ-ray source in the third shelf position and place the empty absorber holder in the second shelf position. Be sure that the source is not moved during the rest of the procedure with γ-rays. For best results let the Geiger tube stabilize for at least 10 minutes before proceeding.

3. Reset the scaler to zero and determine the number of counts from the source with no absorber in the holder. Count for a period of 60 s. Record this count with no absorber in Data Tables 1 and 2 in the space labeled $x = 0$. In principle, a background subtraction should be made for each of the counts. The background should be on the order of less than 10 counts per 60-s counting period and thus will be assumed negligible. Check to see that the background is very low and if for some reason it is not, make the proper adjustments in the procedure to account for the background.

4. Using the lead absorbers, determine the intensity (number of counts) in a 60-s period for the following values of absorber thickness: 0.079, 0.159. 0.318, 0.635, and 0.953 cm. These choices assume the use of a standard set of absorbers 1/32, 1/16, 1/8, and 1/4 in. thick. Any known values of thickness that cover the stated range are satisfactory. Record in Data Table 1 the results of the count for each absorber and the thickness of the absorber.

5. Using the polyethylene absorbers determine the number of counts in a 60-s period for the following values of absorber thickness: 0.051, 0.076, 0.159, 0.318, 0.635, 0.953, and 1.27 cm. Again, a standard set of absorbers are assumed, but any known values of thickness that cover the range are satisfactory. Record in Data Table 2 the count for each absorber and the thickness of the absorber.

β- Absorption

1. Remove the γ-ray source and place a ^{90}Sr β-ray source in the third shelf position. With the empty absorber holder in the second shelf position, determine the number of counts in 60 s and record the results in Data Tables 3 and 4.

2. Using the lead absorbers, determine the number of counts in 60 s for absorbers for the following values of absorber thickness: 0.079, 0.159, and 0.318 cm. Record in Data Table 3 the count for each absorber and the absorber thickness.

3. Using the polyethylene absorbers, determine the number of counts in 60 s for the following values of thickness: 0.010, 0.020, 0.051, 0.076, 0.159, 0.318, and 0.635 cm. Record in Data Table 4 the count for each absorber and the absorber thickness.

CALCULATIONS

1. Although the ^{60}Co source has been referred to as a γ-ray source, it has some β-ray activity as well. In the data taken with the lead absorber, the β-rays from ^{60}Co are completely absorbed by the first thickness of lead used. The rest of the data for count versus absorber thickness represents the absorption of γ-rays alone. Let the count for the 0.079-cm-thick lead absorber represent the initial intensity I_0 of γ-rays alone. Record that intensity in the Calculations Table as I_0. Calculate the increase in thickness that each absorber represents relative to the first absorber. Call this increase in thickness x_1, where $x_1 = x - 0.079$. For each of the absorbers beyond the first record the values of I and x_1 in Calculations Table 1.

2. According to equation 4 the quantity $\ln(I_0/I)$ should be proportional to x_1 for the assumption made in defining x_1 above. Perform a linear least squares fit with $\ln(I_0/I)$ as the ordinate and x_1 as the abscissa. The slope of the straight line obtained is the absorption coefficient μ. Determine the slope μ and the correlation coefficient r and record them in Calculations Table 1.

GRAPHS

1. On semilog graph paper, graph the intensity (counts) versus x for the ^{60}Co data from Data Tables 1 and 2. Use the linear scale for x. According to the theory, the part of the intensity due to γ-rays should be linear on this semilog plot. This means that the linear portion will only occur after the β-rays are absorbed. For the lead absorbers, this will occur after the first absorber. For the polyethylene absorbers, it will take several of the absorbers to remove the β-particles. A typical set of data for ^{60}Co using lead absorbers is shown in Figure 47.2. Note that the zero absorption is not on the straight line drawn by hand.

2. On semilog graph paper, graph the intensity (counts) versus x for the ^{90}Sr data for β-rays on polyethylene from Data Table 4. Although it is only an approximation, these data may be somewhat linear.

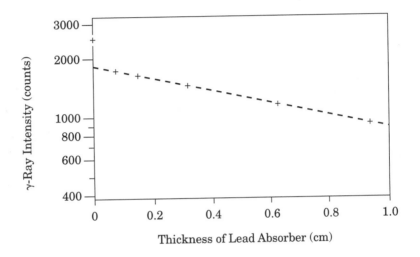

Figure 47.2 Typical data for ^{60}Co using lead absorbers.

LABORATORY REPORT

Data Table 1
γ-Rays on Lead

x (cm)	I (counts)
0	

Data Table 3
β-Rays on Lead

x (cm)	I (counts)
0	

Data Table 2
γ-Rays on Polyethylene

x (cm)	I (counts)
0	

Data Table 4
β-Rays on Polyethylene

x (cm)	I (counts)
0	

SAMPLE CALCULATIONS

Calculations Table 1

	$I_0 =$	Counts	
x (cm)	I (counts)	x_1 (cm)	$\ln (I_0/I)$
$\mu =$	(cm^{-1})	$r =$	

SAMPLE CALCULATIONS

QUESTIONS

1. Comment on the semilog graph of the data from Data Table 1 for the intensity of ^{60}Co radiation versus thickness of lead absorber. At what thickness are all the β-rays absorbed? After the β-rays are absorbed, does the graph of absorption of γ-rays only show a linear behavior?

2. Comment on the semilog graph of the data from Data Table 2 for the intensity of ^{60}Co radiation versus thickness of polyethylene absorber. Does the absorption of the β-rays take place over several absorbers? At what thickness are all of the β-rays absorbed? After the β-rays are absorbed, does the graph of absorption of γ-rays only show a linear behavior?

3. Comment on the data in Data Table 3 for the intensity of ^{90}Sr radiation versus the thickness of lead absorber. What is your conclusion about the absorption of β-rays in lead?

4. Comment on the semilog graph of the data from Data Table 4 for the intensity of ^{90}Sr radiation versus thickness of polyethylene absorber. Is the graph approximately linear? If it is not linear over the whole range, is it at least linear over some portion of the range?

5. Compare the slope of the graphs for ^{60}Co in the two absorbers. Using that as a basis for comparison, estimate how much more strongly the lead absorbs γ-rays than does the polyethylene.

6. For γ-rays of the approximate energy of the ^{60}Co γ-rays in lead, the approximate value of the absorption coefficient is $\mu = 0.65$ cm^{-1}. Considering this as the accepted value, calculate the accuracy of your measurement of μ.

Appendix I

Correlation Coefficients*

This table shows the probability to obtain a given correlation coefficient r for two variables for which there is in fact no correlation. This probability is a strong function of the number of data pairs n. As an illustration of the table, consider the case of $n=12$. There is a 10% probability of obtaining a value of $r \geq 0.497$, a 2% probability of obtaining a value of $r \geq 0.658$, and a 0.1% probability of obtaining a value of $r \geq 0.823$ for data for which no actual correlation exists. For many cases in this laboratory manual, you will take data which produces values of r greater than the 0.1% probability for the particular value of n. In those cases one can conclude that the data is strong evidence for a linear relationship between the variables.

Probability (%)

n	10	5	2	1	0.1
3	0.988	0.997	0.999	1.000	1.000
4	0.900	0.950	0.980	0.990	0.999
5	0.805	0.878	0.934	0.959	0.992
6	0.729	0.811	0.882	0.917	0.974
7	0.669	0.754	0.833	0.874	0.951
8	0.621	0.707	0.789	0.834	0.925
9	0.582	0.666	0.750	0.798	0.898
10	0.549	0.632	0.716	0.765	0.872
11	0.521	0.602	0.685	0.735	0.847
12	0.497	0.576	0.658	0.708	0.823
15	0.441	0.514	0.592	0.641	0.760
20	0.378	0.444	0.516	0.561	0.679

*This table is adapted from Table VI of Fisher and Yates, "Statistical Tables for Biological, Agricultural, and Medical Research," published by Oliver & Boyd, Ltd., Edinburgh, by permission of the authors and publishers.

Appendix II

Properties of Materials

Table II A Density of Substances (kg/m³)

Substance	Density	Substance	Density
Aluminum	2.7×10^3	Cork	$0.22\text{-}0.26 \times 10^3$
Brass	8.4×10^3	Oak wood	$0.60\text{-}0.90 \times 10^3$
Copper	8.9×10^3	Maple wood	$0.62\text{-}0.75 \times 10^3$
Gold	19.3×10^3	Pine wood	$0.35\text{-}0.50 \times 10^3$
Iron	7.85×10^3	Alcohol, ethyl	0.79×10^3
Lead	11.3×10^3	Alcohol, methyl	0.81×10^3
Nickel	8.7×10^3	Mercury	13.6×10^3
Steel	7.8×10^3	Pure water	1.000×10^3
Zinc	7.1×10^3	Sea water	1.025×10^3

Table II B Specific Heats (Calories/gram-⁰C)

Substance	Specific heat	Substance	Specific heat
Aluminum	0.22	Mercury	0.033
Brass	0.092	Steel	0.12
Copper	0.093	Tin	0.054
Iron	0.11	Water	1.000
Lead	0.031	Zinc	0.093

Table II C Thermal Coefficients of Expansion$(^0C)^{-1}$

Substance	α	Substance	α
Aluminum	24×10^{-6}	Brass and bronze	19×10^{-6}
Copper	17×10^{-6}	Lead	29×10^{-6}
Pyrex glass	3.2×10^{-6}	Ordinary glass	9×10^{-6}
Steel	11×10^{-6}	Concrete	12×10^{-6}
Gold	14×10^{-6}	Tin	27×10^{-6}

Table II D   Resistivities and Temperature Coefficients

Substance	Resistivity (Ω-m)	Temperature Coefficient $(^0C)^{-1}$
Aluminum	2.82×10^{-8}	3.9×10^{-3}
Copper	1.72×10^{-8}	3.9×10^{-3}
Silver	1.59×10^{-8}	3.8×10^{-3}
Gold	2.44×10^{-8}	3.4×10^{-3}
Nickel-silver	$33 \quad \times 10^{-8}$	0.4×10^{-3}
Tungsten	$5.6 \quad \times 10^{-8}$	4.5×10^{-3}
Iron	$10 \quad \times 10^{-8}$	5.0×10^{-3}
Lead	$22 \quad \times 10^{-8}$	3.9×10^{-3}
Carbon	$3.5 \quad \times 10^{-5}$	-0.5×10^{-3}

Some Fundamental Constants[*]

Quantity	Symbol	Value[†]
Atomic mass unit	u	$1.660\ 540\ 2(10) \times 10^{-27}$ kg $931.434\ 32(28)$ MeV$/c^2$
Avogadro's number	N_A	$6.022\ 136\ 7(36) \times 10^{23}$ (mol)$^{-1}$
Bohr magneton	$\mu_B = \frac{e\hbar}{2m_e}$	$9.274\ 015\ 4(31) \times 10^{-24}$ J/T
Bohr radius	$a_0 = \frac{\hbar^2}{m_e e^2 k_e}$	$0.5929\ 177\ 249(24) \times 10^{-10}$ m
Boltmann's constant	$k_B = R/N_A$	$1.380\ 658(12) \times 10^{-23}$ J/K
Compton wavelength	$\lambda_C = \frac{h}{m_e c}$	$2.426\ 310\ 58(22) \times 10^{-12}$ m
Deuteron mass	m_d	$3.343\ 586\ 0\ (20) \times 10^{-27}$ kg $2.013\ 553\ 214(24)$ u
Electron mass	m_e	$9.109\ 389\ 7(54) \times 10^{-31}$ kg $5.485\ 799\ 03(13) \times 10^{-4}$ u $0.510\ 999\ 06(15)$ MeV$/c^2$
Electron-volt	eV	$1.602\ 177\ 33\ (49) \times 10^{-19}$ J
Elementary charge	e	$1.602\ 177\ 33\ (49) \times 10^{-19}$ C
Gas Constant	R	$8.314\ 510(70)$ J/K $\cdot$ mol
Gravitational constant	G	$6.672\ 59(85) \times 10^{-11}$ N $\cdot$ m^2/kg^2
Hydrogen ground state	$E_0 = \frac{m_e e^4 k_e^2}{2\hbar^2} = \frac{e^2 k_e}{2a_0}$	$13.605\ 698(40)$ eV
Josephson frequency-voltage ratio	$2e/h$	$4.835\ 976\ 7(14) \times 10^{14}$ Hz/V
Magnetic flux quantum	$\Phi_0 = \frac{h}{2e}$	$2.067\ 834\ 61(61) \times 10^{-15}$ Wb
Neutron mass	m_n	$1.674\ 928\ 6(10) \times 10^{-27}$ kg $1.008\ 664\ 904(14)$ u $939.565\ 63(28)$ MeV$/c^2$
Nuclear magneton	$\mu_n = \frac{e\hbar}{2m_p}$	$5.050\ 786\ 6(17) \times 10^{-27}$ J/T
Permeability of free space	μ_0	$4\pi \times 10^{-7}$ N/A^2 (exact)
Permittivity of free space	$\varepsilon_0 = 1/\mu_0 c^2$	$8.854\ 187\ 817 \times 10^{-12}$ C^2/N $\cdot$ m^2 (exact)
Planck's constant	h	$6.626\ 075(40) \times 10^{-34}$ J $\cdot$ s
	$\hbar = h/2\pi$	$1.054\ 572\ 66(63) \times 10^{-34}$ J $\cdot$ s
Proton mass	m_p	$1.672\ 623(10) \times 10^{-27}$ kg $1.007\ 276\ 470(12)$ u $938.272\ 3(28)$ MeV$/c^2$
Quantized Hall resistance	h/e^2	$25812.805\ 6(12)$ Ω
Rydberg constant	R_H	$1.097\ 373\ 153\ 4(13) \times 10^7$ m^{-1}
Speed of light in vacuum	c	$2.997\ 924\ 58 \times 10^8$ m/s (exact)

[*] These constants are the values recommended in 1986 by CODATA, based on a least squares adjustment of data from different measurements. For a more complete list, see Cohen, E. Richard, and Barry N. Taylor, *Rev. Mod. Phys.* **59**:1121, 1987

[†] The numbers in parentheses for the values below represent the uncertainties in the last decimal places.

Solar System Data

Body	Mass (kg)	Mean Radius (m)	Period (s)	Distance from Sun (m)
Mercury	3.18×10^{23}	2.43×10^6	7.60×10^6	5.79×10^{10}
Venus	4.88×10^{24}	6.06×10^6	1.94×10^7	1.08×10^{11}
Earth	5.98×10^{24}	6.37×10^6	3.156×10^7	1.496×10^{11}
Mars	6.42×10^{23}	3.37×10^6	5.94×10^7	2.28×10^{11}
Jupiter	1.90×10^{27}	6.99×10^7	3.74×10^8	7.78×10^{11}
Saturn	5.68×10^{26}	5.85×10^7	9.35×10^8	1.43×10^{12}
Uranus	8.68×10^{25}	2.33×10^7	2.64×10^9	2.87×10^{12}
Neptune	1.03×10^{26}	2.21×10^7	5.22×10^9	4.50×10^{12}
Pluto	$\approx 1.4 \times 10^{22}$	$\approx 1.5 \times 10^6$	7.82×10^9	5.91×10^{12}
Moon	7.36×10^{22}	1.74×10^6	—	—
Sun	1.991×10^{30}	6.96×10^8	—	—

Physical Data Often Used*

Average Earth-Moon distance	3.84×10^8 m
Average Earth-Sun distance	1.496×10^{11} m
Average radius of the Earth	6.37×10^6 m
Density of air (20°C and 1 atm)	1.20 kg/m^3
Density of water (20°C and 1 atm)	1.00×10^3 kg/m^3
Free-fall acceleration	9.80 m/s^2
Mass of the Earth	5.98×10^{24} kg
Mass of the Moon	7.36×10^{22} kg
Mass of the Sun	1.99×10^{30} kg
Standard atmospheric pressure	1.013×10^5 Pa

* These are the values of the constants as used in the text.

Some Prefixes for Powers of Ten

Power	Prefix	Abbreviation	Power	Prefix	Abbreviation
10^{-18}	atto	a	10^1	deka	da
10^{-15}	femto	f	10^2	hecto	h
10^{-12}	pica	p	10^3	kilo	k
10^{-9}	nano	n	10^6	mega	M
10^{-6}	micro	μ	10^9	giga	G
10^{-3}	milli	m	10^{12}	tera	T
10^{-2}	centi	c	10^{15}	peta	P
10^{-1}	deci	d	10^{18}	exa	E

Standard Abbreviations and Symbols for Units

Symbol	Unit	Symbol	Unit
A	ampere	in.	inch
Å	angstrom	J	joule
u	atomic mass unit	K	kelvin
atm	atmosphere	kcal	kilocalorie
Btu	British thermal unit	kg	kilogram
C	coulomb	kmol	kilomole
°C	degree Celsius	lb	pound
cal	calorie	m	meter
deg	degree (angle)	min	minute
eV	electron volt	N	newton
°F	degree Fahrenheit	Pa	pascal
F	farad	rev	revolution
ft	foot	s	second
G	gauss	T	Tesla
g	gram	V	volt
H	henry	W	watt
h	hour	Wb	weber
hp	horsepower	μm	micrometer
Hz	hertz	Ω	ohm

Mathematical Symbols Used in the Text and Their Meaning

Symbol	Meaning
$=$	is equal to
$\equiv$	is defined as
$\neq$	is not equal to
α	is proportional to
$>$	is greater than
$>$	is less than
$\gg (\ll)$	is much greater (less) than
$\approx$	is approximately equal to
Δx	the change in x
$\sum_{i=1}^{n} x_i$	the sum of all quantities x_i from $i = 1$ to $i = n$
$\lvert x \rvert$	the magnitude of x (always a nonnegative quantity)
$\Delta x \to 0$	Δx approaches zero
$\frac{dx}{dt}$	the derivative of x with respect to t
$\frac{\partial x}{\partial t}$	the partial derivative of x with respect to t
$\int$	integral

Conversions

Length
1 in. = 2.54 cm
1 m = 39.37 in. = 3.281 ft
1 ft = 0.3048 m
12 in. = 1 ft
3 ft = 1 yd
1 yd = 0.9144 m
1 km = 0.621 mi
1 mi = 1.609 km
1 mi = 5280 ft
$1 \text{ Å} = 10^{-10}$ m
$1 \ \mu m = 1\mu = 10^{-6}$ m $= 10^4$ Å
1 lightyear $= 9.461 \times 10^{15}$ m

Area
$1 \text{ m}^2 = 10^4 \text{ cm}^2 = 10.76 \text{ ft}^2$
$1 \text{ ft}^2 = 0.0929 \text{ m}^2 = 144 \text{ in}^2$
$1 \text{ in}^2 = 6.452 \text{ cm}^2$

Volume
$1 \text{ m}^3 = 10^6 \text{ cm}^3 = 6.102 \times 10^4 \text{ in}^3$
$1 \text{ ft}^3 = 1728 \text{ in}^3 = 2.83 \times 10^{-2} \text{ m}^3$
$1 \text{ liter} = 1000 \text{ cm}^3 = 1.0576 \text{ qt} = 0.0353 \text{ ft}^3$
$1 \text{ ft}^3 = 7.481 \text{ gal} = 28.32 \text{ liters} = 2.832 \times 10^{-2} \text{ m}^3$
$1 \text{ gal} = 3.786 \text{ liters} = 231 \text{ in}^3$

Mass
1000 kg = 1 t (metric ton)
1 slug = 14.59 kg
$1 \text{ u} = 1.66 \times 10^{-27}$ kg

Force
$1 \text{ N} = 10^5 = \text{dyne} = 0.2248 \text{ lb}$
1 lb = 4.448 N
$1 \text{ dyne} = 10^{-5} = \text{N} = 2.248 \times 10^{-6} \text{ lb}$

Velocity
1 mi/h = 1.47 ft/s = 0.447 m/s = 1.61 km/h
1 m/s = 100 cm/s = 3.281 ft/s
1 mi/min = 60 mi/h = 88 ft/s

Acceleration
$1 \text{ m/s}^2 = 3.28 \text{ ft/s}^2 = 100 \text{ cm/s}^2$
$1 \text{ ft/s}^2 = 0.3048 \text{ m/s}^2 = 30.48 \text{ cm/s}^2$

Pressure
$1 \text{ bar} = 10^5 \text{ N/m}^2 = 14.50 \text{ lb/in}^2$
1 atm = 760 mm Hg = 76.0 cm Hg
$1 \text{ atm} = 14.7 \text{ lb/in}^2 = 1.013 \ 3 \ 10^5 \text{ N/m}^2$
$1 \text{ Pa} = 1 \text{ N/m}^2 = 1.45 \times 10^{-4} \text{ lb/in}^2$

Time
$1 \text{ year} = 365 \text{ days} = 3.16 \times 10^7 \text{ s}$
$1 \text{ day} = 24 \text{ h} = 1.44 \times 10^3 \text{ min} = 8.64 \times 10^4 \text{ s}$

Energy
$1 \text{ J} = 0.738 \text{ ft} \cdot \text{lb} = 10^7 \text{ erg}$
1 cal = 4.186 J
$1 \text{ Btu} = 252 \text{ cal} = 1.054 \times 10^3 \text{ J}$
$1 \text{ eV} = 1.6 \times 10^{-19} \text{ J}$
$931.5 \ \frac{\text{MeV}}{c^2}$ is equivalent to 1 u
$1 \text{ kWh} = 3.60 \times 10^6 \text{ J}$

Power
1 hp = 550 ft · lb/s = 0.746 kW
1 W = 1 J/s = 0.738 ft · lb/s
1 Btu/h = 0.293 W

The Greek Alphabet

Alpha	A	α	Iota	I	ι	Rho	P	ρ
Beta	B	β	Kappa	K	κ	Sigma	Σ	σ
Gamma	Γ	γ	Lambda	Λ	λ	Tau	T	τ
Delta	Δ	δ	Mu	M	μ	Upsilon	Υ	υ
Epsilon	E	ε	Nu	N	ν	Phi	Φ	ϕ
Zeta	Z	ζ	Xi	Ξ	ξ	Chi	X	χ
Eta	H	η	Omicron	O	o	Psi	Ψ	ψ
Theta	Θ	θ	Pi	Π	π	Omega	Ω	ω